Biotechnology and Safety Assessment
Second Edition

T0225654

Biotechnology and Safety Assessment
Second Edition

Editor

John A. Thomas
University of Texas Health Science Center
San Antonio, Texas

CRC Press
Taylor & Francis Group
Boca Raton London New York

CRC Press is an imprint of the
Taylor & Francis Group, an **informa** business
A TAYLOR & FRANCIS BOOK

BIOTECHNOLOGY AND SAFETY ASSESSMENT, 2/E

CRC Press
Taylor & Francis Group
6000 Broken Sound Parkway NW, Suite 300
Boca Raton, FL 33487-2742

First issued in paperback 2020

© 1999 by Taylor & Francis Group, LLC
CRC Press is an imprint of Taylor & Francis Group, an Informa business

No claim to original U.S. Government works

ISBN-13: 978-0-367-45574-3 (pbk)
ISBN-13: 978-1-56032-721-9 (hbk)

Visit the Taylor & Francis Web site at
http://www.taylorandfrancis.com

and the CRC Press Web site at
http://www.crcpress.com

Composition and editorial services by TechBooks.

A CIP catalog record for this book is available from the British Library.

Library of Congress Cataloging-in-Publication Data

Biotechnology and safety assessment / editor, John A. Thomas.—2nd ed.
    p.   cm.
    Includes bibliographical references and index.
    ISBN 1-56032-721-9 (cloth : alk. paper)
    1. Biotechnology—Safety measures.   I. Thomas J. A. (John A.),
1933– ,
TP248.2.B55126   1999
660.6′0289—dc21   98-21963
                           CIP

Acknowledgment

The editor wishes to express his sincere appreciation to Nancy Ibach for her diligence in processing and collating *Biotechnology and Safety Assessment*. Her dedication to administrative detail and her professional manner greatly expedited the assembling of this monograph.

Contents

Contributing Authors

Jim D. Astwood, *Monsanto Company, 800 North Lindbergh Blvd., Saint Louis, MO 63167*

James R. Beall, *U.S. Department of Energy, 1000 Independence Ave., S.W., Washington, DC 20585*

Shuo Chen, *University of Texas Health Science Center, 7703 Floyd Curl Drive, San Antonio, TX 78284*

Stanley T. Crooke, *Isis Pharmaceuticals, Inc., 2280-B Faraday Avenue Carlsbad, CA 92008*

Jacques Descotes, *INSERM U98-X, Claude Bernard University, Lyon France F-69008*

Roy L. Fuchs, *Monsanto Company, 800 North Lindbergh Blvd., St. Louis, MO 63167*

Richard S. Geary, *Isis Pharmaceuticals, Inc., 2280-B Faraday Avenue, Carlsbad, CA 92008*

Mark E. Groth, *Monsanto Company, 800 North Lindbergh Blvd., Saint Louis, MO 63167*

Bruce G. Hammond, *Monsanto Company, 800 North Lindbergh Blvd., St. Louis, MO 63167*

Scott P. Henry, *Isis Pharmaceuticals, Inc., 2280-B Faraday Avenue, Carlsbad, CA 92008*

Robert V. House, *IIT Research Institute, 10 West 35th Street, Chicago, FL 60616*

Yan Lavrovsky, *University of Texas Health Science Center, 7703 Floyd Curl Drive, San Antonio, TX 78284*

Thomas C. Lee *Monsanto Company, 800 North Lindbergh Blvd., Saint Louis, MO 63167*

Janet M. Leeds, *Isis Pharmaceuticals, Inc., 2208-B Faraday Avenue, Carlsbad, CA 92008*

Samuel B. Lehrer, *Tulane University Medical Center, 1700 Perdido Street, New Orleans, LA 70112*

Arthur A. Levin, *Isis Pharmaceuticals, Inc., 2208-B Faraday Avenue, Carlsbad, CA 92008*

Christine McCullum, *Cornell University, Comstock Hall, Ithaca, NY 14853*

Henry I. Miller, *Stanford University, Hoover Institution, Stanford, CA 94305*

David K. Monteith, *Isis Pharmaceuticals, Inc., 2208-B Faraday Avenue, Carlsbad, CA 92008*

Maurizio G. Paoletti, *Padova University, Via Treste 75, Padova, Italy I-35122*

David Pimentel, *Cornell University, Comstock Hall, Ithaca, NY 14853*

Gerald Reese, *Tulane University Medical Center, 1700 Perdido Street, New Orleans, LA 70112*

Arun K. Roy, *University of Texas Health Science Center, 7703 Floyd Curl Drive, San Antonio, TX 78284*

Patricia R. Sanders, *Monsanto Company, 800 North Lindbergh Blvd., Saint Louis, MO 63167*

Marvin Stodolsky, *U.S. Department of Energy, 1000 Independence Ave., S.W., Washington, DC 20585*

James E. Talmadge, *University of Nebraska Medical Center, 600 South 42nd Street, Omaha, NE 68198*

Mike V. Templin, *Isis Pharmaceuticals, Inc., 2208-B Faraday Avenue, Carlsbad, CA 92008*

John A. Thomas, *University of Texas Health Science Center, 7703 Floyd Curl Drive, San Antonio, TX 78284*

Thierry Vial, *INSERM U98-X, Claude Bernard University, Lyon, France F-69008*

Rosie Yu, *Isis Pharmaceuticals, Inc., 2280-B Faraday Avenue, Carlsbad, CA 92008*

Preface

The second edition of *Biotechnology and Safety Assessment* includes a combination of biotechnology-related topics. Some chapters emphasize the complexities of advanced molecular biology, including antisense therapeutics, the biological features of the cytokines, and triplex and ribozyme technology controlling deleterious gene expression. Important pharmacological applications of new agents may lead to further therapeutic success in the management of various disease states.

The safety assessment of biotechnology-derived agents represents a challenge to the toxicologist. Although the immunotoxicity of a potential agent must be fully evaluated, many other preclinical toxicology tests must be undertaken. A host of safety considerations are pertinent in evaluating biopharmaceuticals, many of which are either large protein molecules or glycoproteins.

Agribiotechnology is also extensively described in this edition. Genetically modified organisms can be used to increase the nutritional value and taste of many food products. Agribiotechnology-derived products can be environmentally friendly, as evidenced by the reduced use of herbicides and pesticides. The incidence of food allergens, whether from naturally derived foods or genetically engineered foods, may necessitate specific safety evaluations.

Regulatory and environmental considerations are of the utmost importance in the safety evaluation of either biotherapeutic agents or genetically engineered food crops. National and international agencies continue to evaluate biotechnology-derived products not only to insure their human safety but also to evaluate any potential impact on the environment. Many of these issues are addressed in *Biotechnology and Safety Assessment*.

Biotechnology and Safety Assessment presents a series of contemporary views on the importance of the advances in mammalian and plant genetics and shows how these advances contribute to better therapeutic agents and more wholesome foods.

Biotechnology and Safety Assessment, 2nd ed.
Edited by John A. Thomas
Copyright © 1998 Taylor & Francis

1

Genomic Information: Frontiers of Toxicology

James R. Beall and Marvin Stodolsky

U.S. Department of Energy, Washington, D.C.

A REVOLUTION IN BIOLOGY

Nearly a decade ago, a small group of biological researchers enabled by the U.S. Department of Energy (DOE) started a revolution in biological research. The revolution created a new paradigm for government support and management of biological research. Previously, most government agencies funded biological research on the basis of individual, investigator-initiated research grants or contracts. The new paradigm evolved from the DOE's long history of sponsoring and managing highly focused, multidisciplinary research intended to accomplish specific goals. This approach to research has a historical basis in DOE. It evolved from the Manhattan Project of World War II in which the target goal was the development of a nuclear bomb. Although subsequently this model was used by DOE in research areas such as nuclear energy and high-energy physics, it had never been used to support biological research. When the research community concluded that DOE should apply this paradigm to biological research and pursue the sequence of the human genome, the revolution started.

Although the revolution is young, it has grown beyond all expectation. It has spawned new industries and opened new doors in medicine, food production, and waste remediation. It is affecting nearly every aspect of biology, including the fundamental understanding of biomolecules, diseases, and toxicity. It will soon begin to alter perceptions of risks affected by these factors. To begin to comprehend the impact that this revolution will have on the field of toxicology, the Society of Toxicology sponsored a symposium on the topic at its 1997 annual meeting (Beall 1997a). This chapter is based in part on presentations made at that symposium.

In 1997, DOE celebrated 50 years of supporting biological and environmental research. The DOE's research programs affect nearly every sector of society. Major DOE contributions include improvements in understanding the hazards and risks associated with the exposure of humans to radiation and chemicals.

In cooperation with Japan since 1947, DOE has been studying the potential long-term effects of radiation on survivors of the World War II atomic bombings at Nagasaki and Hiroshima. The people of concern include the survivors of those bombings, their children, and future generations. The studies involve both radiation dosage analyses and extensive use of model animal systems. Through the years, each gain in new knowledge about radiation, chemicals, and biology has been based on advances in new technology.

In 1984, as part of the joint U.S. and Japanese program, a DOE representative, Dr. Mort Mendleson from Lawrence Livermore National Laboratory (LLNL), attended a meeting in Hiroshima. The meeting summarized current knowledge about the effects of radiation and identified future cooperative research priorities. The people who were exposed to radiation from atomic bombs in World War II are aging. If cell lines from these individuals are not soon preserved, an irreplaceable resource will be forever lost. That fact and changes in modern biotechnology development stimulated a reassessment of research priorities at the 1984 meeting. Meeting participants agreed that developing new DNA analytical tools was the second highest priority for human mutation research, after establishing and preserving cell lines from the bombing survivors (Cook-Deegan 1989).

A follow-up meeting to assess the status of DNA technologies was organized by Drs. David Smith and Mort Mendleson and others. It was sponsored by the DOE and the International Commission for Protection Against Environmental Mutagens and Carcinogens and was held in Alta, Utah. Scientists on the front lines of DNA technology were asked; Is the state of the art of DNA technologies adequate for detecting exogenously induced mutations in humans? This capability was needed to determine if any increase had occurred in the mutation rate among the atomic bombing survivors of Hiroshima and Nagasaki (Cook-Deegan 1989).

By the last day of the Alta meeting, participants had decided that current methods were not adequate to quantify induced mutations in humans. They agreed that a program to develop such methods was desirable and possible. However, some speculated that it would never be possible to measure induced mutations in humans unless the DNA sequence of a normal human genome was available as a control. Those five days at Alta inspired the beginning of the Human Genome Program (HGP) (Segal 1996).

An observer from the Office of Technology Assessment (OTA) of the U.S. Congress attended the meeting in Alta. He understood the power of the newly identified DNA research directions that were suggested at that meeting. The Office of Technology Assessment also knew the potential of DOE to implement and manage such research. Soon after the Alta meeting, an OTA report suggested that DOE should take the leadership role in pursuing new DNA technology development (U.S. OTA 1986). Charles DeLisi, then DOE's Associate Director in the Office of Health and Environmental Research, read the OTA report and responded to the research challenge it posed for DOE.

By 1984, DOE's national laboratories had already achieved unique capabilities to sort chromosomes physically. This ability allowed scientists to subdivide the tremendous tasks of analyzing the entire genome into more manageable tasks of working with the 24 chromosomes individually. Knowing this, DeLisi commissioned a 1986 conference in Santa Fe, New Mexico, to analyze the feasibility of an initiative in the research areas described in the OTA report.

Scientists attending the 1986 Santa Fe conference reported that an intensive program of resource and technology development could achieve the highly desirable goals of developing a high-resolution map of the human chromosomes and subsequently determining the complete DNA sequencing of the ordered DNA clones that make up the map. The researchers also suggested that interpretation of the genome code itself would proceed concurrently with determination of the code. Developing the code was expected to remain a productive research endeavor well into the twenty-first century. DeLisi then initiated a DOE Human Genome Project by redirecting 5.5 million dollars from the DOE Health Effects research program. The funding was to be used to develop critical new resources and technologies.

In 1986, DOE requested money from Congress to support the human genome project (Fig. 1). Congress sanctioned the project by creating a separate line item for a Human Genome Project (HGP) in DOE's fiscal year (FY) 1987 budget (Fig. 1). The following year the National Institutes of Health (NIH) consolidated several research grants involving genetic diseases and DNA technologies and thus started the NIH HGP. Scientists and managers of both agencies felt the dawn of a new era of biomedical research and soon agreed to coordinate their human genome research programs.

The DOE's initiative stimulated much debate in the biomedical and biotechnology communities. An influential National Research Council report recommended that the HGP not only proceed but that mapping and sequencing of the genomes of several model organisms also proceed. The report recommended that funding for the HGP progressively rise to an annual sustained level of $200 million. Figure 1 shows the actual DOE and NIH funding through FY 1997; it excludes support by other agencies.

The HGP rapidly attained international recognition and support. The international Human Genome Organization (HUGO) was soon formed as a nongovernmental coordinating body. The federal research commitment stimulated private sector investments in the supporting resources and technologies. These created new opportunities in related sciences, including areas such as bioengineering, protein-based pharmaceuticals, instrumentation, specialty research reagents, software development, and medicine (Fig. 2). The promise was so great that investments by the pharmaceutical industry alone soon exceeded federal HGP expenditures (Fig. 3).

As with the broader field of biology, the HGP data are revolutionizing the way research in toxicology is conducted. They are also revolutionizing the technologies that are needed to conduct the research. The data and technologies are conducive to designing better animal models, to harvesting new information from older studies, and to building bridges among species and across disciplines (Beall 1997a, b). The new tools that create cloned gene resources are transforming research in several areas of toxicology. These include mutagenicity, teratology, and the knowledge of individual susceptibilities to toxic agents and diseases (Beall 1997b). Soon HGP data and technologies will impact the ways in which hazards and risks are determined and perceived. The examples in the following sections represent but a small selection of many that could be chosen to illustrate the HGP contributions to clearer understandings of toxicology and disease and to the promise of improved therapies.

U.S. Genome Funding

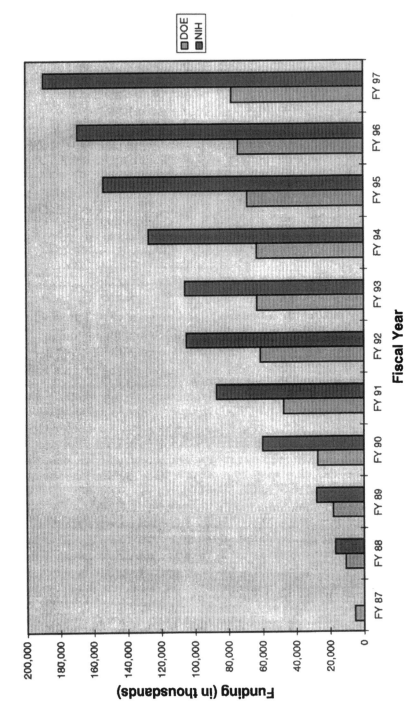

FIG. 1. The Department of Energy's and the National Institutes of Health's financial support of the Human Genome Program by fiscal year from 1986 to 1997. Amounts are in thousands of dollars.

U.S. Biotechnology Impacts Diverse Markets
(1995 Sales)

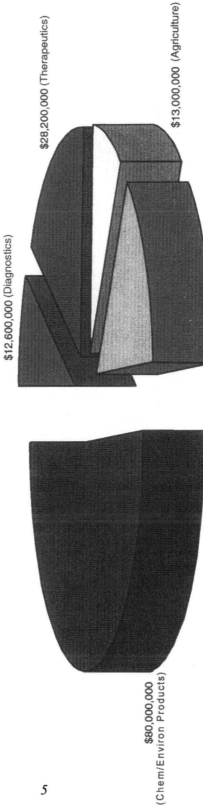

$28,200,000 (Therapeutics)

$13,000,000 (Agriculture)

$12,600,000 (Diagnostics)

$24,000,000 (Supplies)

$80,000,000
(Chem/Environ Products)

FIG. 2. Commercial investments in genomic biotechnology research by the pharmaceutical industry from 1993 to 1996.

5

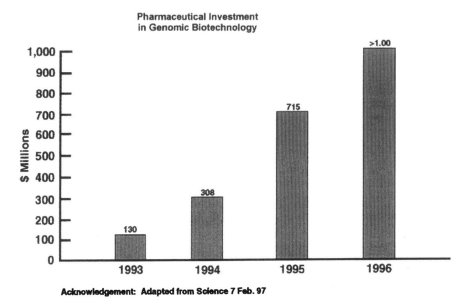

Pharmaceutical Investment
in Genomic Biotechnology

Acknowledgement: Adapted from Science 7 Feb. 97

FIG. 3. Dollar value of biotechnology products in different market sectors during 1995.

Mutagenicity

Subtle mutations, such as point mutations, may affect protein structure and function. Greater damage to chromosomes, such as breaks and deletions of chromosomal material, often leads to cell death. Although both types of alterations in chromosomes are valuable in understanding toxicity, point mutations that occur in embryos or germinal cells are often more informative because they illuminate the role of genes in toxicological responses. This is because the affected cells and embryos often survive and subsequently reproduce. Cellular and embryonic survival permits the effects of the change on normal function or behavior to be studied in greater detail.

A continuing goal in mutation research is finding more economical ways to produce and identify mutations. The need for economy is easily seen when one considers that annual costs of maintaining large animal colonies to study in vivo mutations, such as DOE's famous "mouse house" at Oak Ridge National Laboratory (ORNL), total millions of dollars.

While researching point mutations at ORNL in the 1970s, William Russell discovered that ethyl nitrosyl urea (ENU) is an extremely potent mutagen during spermatogenesis. It produced a tenfold increase in mutagenicity as compared with other agents that generated germline point mutations in the mouse. Russell found that male mice that were given an optimized ENU dosage harbored a large number of point mutations; they occurred in about 1 in 600 genes.

An ENU-treated male can give rise to about 800 offspring per year. For every 10 mice treated, ENU produces approximately 1 mutation per year, per locus, per gene. Thus, treating 100 male mice per year yields about 10 different mutations per year, per locus, per gene. Because of this high rate of inducing mutations, the development of the ENU "super mutagen" system for the mouse was economical enough to permit "saturation" mutagenesis. The goal of saturation mutagenesis is to reveal every gene in a target chromosomal region as a mutation. Before Russell's achievement, saturation mutagenesis was economically feasible only for simpler animal models up to the genomic complexity of the fruit fly (*Drosophila melanogaster*).

The capacity to localize mutations under deletions within a single generation of mice bypasses the need to do extensive and expensive crossbreeding of animals, as is done when in vivo genetic linkage methods are used. The deletion mapping methods do not incur the greater attendant costs of housing and managing a large breeding population of mice.

Complementary and efficient systems were soon developed to recognize chromosomal regions in which the point mutations occurred. This was a preliminary step toward cloning the gene itself and provided a detailed characterization of the region through DNA sequencing. Mouse mutants with large deletions are very useful for identifying the approximate location of genes in chromosomes. An ENU-induced mutation will often not be physiologically manifested in a heterozygote. But if the F_1 generation is cross-bred with a mouse having a large deletion within a chromosome, a recessive mutation does become manifested when it is located within the genetic span of the deletion on the paired chromosome. Once a mutation is roughly localized under a large deletion, it is possible to refine the localization by cross-breeding with mouse strains having smaller deletions in that chromosomal region.

The capacity to do such cross-breeding within a generation, and against mice having such deletions, reduces the need to do the more traditional and expensive multiple cross-breeding within a generation to locate a mutation by classical genetic linkage methods. The new methods do not incur the greater attendant costs of housing and managing a large breeding population of mice.

Before the advent of genetic engineering technologies, X rays or radiometric agents were used to produce deletion mutations at random loci. More recently, highly efficient engineering techniques have been developed for producing clusters of overlapping deletions around a chosen locus. One approach is based on the *Cre* and *lox* system of the bacterial virus P1 (Chambers 1994). In this system, the *Cre* protein acts at chromosomal sites called *lox*, and when two *lox* sites are in tandem, the DNA between the two sites is deleted by the *Cre* protein's action. Introducing a single *lox* site into the mouse genome is generally harmless, and the position of such added foreign DNA on mouse chromosomes can readily be determined. When the *Cre* protein is present, subsequent deletions beginning at the first *lox* site are induced whenever a second *lox* is integrated into the same chromosome (Baubonis and Sauer 1993). In this way, clustered deletion mutations are obtained easily and economically. An advantage of producing deletions in this manner, in contrast to the older X-ray inductions, is that one knows where the deletions will occur. Deletion families produced in this manner are very useful for refined gene and mutation mapping.

Efficient mapping of mutations depends on simple means of recognizing when a mutation is present. Several methods have been devised. One such system uses genes governing coloration patterns in the coats of mice. Several genes affecting coloration now have been well characterized and can be combined to produce coat colors ranging from dark brown, to yellow (agouti), to albino. Through the appropriate combination of coloration genes, the presence of a mutation under a deletion locus is simply manifested as a change in the coat color among the offspring.

An important feature of this method of studying mutations is that if a particular yellow class of mouse (agouti) is not produced from crossing members of the F_1 generation with the mutation, it is because an embryonic lethal gene caused its premature death. Once this is recognized, follow-up cross-breeding of mice that produced the homozygous lethal genes can be done to characterize the embryonic defects caused by the mutations physiologically.

The older ENU mutagenesis techniques, which have now been complemented by the newer genomic tools, are expected to revolutionize the capacity to scan mice for mutations economically. As more mice are scanned, it is expected that many new and interesting developmental mutations will be found. In toxicology for example, having access to extensive new mice genotypes will allow teratologists to learn if a given teratogen is acting on particular gene products that are crucial for the normal development of a specific organ, tissue, or function.

Teratology

Teratology, the study of developmental abnormalities, is a particularly difficult area of toxicology to understand at the mechanistic level. This is partly because embryonic differentiation and growth require genes, enzyme systems, hormones, and nearly every other genetic and molecular control to be activated and deactivated in specific chronological orders.

Traditional teratology studies for regulatory purposes involve administering an experimental substance of interest to pregnant laboratory animals and detecting malformed offspring. These studies are empirical by design and thus produce largely empirical results. They contribute few answers to questions about mechanisms involved in normal or abnormal development. Novel genetic technologies and information from the HGP are providing new ways to study the mechanisms of abnormal development.

By the late 1980s, researchers were finding important genetically inducible mutations that affect limb development. For example, scientists at the University of Cincinnati found a gene that produced completely truncated hindlimbs but left the mice with normally developed forelimbs (McNeish et al. 1988). Findings such as these were bringing the fields of teratology and genetics closer together.

In the early 1990s, Jack Bishop (National Institute of Environmental Health Sciences) organized a series of workshops to bring developmental geneticists and teratologists together to focus on understanding developmental mechanisms (Bishop and Kimmel 1997). The workshops recognized the value of the outputs from the HGP to the field of teratology and catalyzed the integration of knowledge between molecular embryology and developmental toxicology. Participating developmental geneticists and genomic researchers discussed ways to find the genes that orchestrate development, and teratologists sought ways to identify genetic targets that alter development.

The workshops produced a recommendation that science develop large-scale, cost-effective means of establishing the function or role of a gene in development along with its genetic code. According to participants, "Such means could include the use of genetic models generated via targeted mutagenesis and gene-trap approached with sequences identified through the Genome Project, as well as exploitation of new and existing mutations" (Bishop and Kimmel 1997). An example of the kind of research that could benefit from the HGP, as suggested by the workshop, includes that conducted by Dr. Rick Woychik's group at ORNL.

Dr. Woychik's group recently recognized a mutation that causes the fore- and hind-limbs of mice to develop abnormally. These animals had fused digits that made the paws appear to be clubbed or hyperflexed (as if the paw turned back on itself) (Woychik 1997).

Skeletal staining revealed that the radius and ulna were fused in the forelimbs, and in the hindlimbs the tibia and fibula appeared as a single triangular bone. Significant disruptions were seen in the carpals, metacarpals, tarsals, and metatarsals. In contrast, the mutant's axial skeleton was similar to that of the normal wild-type littermate.

Woychik's group searched for the gene that caused the abnormalities. Once the gene was mapped to a chromosomal locus, researchers were able to clone it and called it the legless (lgl) gene. This gene attracted considerable attention from molecular geneticists and toxicologists because it was one of the first examples of a specific gene mutation causing limb abnormalities.

The animals produced by Bishop's and Woychik's groups of researchers provide excellent model systems to study normal and abnormal development. The *lgl* insertional mutation is a useful tool for analyzing specific events in limb development. Retinoic acid is a well known teratogen. In the *lgl* system, the mutant embryos are quite sensitive to the effects of retinoic acid, which indicates that they may make an excellent system for studying other teratogens that affect limb development and other developmental systems (Singh et al. 1991). Mice with the *lgl* gene also provide a useful gene-target system for studying mechanisms of teratogenicity for agents such as thalidomide.

More recently, ORNL scientists identified another mutant gene that causes structural alterations in both the fore- and hindlimbs. The manifestations of its expression include the equivalent of a double thumb, double big toe, and detached fibula.

In the preceding examples, mutations were produced by the process called insertional mutagenesis in which the foreign DNA is injected into pluripotent cells that will become part of a developing embryo. The inserted DNA may disrupt genes to produce mutations with measurable effects (manifested). In addition, the inserted DNA later serves as a "hook" for retrieval of the adjacent normal DNA. Genes that are identified by this method are readily cloned, and the DNA therein can easily be sequenced. Point mutations that have long been studied in relation to teratogenic mechanisms give rise to interesting genotypes, but unlike inserted mutations they do not provide direct information about, or access to, the affected DNA or a particular gene.

To elucidate the roles of a particular gene, the insertional DNA ends can be designed so that integration occurs preferentially at homologous sites in chromosomes. During integration at a homologous locus, the corresponding segment of normal DNA is replaced by a nonfunctioning segment that in essence produces a "knockout" mutation eliminating the useful DNA in the locus. In "knockout" protocols, the role of the "knocked out" parts of a gene is subsequently deduced by the phenotypes of the progeny. There are ways to gain even more information about the affected portion of DNA.

The "knockout" procedure can be designed to be accompanied by a "knock-in" of useful DNA structures. In particular, a reporter gene such as β-galactosidase can thus be introduced with expression under control of the normal gene's promoter for RNA transcription. In a developing embryo, the tissue that is expressing the reporter gene turns blue when X-gal reagent is added, whereas at the same time the complementing chromosome supports normal physiology (Rossant and Nagy 1995). "Knock-in" reporter gene experiments are elucidating the gene's timing and function in development mechanisms (Nagy and Rossant 1996). This system helps illuminate genetic damage done by teratogens that affect normal development.

With these technologies, an important gene involved in early neural embryonic development was found. Crossbreeding of the heterozygotes of mice with the mutations yielded some "knock-in" homozygotes. They had a disorganized neural tube and ex-encephaly (Woychik 1997). Another "knock-in" heterozygote assayed as 7.5 embryos gave a distinctive pattern of staining in the node, somites, foreplate, and the limb buds.

The reporting "knock-ins" are an extremely useful way of making and characterizing mutations. In Woychik's experiment, for example, the afflicted homozygote expressed the mesenchyme that was not undergoing the normal apoptosis during development. In this case, the apoptosis was necessary for forming normal digits. The HGP has greatly advanced these approaches by rapidly providing the sequence information on which these "designer" methodologies depend.

In another study at ORNL, a gene was found to have 10 different repeated sequence motifs. One such repeated motif has homologues in lower eukaryotes in genes that control the cell cycle. That information alone does not reveal enough about the gene's purpose or function in humans, but it certainly is intriguing. This finding also illustrates the way in which genomic information from diverse species brings synergistic benefits. From a toxicological perspective, it is important to know the roles of the gene and its motifs in normal development as well as the effects of toxic agents that alter or affect the gene.

A next step in the process of making more extensive use of the genomic data is to implement faster throughput methods of designer mutagenesis. Faster throughput methods will help make it more economical to determine the genes and subsections wherein mutations give rise to particular developmental phenotypes.

Individual Susceptibility and Response to Toxic Agents and Disease

For five decades, research on the biological and medical implications of the genetic code has monopolized biophysics, biochemistry, and molecular biology (Murphy and Pyeritz 1995). That research has produced remarkable progress and great successes. Still, understanding the roles of genetics in complex diseases such as cancer remains a difficult challenge. As more and more cancer-related genes are identified in the HGP, the mechanisms of cancer formation and mitigation become clearer. In a broader sense, as disease-related genes are recognized, comparisons across the human population can be made to begin correlation of disease phenotypes and particular gene mutations.

One research goal is that individual variations in DNA will soon be used as indicators of susceptibility to toxicity and disease. The general concept is that mutations in some specific genes will affect individual response to specific toxic agents. The consequences of mutations are not always clear or predictable. For example, LLNL scientists found significant variability in several closely related DNA repair genes but have thus far not found a correlation of changes in the amino acids of the associated proteins with specific phenotypes (Thompson 1997).

Because of their importance in maintaining genome integrity and preventing progressions to cancer, DNA repair genes are of particular interest in mutation research. Dr. Larry Thompson and others at LLNL have cloned several genes that function in DNA repair pathways. These genes have been mapped, the cDNAs have been isolated, and, in many cases, the genes have been sequenced (Thompson 1997). The protein functions that were initially recognized as contributing to chromosomal repair include helicases, endonucleases, exonucleases, polymerases, and ligases. Other enzymes are involved in cell cycle staging. The functions of many others remain unknown.

About 25% of individuals examined have protein-altering variations in the three DNA repair genes studied: *XRCC1*, *ERCC2*, and *XRCC3* (Thompson 1977). Therefore, there may be a sizable population for studying the meaning of these variations. Can science now link this information to people or subpopulations who have increased susceptibility to certain chemicals, diseases, or other factors? Those are the key questions, and they offer a next step in the research.

The gene *ERCC2* provides an interesting example of a gene that, when altered in various ways, imparts variable expressions in different individuals. Disorders now traced to this gene are xeroderma pigmentosum, Cockayne's syndrome, and trichothiodystrophy (Takayama et al. 1996a). Most severe is xeroderma pigmentosum Group D (XP-D), that is characterized by high sensitivity to ultraviolet light, a high cancer incidence, and some neurological disorders. Cockayne's syndrome is marked by slight photosensitivity without a high cancer incidence. Cockayne's syndrome also includes more severe neurological defects than those of XP-D (Takayama et al.

1995). Trichothiodystrophy is characterized by a defect in sulfur metabolism. Because of low sulfur content, the hair of these individuals is brittle, and the skin is pale. The condition is manifested by slight photosensitivity in about 50% of patients and some minor neurological defects, but its severity varies (Takayama et al. 1996).

Chris Webber and collaborators at LLNL identified ERCC2 mutations in afflicted individuals. They found mutated sites throughout the gene. Some mutations had a more severe effect than did others. Multiple mutations in a given region tend to cause the most severe diseases. However, there tends not to be a specific mutation clustering along the region of the gene. The ERCC2 protein is part of a transcription factor complex made of several proteins. How each protein folds and interacts in that complex may affect the transcription specificity of the complex and the phenotypes of the three disorders.

Migraine is another example of a disease that is elicited by exogenous and endogenous factors and involves individual susceptibilities to these factors. About 24% of females and 12% of males are affected by migraines. Some believe that the condition is the largest reason for people in the United States to seek outpatient health care. Those who suffer from migraines know that certain foods, alcohol, and environmental stresses initiate attacks and have learned to avoid these factors. Although this disease has both strong genetic and environmental components, neither component is well understood. However, the HGP is providing new insights into the genetic component of some forms of migraine.

The hemiplegic form of migraine is often associated with a distinct disorder called episodic ataxia. Some forms of episodic ataxia are associated with calcium channel genes. A clue found in searching for this defect and its gene was that familial hemiplegic migraine is associated with episodic ataxia. Thus, this form of migraine may be associated with an altered calcium channel gene (Ophoff 1996).

Scientists at LLNL in cooperation with researchers in the Netherlands used genetic linkage analysis to implicate a region of chromosome 19 with alterations in calcium channels. Within the deep contigs of chromosome 19 DNA clones, LLNL scientists identified a candidate calcium channel gene. It was located within the 1–3-megabase long region associated with familial hemiplegic migraine. The calcium channel gene encoded the alpha-1 subunit of that gene. The alpha-1 protein is present in four copies in the calcium channel complex. Each subunit has six alpha helical turns spanning the membrane and a pore section. The pore section controls and forms a channel in the membrane.

Tissue samples were obtained from normal and afflicted family members, and their gene for the alpha-1 subunit was sequenced. Mutations associated with both hemiplegic migraine and episodic ataxia were found in this single gene. For familial hemiplegic migraine, several mutations were within the alpha helical units as well as with the pore unit. Mutations associated with episodic ataxia in the same region correlated closely with the hemiplegic migraine in these families. This represents an example of an understood but complex genetic defect. It is also remarkable that the same calcium channel is involved in 5-hydroxytryptamine (5HT) release. Studies of the pharmacological role of 5HT in migraine would be informative. Some environmental factors seem to trigger the onset of migraine. Another important area for study is the relationship between those environmental factors and alterations in the gene. When completed, research in these areas will help provide an understanding of environmental toxicity and create new options for preventing migraine.

In a recent book, *Principles of Nuclear Medicine*, Dr. Tony Murphy stated, "The ubiquity of physiological homeostatic processes explains why classical and population genetics have done so little to clarify the causes of human disease and may have even obscured genetic components" (Murphy and Pyeritz 1995). Homeostatic processes operate in all diseases and can mask the disease if the homeostatic process itself is not measured, which we can now do at the molecular level with these radioactive tracers. Murphy illustrated this point with an example involving drug toxicity.

As a result of exposure to a toxic chemical, adriamycin in this example, the *MDR*1 gene produces the p-glycoprotein, which functions by extruding the chemical out of the cell. This fact was discovered when physicians noticed that cancers that developed resistance to a single drug such as adriamycin would automatically have resistance to other drugs in the chemotherapeutic arsenal (Wagner 1997).

One can produce a phenotypic marker of the expression of the *DMR*1 gene with a widely used radiopharmaceutical called Tc 99m sestamibi. It is labeled with the radionuclide technetium 99m (Tc) (Piwnica-Worms et al. 1995). This agent was originally conceived to be used in single-photon-emission computed tomography to study the heart. There are several lipophilic cation markers of this type labeled with technetium 99m.

As a practical application of the marker in oncology, uptake of the Tc 99m sestamibi tracer in tumors is measured. If the tumor contains a high concentration of p-glycoprotein, the Tc 99m sestamibi concentration will be low (Piwnica-Worms et al. 1993). By this method, it can be predicted whether the patient is likely to have a genetically based resistance to the beneficial effects of the chemotherapeutic agent (Del Vecchio et al. 1997).

After the patient is on treatment, one can determine if the p-glycoprotein is being expressed by its effect on the time course of the Tc 99m sestamibi tracer as it leaves the lesion (Luker et al. 1997; Chen et al. 1987). The effect is so predictable that the system has been suggested for use as a measure of response to chemotherapy (Bom et al. 1997).

A high uptake of Tc 99m sestamibi by tumor tissue meant that p-glycoprotein activity was low. Patients with low p-glycoprotein activities were found to have a complete remission. Conversely, when p-glycoprotein activity was high, Tc 99m sestamibi (as a surrogate marker for a chemotherapeutic agent) was pumped rapidly out of the cell, and patients were found to have no response to the chemotherapy.

Whole-body dynamic scintigraphy after injection with Tc 99m sestamibi tracer reveals that the tracer accumulates in the liver and is gradually excreted into the gall bladder. Using this time-activity curve, and doing a quantitative analysis using mathematical models, one can get an idea of the activity of the p-glycoprotein, which in turn is reflected in the activity of the *MDR*1 gene.

Dr. Henry Wagner studied whether this quantifiable measure of the detoxification process would be distributed or clustered among people. Although studies are preliminary, he found only 1 chance in 10,000 of following a single normal distribution; some normal people had a very rapid clearance of the tracer, and others had a very prolonged clearance. This may be due to some unknown common environmental factors, but results suggest genetic differences in p-glycoprotein expression may be related to susceptibility to the effects of the toxic chemical.

A fundamental shift is occurring in the practice of medicine and thus in therapeutic agents and their potential for toxicity. Most medicine is based on the combination of history, physical examination, and laboratory findings with a foundation in pathological anatomy. Although these approaches will continue to be important, modern medicine now is looking also at molecular homeostasis related to a foundation in genetic pathology.

An example of the transition in the understanding of diseases characterized by a movement from seeing the disease phenotype to understanding its genotype involves Alzheimer's disease. In about 10% of the patients, Alzheimer's disease is probably related to a mutation of the *ApoE4* gene. The phenotypic marker, which can be used noninvasively in human beings and animals, measures regional cerebral blood flow or regional cerebral glucose utilization. A recent study by UCLA investigators shows a correlation between positron emission tomography (PET) imaging of a defect in glucose utilization and the *ApoE4* gene mutation. This provides a noninvasive way of looking at an abnormality that is related to a gene mutation in at least 10% of cases.

Molecular markers, when measured on a regional basis, can improve fundamental studies. When conventional neuropsychological tests are used to identify patients as having Alzheimer's disease, there may be a 35% error rate in diagnosis. But when a combination of anatomical imaging methods are used, such as computed tomography (CT) combined with nuclear imaging of regional cerebral blood flow, the error rate can be reduced to about 3%. Testing a potentially useful drug for Alzheimer's disease would be difficult if 35% of the patient study group did not have the disease. Thus, here we see the homogenization of patients for either genetic linkage studies, pharmacological studies, or diagnostic purposes.

More sophisticated techniques consist of having an image database of normal people, for example, in distribution of glucose utilization. In early Alzheimer's disease, these images can be subtracted from the database of normal people, and differences in the images can thus be obtained. When this was done in a study from the University of Michigan, the posterior singlet gyrus was found to be the most sensitive indicator of Alzheimer's disease, not only in diagnosed patients but also in their family members. Going beyond measurements of blood flow and glucose utilization is the study of specific enzyme patterns such as acetylcholine esterase.

Studies of Alzheimer's patients with particular tracers that reflect muscarinic cholinergic neurotransmission have not been very useful. However, with nicotinic acetylcholine receptors, there are gross abnormalities. These tracers have just been developed. They show how studies can progress from blood flow to energy metabolism to the neurotransmission process itself.

A powerful way in which genome reagents can be used for the toxicology experiments is in obtaining differential gene-expression profiles. This approach has long been in use at the protein level (i.e., comparisons of two-dimensional protein fractions), but this method's beneficial extension to the messenger RNA level is perhaps more powerful. Before continuing with mRNA, however, a brief discussion of cDNA technology may be worthwhile.

The great value of the protein coding information of genes has long been recognized. This information can most quickly be obtained from sequencing the stable cDNAs derived in vitro from the cell's unstable messenger RNAs. Once a protein's code is displayed, diagnostic applications and enhanced genetic engineering are more easily done.

There are numerous coordination and technical difficulties in operating a cost-efficient cDNA research program. Following a series of planning workshops, DOE initiated a cDNA mapping and sequencing sector of the HGP in 1990. The economic value of knowing the most commercially important cDNA "jewels" soon caught the attention of the investors in venture capital. Investments by private companies quickly exceeded those of the public sector HGP. Merck, Inc., and the Howard Hughes Medical Foundation have funded the systematic acquisition of cDNA sequences for public databases for humans and the mouse, respectively. (Current information on these projects can be accessed through the World Wide Web at http://genome.wustl.edu:80/est/ esthmpg.html and at http://genome.wustl.edu/est/mouse_esthmpg.htm)

The Institute for Genomic Research (TIGR) made public a large cDNA data set. Even without having the full (and much more expensive) sequence of a cDNA, a short sequence can serve as an expressed sequence tag (EST) identifier to discriminate one cDNA from other cDNAs. Craig Venter of TIGR first championed the use of short cDNA sequences that are the ESTs (Venter 1997).

Today some 80% of the estimated 60–80,000 human genes are represented as ESTs. Dedicated research programs to complete this representation are in place. The locations of a quarter of the human cDNAs have now been mapped to specific chromosomal loci that represent the sources of their progenitor mRNAs. The EST-distinguished cDNAs are being compactly arrayed by the thousands in an area no larger than a microscope slide. The specific binding of related nucleic acids to the tags can be disclosed by fluorescent microscopy technologies.

Toxicological Applications of cDNA Chips and RNA

Experimental animals are treated with the experimental agent, or a placebo, and their RNAs are collected and prepared at specific intervals from appropriate organs, tissues, or fluids and processed into fluorescence-labeled cDNA. These cDNAs represent the subpopulation of expressed genes that are of interest. After selective binding to their homologues on the cDNA chips, the spatially resolved binding sites are quantitated. Variations in gene expression due to treatments are seen (manifested) in massively parallel fashion as binding charges on cDNA chips representing all genes of concern.

The greatly reduced number of relevant genes can then be studied in more detail to ascertain the toxicological or pharmacological mechanisms of chemical effects better.

Genomic Computational Power as a Resource in Toxicology

In gene sequencing, as in other areas of science, data support, analytical systems, and computational powers are evolving rapidly as new analytical needs lead to technological improvements. Some of the new computational power flows from massive capabilities for doing cDNA analyses and provides new ways of understanding the products of altered DNA sequence. These capabilities in data analysis will provide new insights into toxicology and new ways to evaluate hazards and assess risks from exposures to chemicals.

Computerized comparisons can be made to identify homologous, already sequenced genes and cDNAs whose functions are known. Genomes of several bacteria and yeast have been completely sequenced. The nematode and the fruit fly genomes will soon follow. Functional identifications of the proteins are proceeding very rapidly for the sequenced genomes of the organisms. Computerized homology searches within and across species are beneficial for assigning candidate protein functions for newly sequenced human and mouse genes or corresponding cDNAs.

Research on colon cancer provides an interesting example of the use of computational EST methods to recognize important genes. Because the incidences of colon cancer vary among countries with significantly different diets, many scientists believe it has an environmental (dietary) basis. It also has a genetic component. Bert Vogelstein (Johns Hopkins University) and colleagues found a genetic basis for, and then cloned, a DNA mismatch repair enzyme that explains about 8% of colon cancer cases. Craig Venter compared its DNA sequences against the EST database (Venter 1997). By use of these databases, over the course of a week related ESTs were found and mapped onto a specific chromosomal locus. A second colon cancer locus was also thus identified. Within another week, the complete coding sequences for the DNA repair genes were identified. Within 6 weeks, Volgelstein's group correlated mutations in these key genes with the development of colon cancer in their patients.

Similar results were also achieved after the finding of the first Alzheimer's disease gene. As soon as a chromosome 14 early-onset Alzheimer's disease gene was found, a quick search of cDNA databases came up with a cDNA sequence homology representing only a few amino acid differences in the two proteins. This messenger RNA sequence was known to originate from a region of chromosome 1. The Darwin Molecular Company in Seattle had isolated genomic DNA clones from this region of chromosome 1. Within 6 weeks of the chromosome 14 gene finding, the Darwin group published a paper on the chromosome 1 Alzheimer's gene.

As the focus on the use of these data turns more to understanding changes in phenotypic expression, genomic data will become even more useful to toxicology. For example, pheochromocytoma cell-culture systems provide useful models for studying neuronal development and the effects of exogenous agents on that development. In this system, significant differences can be seen between untreated cells and cells that were treated with nerve growth factor that causes neurite growth. Among the thousands of sequenced cDNAs from libraries representing many tissues, Venter's group was able to locate the growth-controlling genes. Their representative cDNAs are abundant in cells with high catecholamine synthesis and storage. Chemicals that affect these cells may also affect neurite growth.

Early planning of the research approach for the genome project was based on the bacteriophage model for sequencing DNA that was in the literature by 1982. The bacteriophage lambda genome with 48,000 base pairs had been sequenced. This was done with a process involving several individual sequence "reads" of only a few hundred subunits, to cover a few kilobase lambda fragment. The model suggested that before actual sequencing was started it would be necessary to map out the serial order of such small fragments to represent the chromosome of interest and thus enable proper assembly of the entire sequence. This mapping prerequisite had less to do with sequencing technology than it did with an imposition created by the existing software. At that time, the software could not manage more than 1000 sequence fragments at a time. Because scientists thought a 40–50-kilobase-long chromosome was all one needed to sequence at a time, more powerful computational software development had not been actively pursued. Later, needs arose to assemble fragments representing source DNAs of sizes of 500,000 and longer, such as those in whole bacterial genomes. Meeting these needs required new algorithms and computational approaches for analyzing and assembling larger fragments. Venter's group pioneered the development of software with the requisite capabilities (Venter 1997).

This example shows that computational biology has developed rapidly in concert with research needs and computer technology. Some scientists predict that desktop computers with the power of Cray systems will likely be available widely in the next few years. As will be discussed in the next section, combining the explosion of genomic data and advances in biotechnologies with new, powerful computer systems will catalyze even greater advances in all fields of biology, including toxicology.

A VIEW OF THE FUTURE

Making transgenic clones of laboratory species is becoming increasingly easy. With knowledge of the genome and these new cloning technologies comes the possibility of having colonies of cloned rats and mice that carry genes for human enzymes and metabolic systems. Such engineered animal models are opening new doors in biology, including new ways of studying diseases and toxicity and producing new therapies.

In 1996, about 5000 human genes had been reported in public databases. These data were from traditional studies that found proteins or isolated and cloned genes based on homologies. Finding genes at a much faster rate is a major justification for the genome project. Aside from discovering genes linked to genetic diseases, without the HGP it would have been well into the end of the next century before all the human genes had been found. Now this will be done in the next few years.

Studies involving animals are generating new genetic understandings in human susceptibilities. In comparing human and mouse responses to exogenous agents or disease processes, it is important to have gene probes that are useful for finding both human and mouse clones. With these it is possible to sequence the clones containing the genes and then make them function biochemically, structurally, or physiologically by using "knockout" and "knock-in" technologies in mice. Many laboratories use these approaches.

The DNA repair gene XRCC1 illustrates a growing understanding of sequence homologies across species (Lamerdin et al. 1995). The sequence of the XRCC1 gene in the human contains 17 exons. There are also 17 highly conserved (95 to 100%) exons in the mouse for this same gene. Interestingly, the noncoding regions are also conserved (Lamerdin et al. 1995). What functions do these highly conserved, noncoding regions perform? Are they regulatory regions? These questions present primary targets for looking at subtle aspects of these particular genes and gene regions. There are other conserved noncoding regions that are also exciting targets for study.

Comparing different genomes will allow scientists to determine a more precise history of evolutionary events that led to all the separate species and genomes. Advances in understanding human physiology and human genes will significantly advance toxicological research.

Having access to the genes and knowledge of their map positions is revolutionizing the way science does functional genomic experiments. But, this cannot be done without automation. Robotic systems have been developed that create the high-density filters that can put 40,000 DNA or cDNA spots representing genomic DNA or cDNA clones on a filter in a few minutes (Carrano 1997). Hundreds of filters can be created automatically in a day. A new probe–cDNA can be used to interrogate the whole filter and display spots of homology. Use of these filters is the fastest way to find related clones and genes. Filters can be mailed anywhere in the world and probed by the receiving laboratories. Positive probe coordinates can be sent back to the laboratory that created the filters, and clones with the genes can then be retrieved (Carrano 1997).

Sequencing technologies are changing dramatically. The rate of DNA sequencing is likely to increase at least ten- to thirtyfold over the next 3 to 4 years. Incremental technologies alone will yield this kind of throughput, and truly revolutionizing approaches are soon to come. For example, technologies are being developed at LLNL that will use microchannel arrays for sequencing and totally automated refilling of reagents (Thompson 1997).

To help understand events at a crime scene, forensic toxicologists will soon be able to use polymerase chain reaction (PCR) technologies in the field. As part of the HGP, scientists at LLNL recently developed a portable PCR machine contained in a small suitcase that can do PCR analyses in 20 min for either a single gene or a complex set of genes. The device's main component is a silicon chamber that was developed at LLNL (Caranno 1997). To take this technology even further, the LLNL scientists built a portable PCR device that has not been commercialized yet. It will have the same cycle time and operate on just four 9-volt batteries. These are only some technologies that are available or will be available in the next few months. Toxicologists will be among the communities of scientists who benefit from these technologies.

Recently developed computer software and more powerful computers are offering new ways of extracting meaningful information from massive amounts of new data. A new system called Spatial Paradigm for Informations Retrieval and Exploration (SPIRE) illustrates such an advance in analytical software. Developed as part of the nation's defense programs, the SPIRE relational analysis system can search tens of thousands of documents, analyze the frequency of word usage, find relational themes in the use of the word in context in the documents, and display in star cluster or terrain maps as groups of documents that are related in theme, hypotheses, or other elements derived from the search. When these new systems are combined with traditional ways of viewing data (numbers, graphs, and charts), the total content of information can be displayed in landscapes of relational themes called "Themescapes" or informational "Galaxies" (Thomas 1997). This technology quickly reveals voids and clusters of related information within massive information spaces in seconds.

Such relational software analyses, when combined with high-level computation (such as permitted by Cray computers), will produce new insights into, and understandings of, the effects of toxic agents on genomes, protein production, and numerous other biological systems and will assemble information from documents. Cray-level computing power in desktop computers will permit this powerful software to be applied to data such as those from cDNA chips containing tens of thousands of ESTs for specific genes (or vast arrays of proteins separated in two-dimensional gel analyses). This technology will permit analysis of genetic sequence data in ways that will form molecular associations between animal and human responses to exogenous agents. When combined, these technologies will yield new understandings of toxicology and produce meaningful risk assessments that will be unbelievably powerful and relevant.

To gain a useful perspective on the ways in which these new advances could affect future health-related research, DOE organized a workshop that was held on 24 July 1996. It was modeled after the Alta meeting held a decade earlier. In the 1996 meeting, participants were asked to address four questions:

1. What are major research opportunities or needs in biological research today that can take advantage of advances in genomics, structural biology, and informatics–the computing power of DOE?
2. What outstanding limiting factors need to be addressed to open up these research opportunities?
3. How can the national laboratories be used to facilitate research?
4. What are the best opportunities for creating partnerships between the national laboratories and the industrial sector to overcome research limitations?

It is easy to imagine that answers to these questions will greatly impact the future of biology and toxicology—perhaps even more than did the answers from the Alta workshop that led to the human genome program.

ACKNOWLEDGMENTS

This chapter is a product of the Department of Energy and is based in part on a Society of Toxicology Symposium entitled Genomic Information as a Frontier of Toxicology, which was chaired by James Beall (Department of Energy) and held March 10, 1997, in Cincinnati, Ohio. Symposium speakers were Craig Venter, Anthony Caranno, Henry Wagner, Rick Woychik, and James Beall. We thank the speakers for use of their symposium materials and Ms. Betty Mansfield for transcribing the symposium proceedings. The valuable suggestions of Benjamin Barnhart, Sandra Morseth, Prem Srivastava, and David Thomassen are greatly appreciated.

REFERENCES

Baubonis, W., and Sauer, B. 1993. Genomic targeting with purified Cre recombinase. *Nucleic Acids Res.* 11(9):2025–2029.

Beall, J. R. 1997a. Genomic information as a frontier of toxicology: Building bridges in biology. In *The Toxicologist. Fundamental and Applied Toxicology*, Supplement V, 36(1, part 2):1 (abstract 1).

Beall, J. R. 1997b. Genomic sequences: Foundations for new directions in toxicology. In *The Toxicologist. Fundamental and Applied Toxicology*, Supplement V, 36(1, part 2):1 (abstract 6).

Bishop, J., and Kimmel, C. A. 1997. Molecular and cellular mechanisms of early mammalian development: An overview of NIEHS/EPA developmental toxicity workshops. *Reprod. Toxicol.* 11(2/3):285–291.

Bom, H. S., Kim, Y. C., Lim, S. C., and Park, K. O. 1997. Dipyridamole modulated Tc 99m sestamibi scintigraphy: A predictor of response to chemotherapy in patients with small cell lung cancer. *J. Nucl. Med.* 38:240.

Caranno, A. V. 1997. Genomic and DNA sequencing technologies: Implications for toxicology and health research in the 21st century. In *The Toxicologist. Fundamental and Applied Toxicology*, Supplement V, 36(1, part 2):1 (abstract 3).

Chambers, C. A. 1994. TKO'ed lox stock and barrel. *Bioassays* 16(12):865–868.

Chen, C. C., Meadows, B., Regis, J., Kalafsky, G., Fojo, T., Carrasquillo, J. A., and Bates, S. E. 1987. Detection of in vivo P-glycoprotein inhibition by PSC 833 using Tc-99m sestamibi. *Clin. Cancer Res.* 3:545–552.

Cook-Deegan, R. M. 1989. The Alta summit December 1984. *Genomic* 5:661–663.

Del Vecchio, S., Ciarmiello, A., Potena, M. I., Carriero, M. V., Mainolfi, C., Botti, G., Thomas, R., Cerra, M., D'Aiuto, G., Tsuruo, T., and Salvatore, M. 1997. In vivo detection of multidrug resistance (MDR1) phenotype by technetium-99m-sestamibi scan in untreated breast cancer patients. *Eur. J. Nucl. Med.* 24:150–159.

Lamerdin, J. E., Montgomery, M. A., Stilwaten, S. A., Scheidecker, L. K., Tebbs, R. S., Brookman, K. W., Thompson, L. H., and Carrano, A. V. 1995. Genomic sequence comparison of the human and mouse XRCC1 DNA repair gene regions. *Genomics* 25:547–554.

Luker, G. D., Fracasso, P. M., Dobkin, J., and Piwnica-Worms, D. 1997. Modulation of the multidrug resistance P-glycoprotein: Detection with Tc-99m-sestamibi in vivo. *J. Nucl. Medi.* 38:369–372.

Murphy, E. A., and Pyeritz, R. E. 1995. Genes and disease. Chap. 3 in *Principles of nuclear medicine*, 2d ed., eds. H. N. Wagner, Z. Szabo and J. Buchanan, 18–27. Philadelphia: W.B. Saunders Co.

Mc Neish, J. D., Scott, W. J., and Potter, S. S. 1988. Legless, a novel mutation found in PHT1-1 transgenic mice. *Science* 241(12):837–839.

Nagy, A., and Rossant, J. 1996. Targeted mutagenesis: analysis of phenotype without germ line transmission. *J. Clini. Invest.* 98(11):S31–35.

Ophoff, R. A., Terwindt, G. M., Vergouwe, M. M., Van Eijk, R., Oefner, J., Hoffman, S. M., Lamerdin, J. E., Bulman, D. E., Ferrari, M., Hann, J., Lindhout, D., Van Ommen, G. B., Hofker, M. H., Ferrari, M. D., and Frants, R. R. 1996. Familial hemiplegic migraine and episodic ataxia type-2 are caused by mutations in the CA2+ channel gene CACNL1A4. *Cell* 87:4543–4552.

Piwnica-Worms, D., Chiu, M. L., Budding, M., Kronauge, J. F., Kramer, R. A., and Croop J. M. 1993. Functional imaging of multidrug resistant P-glycoprotein with an organotechnetium complex. *Cancer Res.* 53:977–984.

Piwnica-Worms, D., Rao, V. V., Kronauge, J. F., and Croop, J. M. 1995. Characterization of multidrug resistance P-glycoprotein transport function with an organotechnetium cation. *Biochem.* 34:12210–12220.

Rossant, J., and Nagy, A. 1995. Genome engineering: The new mouse genetics. *Natural Med.* 1(6):592–594.

Segal, Y. 1996. *The human genome project*, 1–31. Jerusalem: The Israel Academy of Sciences and Humanities.

Singh, G., Supp, D. M., Schreiner, C., McNeish, J., Merker, H. J., Copeland, N. G., Jenkins, N. A., Potter, S., and Scott, W. 1991. *Legless* insertional mutation: Morphological, molecular, and genetic characterization. *Genes Dev.* 5:2245–2255.

Takayama, K., Salazar, E. P., Broughton, B. C., Lehmann, A. R., Sarasin, A., Thompson, L. H., and Weber, C. A. 1996. Defects in the DNA repair and transcription gene ERCC2(XPD) in trichothiodystrophy. *Am. J. Human Genetics* 58:263–270.

Takayama, K., Salazar, E. P., Lehmann, A. R., Stefanini, M., Thompson, L. H., and Weber, C. A. 1995. Defects in the DNA repair and transcription gene ERCC2 in the cancer-prone disorder xeroderma pigmentosum group D. *Cancer Res.* 55:5656–5663.

Thomas, J. J. 1997. Computer technology for mining, synthesis, and visualization of textual data. Paper presented at the Symposium Predicting Chemical Carcinogenicity, 17–18 July 1997, at Washington, D.C.

U.S. Congress Office of Technology Assessment. 1986. *Technologies for detecting heritable mutations in human beings*, Report No. OTA-H-298. Washington, DC: U.S. Government Printing Office.

Venter, J. C. 1997. Gene sequence data as a rational basis for understanding biological effects and human health. In *The Toxicologist. Fundamental and Applied Toxicology*, Supplement V, 36(1, part 2):1 (abstract 2).

Wagner, H. 1997. Molecular nuclear medicine: From genotype to phenotype. In *The Toxicologist. Fundamental and Applied Toxicology*, Supplement V, 36(1, part 2):1 (abstract 4).

Woychik, R. P., 1997. Building genetic bridges across biological barriers. In *The Toxicologist. Fundamental and Applied Toxicology*, Supplement V, 36(1, part 2):1 (abstract 5).

Biotechnology and Safety Assessment, 2nd ed.
Edited by John A. Thomas
Copyright © 1998 Taylor & Francis

2

Antisense Therapeutics

Stanley T. Crooke

Isis Pharmaceuticals, Inc., Carlsbad, California

During the past few years, interest in developing antisense technology and in exploiting it for therapeutic purposes has been intense. Although progress has been gratifyingly rapid, the technology remains in its infancy, and the questions that remain to be answered still outnumber the questions for which there are answers. Appropriately, considerable debate continues about this approach's breadth of utility and the type of data required to "prove that a drug works through an antisense mechanism."

The objectives of this chapter are to provide a summary of recent progress, to assess the status of the technology, to place the technology in the pharmacological context in which it is best understood, and to deal with some of the controversies with regard to the technology and the interpretation of experiments.

PROOF OF MECHANISM

Factors That May Influence Experimental Interpretations

Clearly, the ultimate biological effect of an oligonucleotide will be influenced by the local concentration of the oligonucleotide at the target RNA, the concentration of the RNA, the rates of synthesis and degradation of the RNA, the type of terminating mechanism, and the rates of the events that result in termination of the RNA's activity. At present, we understand essentially nothing about the interplay of these factors.

Oligonucleotide Purity

Currently, phosphorothioate oligonucleotides can be prepared consistently and with excellent purity (Crooke et al. 1993a). However, this has only been the case for the past 3 to 4 years. Before that time, synthetic methods were evolving, and analytical methods were inadequate. In fact, our laboratory reported that different synthetic and purification procedures resulted in oligonucleotides that varied in cellular toxicity (Crooke 1991) and that potency varied from batch to batch. Though there are no longer synthetic problems with phosphorothioates, they undoubtedly complicated earlier studies. More importantly, with each new analog class, new synthetic, purification, and analytical challenges are encountered.

Oligonucleotide Structure

Antisense oligonucleotides are designed to be single-stranded. We now understand that certain sequences (e.g., stretches of guanosine residues) are prone to adopt more complex structures (Wyatt et al. 1994). The potential to form secondary and tertiary structures also varies as a function of the chemical class. For example, higher affinity 2'-modified oligonucleotides have a greater tendency to self-hybridize, resulting in more stable oligonucleotide duplexes than would be expected based on rules derived from oligodeoxynucleotides (Freier, unpublished results).

RNA Structure

Ribonucleic acid (RNA) is structured. The structure of the RNA has a profound influence on the affinity of the oligonucleotide and on the rate of binding of the oligonucleotide to its RNA target (Ecker 1993; Freier 1993). Moreover, RNA structure produces asymmetrical binding sites that then result in very divergent affinity constants depending on the position of oligonucleotide in that structure (Ecker et al. 1992; Lima et al. 1992; Ecker 1993). This in turn influences the optimal length of an oligonucleotide needed to achieve maximal affinity. We understand very little about how RNA structure and RNA protein interactions influence antisense drug action.

Variations in In Vitro Cellular Uptake and Distribution

Studies in several laboratories have clearly demonstrated that cells in tissue culture may take up phosphorothioate oligonucleotides via an active process and that the uptake of these oligonucleotides is highly variable on the basis of many conditions (Crooke 1991; Crooke et al. 1994). Cell type has a dramatic effect on total uptake, kinetics of uptake, and pattern of subcellular distribution. At present, there is no unifying hypothesis to explain these differences. Tissue culture conditions, such as the type of medium, degree of confluence, and the presence of serum can all have enormous effects on uptake (Crooke et al. 1994). Oligonucleotide chemical class obviously influences the characteristics and mechanism of uptake. Within the phosphorothioate class of oligonucleotides, uptake varies as a function of length but not linearly. Uptake varies as a function of sequence and stability in cells and is also influenced by sequence (Crooke et al. 1994; Crooke et al. 1995).

Given the foregoing, it is obvious that conclusions about in vitro uptake must be very carefully made and that generalizations are virtually impossible. Thus, before an oligonucleotide can be said to be inactive in vitro, it should be studied in several cell lines. Furthermore, although it may be absolutely correct that receptor-mediated endocytosis is a mechanism of uptake of phosphorothioate oligonucleotides (Loke et al. 1989), it is obvious that a generalization that all phosphorothioates are taken up by all cells in vitro primarily by receptor-mediated endocytosis is simply unwarranted.

Finally, extrapolations from in vitro uptake studies to predictions about in vivo pharmacokinetic behavior are entirely inappropriate and, in fact, there are now several lines of evidence in animals and man that, even after careful consideration of all in vitro uptake data, one cannot predict in vivo pharmacokinetics of the compounds (Cossum et al. 1993; Cossum et al. 1994; Crooke et al. 1994; Sands et al. 1995).

The Binding to and Effects of Binding to Nonnucleic Acid Targets

Phosphorothioate oligonucleotides tend to bind to many proteins, and these interactions are influenced by many factors. The effects of binding can influence cell uptake, distribution, metabolism, and excretion. They may induce nonantisense effects that can mistakenly be interpreted as antisense or complicate the identification of an antisense mechanism. By inhibiting RNase H, protein binding may inhibit the antisense activity of some oligonucleotides. Finally, binding to proteins can certainly have toxicological consequences.

In addition to proteins, oligonucleotides may interact with other biological molecules, such as lipids or carbohydrates, and such interactions, like those with proteins, will be influenced by the chemical class of oligonucleotide studied. Unfortunately, essentially no data bearing on such interactions are currently available.

An especially complicated experimental situation is encountered in many in vitro antiviral assays. In these assays, high concentrations of drugs, viruses, and cells are often coincubated. The sensitivity of each virus to nonantisense effects of oligonucleotides varies, depending on the nature of the virion proteins and the characteristics of the oligonucleotides (Azad et al. 1993; Cowsert 1993). This has resulted in considerable confusion. In particular for HIV, herpes simplex viruses, cytomegaloviruses, and influenza virus, the nonantisense effects have been so dominant that identifying oligonucleotides that work via an antisense mechanism has been difficult. Given the artificial character of such assays, it is difficult to know whether nonantisense mechanisms would be as dominant in vivo or result in antiviral activity.

Terminating Mechanisms

It has been amply demonstrated that oligonucleotides may employ several terminating mechanisms. The dominant terminating mechanism is influenced by RNA receptor site, oligonucleotide chemical class, cell type, and probably many other factors (Crooke 1995a). Obviously, because variations in terminating mechanism may result in significant changes in antisense potency and studies have shown significant variations from cell type to cell type in vitro, it is essential that the terminating mechanism be well understood. Unfortunately, at present our understanding of terminating mechanisms remains rudimentary.

Effects of "Control Oligonucleotides"

Several types of control oligonucleotides have been used, including randomized oligonucleotides. Unfortunately, we know little to nothing about the potential biological effects of such "controls," and the more complicated a biological system and test the more likely that "control" oligonucleotides may have activities that complicate interpretations. Thus, when a control oligonucleotide displays a surprising activity, the mechanism of that activity should be explored carefully before concluding that the effects of the "control oligonucleotide" prove that the activity of the putative antisense oligonucleotide are not due to an antisense mechanism.

Kinetics of Effects

Many rate constants may affect the activities of antisense oligonucleotides (e.g., the rate of synthesis and degradation of the target RNA and its protein); the rates of uptake into cells; the rates of distribution, extrusion, and metabolism of an oligonucleotide in cells; and similar pharmacokinetic considerations in animals. Despite this, relatively few time courses have been reported, and in vitro studies have been reported that range from a few hours to several days. In animals, we have a growing body of information on pharmacokinetics, but in most studies reported to date, the doses and schedules were chosen arbitrarily and, again, little information on duration of effect and onset of action has been presented.

Clearly, more careful kinetic studies are required, and rational in vitro and in vivo dose schedules must be developed.

Recommendations

Until more is understood about how antisense drugs work, it is essential to demonstrate effects consistent with an antisense mechanism positively. For RNase H activating oligonucleotides, Northern blot analysis showing selective loss of the target RNA is the best choice, and many laboratories are publishing reports in vitro and in vivo of such activities (Chiang et al. 1991; Dean et al. 1994; Hijiya et al. 1994; Skorski et al. 1994). Ideally, a demonstration that closely related isotypes are unaffected should be included. In brief, then, for proof of mechanism, the following steps are recommended:

- Perform careful dose-response curves in vitro using several cell lines and methods of in vitro delivery.
- Correlate the rank-order potency in vivo with that observed in vitro after thorough dose response curves are generated in vivo.
- Perform careful "gene walks" for all RNA species and oligonucleotide chemical classes.

- Perform careful time courses before drawing conclusions about potency.
- Directly demonstrate proposed mechanism of action by measuring the target RNA, protein, or both.
- Evaluate specificity and therapeutic indices via studies on closely related isotypes and with appropriate toxicological studies.
- Perform sufficient pharmacokinetics to define rational dosing schedules for pharmacological studies.
- When control oligonucleotides display surprising activities, determine the mechanisms involved.

CHARACTERISTICS OF PHOSPHOROTHIOATE OLIGODEOXYNUCLEOTIDES

Of the first-generation oligonucleotide analogs, the class that has resulted in the broadest range of activities and about which the most is known is the phosphorothioate class. Phosphorothioate oligonucleotides were first synthesized in 1969 when a poly rI-rC phosphorothioate was synthesized (De Clercq et al. 1969). This modification clearly achieves the objective of increased nuclease stability. In this class of oligonucleotides, one of the oxygen atoms in the phosphate group is replaced with a sulfur. The resulting compound is negatively charged, is chiral at each phosphorothioate, and is much more resistant to nucleases than the parent phosphorothioate (Cohen 1993).

Hybridization

The hybridization of phosphorothioate oligonucleotides to DNA and RNA has been thoroughly characterized (Crooke 1992; Crooke 1993; Crooke et al. 1993). The melting point T_m of a phosphorothioate oligodeoxynucleotide for RNA is approximately 0.5 °C less per nucleotide than for a corresponding phosphodiester oligodeoxynucleotide. This reduction in T_m per nucleotide is virtually independent of the number of phosphorothioate units substituted for phosphodiesters. However, sequence context has some influence, for the ΔT_m can vary from -0.3–1.0 °C, depending on sequence. Compared with RNA and RNA duplex formation, a phosphorothioate oligodeoxynucleotide has a T_m approximately -2.2 °C lower per unit (Freier 1993). This means that to be effective in vitro, phosphorothioate oligodeoxynucleotides must typically be 17–20 nucleotides in length and that invasion of double-stranded regions in RNA is difficult (Vickers et al. 1991; Lima et al. 1992; Monia et al. 1992; Monia et al. 1993).

Association rates of phosphorothioate oligodeoxynucleotide to unstructured RNA targets are typically 10^6–10^7 m^{-1} s^{-1} independent of oligonucleotide length or sequence (Lima et al. 1992; Freier 1993). Association rates to structured RNA targets can vary from 10^2–10^8 m^{-1} s^{-1}, depending on the structure of the RNA, site of binding in the structure, and other factors (Freier 1993). Said another way, association rates for oligonucleotides that display acceptable affinity constants are sufficient to support biological activity at therapeutically achievable concentrations.

The specificity of hybridization of phosphorothioate oligonucleotides is, in general, slightly greater than phosphodiester analogs. For example, a T–C mismatch results in a 7.7 or 12.8 °C reduction in T_m, respectively, for a phosphodiester or phosphorothioate oligodeoxynucleotide 18 nucleotides in length with the mismatch centered (Freier 1993). Thus, from this perspective, the phosphorothioate modification is quite attractive.

Interactions with Proteins

Phosphorothioate oligonucleotides bind to proteins. The interactions with proteins can be divided into nonspecific, sequence-specific, and structure-specific binding events, each of which may have different characteristics and effects. Nonspecific binding to a wide variety of proteins has been demonstrated. Exemplary of this type of binding is the interaction of phosphorothioate oligonucleotides with serum albumin. The affinity of such interactions is low. The binding constant K_d for albumin is approximately 200 μM and is thus in a similar range with aspirin or penicillin (Joos et al. 1969; Crooke et al. 1996). Furthermore, in this study, no competition between phosphorothioate oligonucleotides and several drugs that bind to bovine serum albumin was observed. In this study, binding and competition were determined in an assay in which electrospray mass spectrometry was used. In contrast, in a study in which an equilibrium dissociation constant was derived from an assay using albumin loaded on a CH-sephadex column, the Michaelis constant K_M ranged from 1×10^{-5} to 5×10^{-5} M for bovine serum albumin and from 2×10^{-4} to 3×10^{-4} M for human serum albumin. Moreover, warfarin and indomethacin were reported to compete for binding to serum albumin (Srinivasan et al. 1995). Clearly, much more work is required before definitive conclusions can be drawn.

Phosphorothioate oligonucleotides can interact with nucleic acid binding proteins such as transcription factors and single-strand nucleic acid binding proteins. However, very little is known about these binding events. Additionally, it has been reported that phosphorothioates bind to an 80-kDa membrane protein that was suggested to be involved in cellular uptake processes (Loke et al. 1989). However, again, little is known about the affinities, sequence, or structure specificities of these putative interactions. More recently, interactions with 30-kDa and 46-kDa surface proteins in T15 mouse fibroblasts were reported (Hawley et al. 1996).

Phosphorothioates interact with nucleases and DNA polymerases. These compounds are slowly metabolized by both endo- and exonucleases and inhibit these enzymes (Crooke 1992; Crooke et al. 1995). The inhibition of these enzymes appears to be competitive, and this may account for some early data suggesting that phosphorothioates are almost infinitely stable to nucleases. In these studies, the oligonucleotide-to-enzyme ratio was very high, and thus the enzyme was inhibited. Phosphorothioates also bind to RNase H when in an RNA–DNA duplex and the duplex serves as a substrate for RNase H (Gao et al. 1992). At higher concentrations, presumably by binding as a single strand to RNase H, phosphorothioates inhibit the enzyme (Crooke 1992; Crooke et al. 1995). Again, the oligonucleotides appear to be competitive antagonists for the DNA–RNA substrate.

Phosphorothioates have been shown to be competitive inhibitors of DNA polymerase α and β with respect to the DNA template and noncompetitive inhibitors of DNA polymerases g and d (Gao et al. 1992). Despite this inhibition, several studies have suggested that phosphorothioates might serve as primers for polymerases and be extended (Agrawal et al. 1991; Stein et al. 1993; Crooke et al. 1995). In our laboratories, we have shown extensions of 2–3 nucleotides only. At present, a full explanation as to why no longer extensions are observed is not available.

Phosphorothioate oligonucleotides have been reported to be competitive inhibitors for HIV-reverse transcriptase and inhibit RT-associated RNase H activity (Majumdar et al. 1989; Cheng et al. 1991). They have been reported to bind to the cell-surface protein CD4 and to protein kinase C (Stein et al. 1991). Various viral polymerases have also been shown to be inhibited by phosphorothioates (Stein et al. 1993). Additionally, we have shown potent, nonsequence-specific inhibition of RNA splicing by phosphorothioates (Hodges et al. 1995).

Like other oligonucleotides, phosphorothioates can adopt a variety of secondary structures. As a general rule, self-complementary oligonucleotides are avoided, if possible, to prevent the occurrence of duplex formation between oligonucleotides. However, other structures that are less well understood can also form. For example, oligonucleotides containing runs of guanosines can form tetrameric structures called G-quartets, and these appear to interact with many proteins having relatively greater affinity than unstructured oligonucleotides (Wyatt et al. 1994).

In conclusion, phosphorothioate oligonucleotides may interact with a wide range of proteins via several types of mechanisms. These interactions may influence the pharmacokinetic, pharmacologic, and toxicologic properties of these molecules. They may also complicate studies on the mechanism of action of these drugs, and may, in fact, obscure an antisense activity. For example, phosphorothioate oligonucleotides were reported to enhance lipopolysaccharide-stimulated synthesis or tumor necrosis factor (Hartmann et al. 1996). This would obviously obscure antisense effects on this target.

Pharmacokinetic Properties

To study the pharmacokinetics of phosphorothioate oligonucleotides, a variety of labeling techniques have been used. In some cases, 3'- or $5'^{32}P$ end-labeled or fluorescently labeled oligonucleotides have been used in in vitro or in vivo studies. These are probably less satisfactory than internally labeled compounds because terminal phosphates are rapidly removed by phosphatases, and fluorescently labeled oligonucleotides have physicochemical properties that differ from the unmodified oligonucleotides. Consequently, either uniformly (Cowsert et al. 1993) S-labeled or base-labeled phosphorothioates are preferable for pharmacokinetic studies. In our laboratories, a tritium exchange method that labels a slowly exchanging proton at the C-8 position in purines was developed and has proved to be quite useful (Graham et al. 1993). Very recently, a method that adds radioactive methyl groups via S-adenosyl methionine has also been used successfully (Sands et al. 1994). Finally, advances in extraction, separation, and detection methods have resulted in methods that provide excellent pharmacokinetic analyses without radiolabeling (Crooke et al. 1996).

Nuclease Stability

The principal metabolic pathway for oligonucleotides is cleavage via endo- and exo-nucleases. Phosphorothioate oligonucleotides, although quite stable to various nucleases, are competitive inhibitors of nucleases (Wickstrom 1986; Campbell et al. 1990; Hoke et al. 1991; Gao et al. 1992; Crooke et al. 1995). Consequently, the stability of phosphorothioate oligonucleotides to nucleases is probably a bit less than initially thought, for high concentrations (that inhibited nucleases) of oligonucleotides were employed in the early studies. Similarly, phosphorothioate oligonucleotides are degraded slowly by cells in tissue culture with a half-life of 12–24 h and are slowly metabolized in animals (Hoke et al. 1991; Cossum et al. 1993; Crooke et al. 1995). The pattern of metabolites suggests primarily exonuclease activity with perhaps modest contributions by endonucleases. However, several lines of evidence suggest that endonucleases play an important role in many cells and tissues in the metabolism of oligonucleotides. For example, 3'- and 5'-modified oligonucleotides with phosphodiester backbones have been shown to be degraded relatively rapidly in cells and after administration to animals (Miyao et al. 1995; Sands et al. 1995). Thus, strategies in which oligonucleotides are modified at only the 3'- and 5'-terminus as a means of enhancing stability have not proven to be successful.

In Vitro Cellular Uptake

Phosphorothioate oligonucleotides are taken up by a wide range of cells in vitro (Crooke 1991; Gao et al. 1992; Crooke 1993b; Neckers 1993; Crooke et al. 1995). In fact, uptake of phosphorothioate oligonucleotides into a prokaryote, *Vibrio parahaemoyticus*, has been reported as has uptake into *Schistosoma mansoni* (Chrisey et al. 1993; Tao et al. 1995). Uptake is time- and temperature-dependent. It is also influenced by cell type, cell-culture conditions, media and sequence, and length of the oligonucleotide (Crooke et al. 1995). No obvious correlation between the lineage of cells, whether the cells are transformed or whether they are virally infected, and uptake has been identified (Crooke et al. 1995), nor are the factors that result in variations in uptake of different sequences of oligonucleotide understood. Although several studies have suggested that receptor-mediated endocytosis may be a significant mechanism of cellular uptake, the data are not yet compelling enough to conclude that receptor-mediated endocytosis accounts for a significant portion of the uptake in most cells (Loke et al. 1989).

Numerous studies have shown that phosphorothioate oligonucleotides become distributed broadly in most cells once taken up (Crooke 1993a; Crooke et al. 1995). Again, however, significant differences in subcellular distribution among various types of cells have been noted.

Cationic lipids and other approaches have been used to enhance uptake of phosphorothioate oligonucleotides in cells that take up little oligonucleotide in vitro (Bennett et al. 1992; Bennett et al. 1993; Quattrone et al. 1994). Again, however, there are substantial variations from cell type to cell type. Other approaches to enhanced intracellular uptake in vitro have included streptolysin D treatment of cells and the use of dextran sulfate and other liposome formulations as well as physical means such as microinjections (Crooke 1995a; Giles et al. 1995; Wang et al. 1995).

In Vivo Pharmacokinetics

Phosphorothioate oligonucleotides bind to serum albumin and α-2 macroglobulin. The apparent affinity for albumin is quite low (200–400 μM) and is comparable to the low-affinity binding observed for several drugs such as aspirin and penicillin (Joos et al. 1969; Srinivasan et al. 1995; Crooke et al. 1996). Serum protein binding, therefore, provides a repository for these drugs and prevents rapid renal excretion. Because serum protein binding is saturable, at higher doses, intact oligomer may be found in urine (Agrawal et al. 1991; Iversen 1991). Studies in our laboratory suggest that in rats, oligonucleotides administered intravenously at doses of 15–20 mg/kg saturate the serum protein binding capacity (Leeds, unpublished data).

Phosphorothioate oligonucleotides are rapidly and extensively absorbed after parenteral administration. For example, in rats, after an intradermal dose 3.6 mg/kg of ^{14}C-ISIS 2105, a 20-mer phosphorothioate, approximately 70% of the dose was absorbed within 4 h, and total systemic bioavailability was in excess of 90% (Cossum et al. 1994). After intradermal injection in man, absorption of ISIS 2105 was similar to that observed in rats (Crooke et al. 1994). Subcutaneous administration to rats and monkeys results in somewhat lower bioavailability and greater distribution to lymph, as would be expected (Leeds, unpublished observations).

Distribution of phosphorothioate oligonucleotides from blood after absorption or intravenous administration is extremely rapid. We have reported distribution half-lives of less than 1 h, and similar data have been reported by others (Agrawal et al. 1991; Iversen 1991; Cossum et al. 1993; Cossum et al. 1994). Blood and plasma clearance is multiexponential and has a terminal elimination half-life from 40–60 h in all species except man. In man, the terminal elimination half-life may be somewhat longer (Crooke et al. 1994).

Phosphorothioates are broadly distributed to all peripheral tissues. Liver, kidney, bone marrow, skeletal muscle, and skin accumulate the highest percentage of a dose, but other tissues display small quantities of drug (Cossum et al. 1993; Cossum et al. 1994). No evidence of significant penetration of the blood–brain barrier has been reported. The rates of incorporation and clearance from tissues vary as a function of the organ studied; the liver accumulates drug most rapidly (20% of a dose within 1–2 h), and other tissues accumulate drug more slowly. Similarly, elimination of drug is more rapid from liver than any other tissue (e.g., terminal half-life from liver: 62 h; from renal medulla: 156 h). The distribution into the kidney has been studied more extensively, and drug has been shown to be present in Bowman's capsule, the proximal convoluted tubule, the bush border membrane, and within renal tubular epithelial cells (Rappaport et al. 1995). The data suggested that the oligonucleotides are filtered by the glomerulus and then reabsorbed by the proximal convoluted tubule epithelial cells. Moreover, the authors suggested that reabsorption may be mediated by interactions with specific proteins in the bush border membranes.

At relatively low doses, clearance of phosphorothioate oligonucleotides is due primarily to metabolism (Iversen 1991; Cossum et al. 1993; Cossum et al. 1994). Metabolism is mediated by exo- and endonucleases that result in shorter oligonucleotides and, ultimately, nucleosides that are degraded by normal metabolic pathways. Although no direct evidence of base excision or modification has been reported, these are theoretical possibilities that may occur. In one study, a larger-molecular-weight radioactive material was observed in urine but not fully characterized (Agrawal et al. 1991). Clearly, the potential for conjugation reactions and extension of oligonucleotides via these drugs serving as primers for polymerases must be explored in more detail. In a very thorough study, 20-nucleotide phosphodiester and phosphorothioate oligonucleotides were administered intravenously at a dose of 6 mg/kg to mice. The oligonucleotides were internally labeled with ^{3}H-CH$_3$ by methylation of an internal deoxycytidine residue using Hha1 methylase and S-(^{3}H) adenosyl methionine (Sands et al. 1994). The observations for the phosphorothioate oligonucleotide were entirely consistent with those made in our studies. Additionally, as reported by Sands and coworkers, autoradiographic analyses showed drug in renal cortical cells (Sands et al. 1994).

One study of prolonged infusions of a phosphorothioate oligonucleotide into human beings has been reported (Bayever et al. 1993). In this study, five patients with leukemia were given 10-day intravenous infusions at a dose of 0.05 mg/kg/h. Elimination half-lives reportedly varied from 5.9 to 14.7 days. Urinary recovery of radioactivity was reported to be 30–60% of the total dose, and 30% of the radioactivity was intact drug. Metabolites in urine included both higher- and lower-molecular-weight compounds. In contrast, when GEM-91 (a 25-mer phosphorothioate oligodeoxynucleotide) was administered to humans as a 2 h intravenous infusion at a dose of 0.1 mg/kg, a peak plasma concentration of 295.8 mg/ml was observed at the cessation of the infusion. Plasma clearance of total radioactivity was biexponential with initial and terminal eliminations half-lives of 0.18 and 26.71 h, respectively. However, degradation was extensive, and intact drug pharmacokinetic models were not presented. Nearly 50% of the administered radioactivity was recovered in urine, but most of the radioactivity represented degradates. In fact, no intact drug was found in the urine at any time (Zhang et al. 1995).

In a more recent study in which the level of intact drug was carefully evaluated using capillary gel electrophoresis, the pharmacokinetics of ISIS 2302, a 20-mer phosphorothioate oligodeoxynucleotide, after a 2-h infusion, were determined. Doses from 0.06 mg/kg to 2.0 mg/kg were studied, and the peak plasma concentrations were shown to increase linearly with dose with the 2 mg/kg dose resulting in peak plasma concentrations of intact drug of approximately 9.5 μg/ml. Clearance from plasma, however, was dose dependent, and the 2 mg/kg dose had a clearance of 1.28 ml min^{-1} kg^{-1}, whereas that of 0.5 mg/kg was 2.07 ml min^{-1} kg^{-1}. Essentially, no intact drug was found in urine.

Clearly, the two most recent studies differ from the initial report in several facets. Although many factors may explain the discrepancies, the most likely explanation is related to the evolution of assay methodology, not differences between compounds. Overall, the behavior of phosphorothioates in the plasma of humans appears to be similar to that in other species.

In addition to the pharmacological effects that have been observed with phosphorothioate oligonucleotides, there are several lines of evidence supporting the notion that these drugs enter cells in various organs. As an example, Figure 1 shows autoradiographic, fluorescent, and immunohistochemical data demonstrating the intracellular location of phosphorothioate oligonucleotides in renal proximal convoluted tubular cells. Similar results have been observed in liver, skin, and bone marrow in similar studies. Using radiolabeled drugs and isolated perfused rat liver cells, uptake into parenchymal and nonparenchymal cells of the liver (Takakura et al. 1996) has been reported.

We have also performed oral bioavailability experiments in rodents treated with an H$_2$ receptor antagonist to avoid acid-mediated depurination or precipitation. In these studies, very limited (<5%) bioavailability was observed (Crooke, unpublished observations). However, it seems likely that the principal limiting factor in the oral bioavailability of phosphorothioates may be degradation in the gut rather than absorption. Studies using everted rat jejunum sacs demonstrated passive transport across the intestinal epithelium (Hughes et al. 1995). Further, studies using more stable 2′-methoxy phosphorothioate oligonucleotides showed a significant increase in oral bioavailability that appeared to be associated with the improved stability of the analogs (Agrawal et al. 1995).

In summary, pharmacokinetic studies of several phosphorothioates demonstrate that they are well absorbed from parenteral sites, distribute broadly to all peripheral tissues, do not cross the blood–brain barrier, and are eliminated primarily by slow metabolism. In short, once-a-day or every-other-day systemic dosing should be feasible. Although the similarities between oligonucleotides of different sequences are far greater than the differences, additional studies are required before determining whether there are subtle effects of sequence on the pharmacokinetic profile of this class of drugs.

Pharmacological Properties

Molecular Pharmacology

Antisense oligonucleotides are designed to bind to RNA targets via Watson–Crick hybridization. Because RNA can adopt a variety of secondary structures via Watson–Crick hybridization, one useful way to think of antisense oligonucleotides is as competitive antagonists for self-complementary regions of the target RNA. Obviously, creating oligonucleotides with the highest affinity per nucleotide unit is pharmacologically important, and a comparison of the affinity of the oligonucleotide with a complementary RNA oligonucleotide is the most sensible comparison. In this context, phosphorothioate oligodeoxynucleotides are relatively competitively disadvantaged, for the affinity per nucleotide unit of oligomer is less than RNA ($> -2.0 °C$ T_m per unit) (Cook 1993). This results in a requirement of at least 15–17 nucleotides in order to have sufficient affinity to produce biological activity (Monia et al. 1992).

Although multiple mechanisms by which an oligonucleotide may terminate the activity of an RNA species to which it binds are possible, examples of biological activity have been reported for only three of these mechanisms. Antisense oligonucleotides have been reported to inhibit RNA splicing, effect translation of mRNA, and induce degradation of RNA by RNase H (Agrawal et al. 1988; Kulka et al. 1989; Chiang et al. 1991). Without question, the mechanism that has resulted in the most potent compounds and is best understood is RNase H activation. To serve as a substrate for RNase H, a duplex between RNA and a "DNA-like" oligonucleotide is required. Specifically, a sugar moiety in the oligonucleotide that induces a duplex conformation equivalent to that of a DNA–RNA duplex and a charged phosphate is required (Mirabelli et al. 1993). Thus, phosphorothioate oligodeoxynucleotides are expected to induce RNase H-mediated cleavage of the RNA when bound. As will be discussed later, many chemical approaches that enhance the affinity of an oligonucleotide for RNA result in duplexes that are no longer substrates for RNase H.

Selection of sites at which optimal antisense activity may be induced in an RNA molecule is complex, is dependent on terminating mechanism, and is influenced by the chemical class of the oligonucleotide. Each RNA appears to display unique patterns of sensitivity sites. Within the phosphorothioate oligodeoxynucleotide chemical class, studies in our laboratory have shown antisense activity can vary from undetectable to 100% by shifting an oligonucleotide by just a few bases in the RNA target (Chiang et al. 1991; Crooke 1992; Bennett et al. 1996). Although significant progress has been made in developing general rules that help define potentially optimal sites in RNA species, to a large extent this remains an empirical process that must be performed for each RNA target and every new chemical class of oligonucleotides.

Phosphorothioates have also been shown to have effects inconsistent with the antisense mechanism for which they were designed. Some of these effects are due to sequence or are structure-specific. Others are due to nonspecific interactions with proteins. These effects are particularly prominent in in vitro tests for antiviral activity because high concentrations of cells, viruses, and oligonucleotides are often coincubated (Azad et al. 1993; Wagner et al. 1993). Human immune deficiency virus (HIV) is particularly problematic, for many oligonucleotides bind to the gp120 protein (Wyatt et al. 1994). However, the potential for confusion arising from the misinterpretation of an activity as being due to an antisense mechanism when, in fact, it is due to nonantisense effects is certainly not limited to antiviral or just in vitro tests (Barton et al. 1995; Burgess et al. 1995; Hertl et al. 1995). Again, these data simply urge caution and argue for careful dose-response curves, direct analyses of target protein or RNA, and inclusion of appropriate controls before drawing conclusions concerning the mechanisms of action of oligonucleotide-based drugs. In addition to protein interactions, other factors, such as overrepresented sequences of RNA and unusual structures that may be adopted by oligonucleotides, can contribute to unexpected results (Wyatt et al. 1994).

Given the variability in cellular uptake of oligonucleotides, the variability in potency as a function of binding site in an RNA target, and potential nonantisense activities of oligonucleotides, careful evaluation of dose-response curves and clear demonstration of the antisense mechanism are required before drawing conclusions from in vitro experiments. Nevertheless, numerous well-controlled studies have been reported in which antisense activity was conclusively demonstrated. Because many of these studies have been reviewed previously, suffice it to say that antisense effects of phosphorothioate oligodeoxynucleotides against a variety of targets are well documented (Crooke 1992; Crooke 1993; Crooke and Lebleu 1993; Nagel et al. 1993; Stein et al. 1993; Crooke 1995b).

In Vivo Pharmacological Activities

A relatively large number of reports of in vivo activities of phosphorothioate oligonucleotides have now appeared documenting activities after local and systemic administration (Table 1) (Crooke 1995b). However, for only a few of these reports have sufficient studies been performed to draw relatively firm conclusions concerning the mechanism of action. Consequently, I will review in some detail only a few reports that provide sufficient data to support a relatively firm conclusion with regard to mechanism of action. Local effects have been reported for phosphorothioate and methylphosphonate oligonucleotides. A phosphorothioate oligonucleotide designed to inhibit c-*myb* production and applied locally was shown to inhibit intimal accumulation in the rat carotid artery (Simons et al. 1992). In this study, a Northern blot analysis showed a significant reduction in c-*myb* RNA in animals treated with the antisense compound but no effect by a control oligonucleotide. In a recent study, the effects of the oligonucleotide were suggested to be due to a nonantisense mechanism (Burgess et al. 1995). However, only one dose level was studied; therefore, much remains to be done before definitive conclusions are possible. Similar effects were reported for phosphorothioate oligodeoxynucleotides designed to inhibit cyclin-dependent kinases (CDC-2 and CDK-2). Again, the antisense oligonucleotide

TABLE 1. *Reported activity of antisense oligonucleotides in animal models*

Target	Route	Species	Reference
Cardiovascular models			
c-*myb*	Topically	Rat	(Simons et al. 1992)
cdc2 kinase	Topically	Rat	(Morishita et al. 1993)
PCNA	Topically	Rat	(Morishita et al. 1993)
cdc2 kinase	Topically	Rat	(Abe et al. 1994)
CDK2	Topically	Rat	(Abe et al. 1994)
Cyclin B_1	Topically	Rat	(Morishita et al. 1994)
PCNA	Topically	Rat	(Simons et al. 1994)
Angiotensin 1 receptor	Intracerebral	Rat	(Gyurko et al. 1993)
Angiotensinogen	Intracerebral	Rat	(Phillips et al. 1994)
c-*fos*	Intracerebral	Rat	(Suzuki et al. 1994)
Inflammatory models			
Type 1 IL-1 receptor	Intradermal	Mouse	(Burch et al. 1991)
ICAM-1	Intravenous	Mouse	(Stepkowski et al. 1994)
ICAM-1	Intravenous	Mouse	(Kumasaka et al. 1996)
ICAM-1	Intravenous	Mouse	(Katz et al. 1995)
ICAM-1	Intravenous	Mouse	(Stepkowski et al. 1995)
ICAM-1	Intravenous	Mouse	(Bennett et al. 1997)
Adenosine A_1 receptor	Aerosol	Rabbit	(Nyce et al. 1997)
Cancer models			
N-*myc*	Subcutaneous	Mouse	(Whitesell et al. 1991)
NF-kB p65	Intraperitoneal	Mouse	(Kitajima et al. 1992)
c-*myb*	Subcutaneous	Mouse	(Ratajczak et al. 1992)
p120 nucleolar antigen	Intraperitoneal	Mouse	(Perlaky et al. 1993)
NK-kB p65	Subcutaneous	Mouse	(Higgins et al. 1993)
protein kinase C-a	Intraperitoneal	Mouse	(Dean et al. 1994)
c-*myb*	Subcutaneous	Mouse	(Hijiya et al. 1994)
Ha-*ras*	Intratumor	Mouse	(Schwab et al. 1994)
BCR-ABL	Intravenous	Mouse	(Skorski et al. 1994)
PTHrP	Intraventricular	Rat	(Akino et al. 1996)
c-raf kinase	Intravenous	Mouse	(Monia et al. 1995)
protein kinase C-a	Intravenous	Mouse	(Dean et al. 1996)
protein kinase C-a	Intravenous	Mouse	(Yazaki et al. 1996)
Neurological models			
c-*fos*	Intracerebral	Rat	(Chiasson et al. 1992)
SNAP-25	Intracerebral	Chicken	(Osen-Sand et al. 1993)
Kinesin heavy chain	Intravitreal	Rabbit	(Amaratunga et al. 1993)
Arginine vasopressin	Intracerebral	Rat	(Flanagan et al. 1993)
c-*fos*	Intracerebral	Rat	(Heilig et al. 1993)
Progesterone receptor	Intracerebral	Rat	(Pollio et al. 1993)
Dopamine D_2 receptor	Intracerebral	Rat	(Zhang and Creese 1993)
Y-Y1 receptor	Intracerebral	Rat	(Wahlestedt et al. 1993)
Neuropeptide Y	Intracerebral	Rat	(Akabayashi et al. 1994)
k-opioid receptor	Intracerebral	Rat	(Adams et al. 1994)
IGF-1	Intracerebral	Rat	(Castro-Alamancos and Torres-Aleman 1994)
c-*fos*	Intraspinal	Rat	(Gillardon et al. 1994)
c-*fos*	Intracerebral	Rat	(Hooper et al. 1994)
c-*fos*	Intraspinal	Rat	(Woodburn et al. 1994)
NMDA receptor	Intracerebral	Rat	(Kindy 1994)

TABLE 1. *Continued.*

Target	Route	Species	Reference
CREB	Intracerebral	Rat	(Konradi et al. 1994)
Delta-opioid receptor	Intracerebral	Mice	(Lai et al. 1994)
Progesterone receptor	Intracerebral	Rat	(Mani et al. 1994)
GAD65	Intracerebral	Rat	(McCarthy et al. 1994)
GAD67	Intracerebral	Rat	(McCarthy et al. 1994)
AT1-angiotensin receptor	Intracerebral	Rat	(Sakai et al. 1995)
Tryptophan hydroxylase	Intracerebral	Mouse	(McCarthy et al. 1995)
AT1-angiotensin receptor	Intracerebral	Rat	(Ambuhl et al. 1995)
CRH_1-corticotropin-releasing hormone receptor	Intracerebral	Rat	(Liebsch et al. 1995)
opiod receptor	Intracerebral	Rat	(Cha et al. 1995)
opiod receptor	Intracerebral	Mouse	(Mizoguchi et al. 1995)
Oxytocin	Intracerebral	Rat	(Morris et al. 1995)
Oxytocin	Intracerebral	Rat	(Neumann et al. 1994)
Substance P receptor	Intracerebral	Rat	(Ogo et al. 1994)
Tyrosine hydroxylase	Intracerebral	Rat	(Skutella et al. 1994)
c-*jun*	Intracerebral	Rat	(Tischmeyer et al. 1994)
D_1 dopamine receptor	Intracerebral	Mouse	(Zhang et al. 1994)
D_2 dopamine receptor	Intracerebral	Mouse	(Zhou et al. 1994)
D_2 dopamine receptor	Intracerebral	Mouse	(Weiss et al. 1993)
D_2 dopamine receptor	Intracerebral	Mouse	(Qin et al. 1995)
Viral models			
HSV-1		Mouse	(Kulka et al. 1989)
Tick-born encephalitis		Mouse	(Vlassov 1989)
Duck hepatitis virus	Intravenous	Duck	(Offensperger et al. 1993)

inhibited intimal thickening and cyclin-dependent kinase activity, whereas a control oligonucleotide had no effect (Abe et al. 1994). Additionally, local administration of a phosphorothioate oligonucleotide designed to inhibit N-*myc* resulted in reduction in N-*myc* expression and slower growth of a subcutaneously transplanted human tumor in nude mice (Whitesell et al. 1991).

Antisense oligonucleotides administered intraventricularly have been reported to induce a variety of effects in the central nervous system. Intraventricular injection of antisense oligonucleotides to neuropeptide-y-y1 receptors reduced the density of the receptors and resulted in behavioral signs of anxiety (Wahlestedt et al. 1993). Similarly, an antisense oligonucleotide designed to bind to NMDA-R1 receptor channel RNA inhibited the synthesis of these channels and reduced the volume of focal ischemia produced by occlusion of the middle cerebral artery in rats (Wahlestedt et al. 1993).

In a series of well-controlled studies, antisense oligonucleotides administered intraventricularly selectively inhibited dopamine type-2 receptor expression, dopamine type-2 receptor RNA levels, and behavioral effects in animals with chemical lesions. Controls included randomized oligonucleotides and the observation that no effects were observed on dopamine type-1 receptor or RNA levels (Weiss et al. 1993; Zhou et al. 1994; Qin et al. 1995). This laboratory also reported the selective reduction of dopamine type-1 receptor and RNA levels with the appropriate oligonucleotide (Zhang et al. 1994).

Similar observations were reported in studies on AT-1 angiotensin receptors and tryptophan hydroxylase. In studies in rats, direct observations of AT-1 and AT-2 receptor densities in various sites in the brain after administration of different doses of phosphorothioate antisense, sense, and scrambled oligonucleotides were reported (Ambuhl et al. 1995). Again, in rats, intraventricular administration of phosphorothioate antisense oligonucleotide resulted in a decrease in tryptophan hydroxylase levels in the brain, whereas a scrambled control did not (McCarthy et al. 1995).

Injection of antisense oligonucleotides to synaptosomal-associated protein-25 into the vitreous body of rat embryos reduced the expression of the protein and inhibited neurite elongation by rat cortical neurons (Osen-Sand et al. 1993).

Aerosol administration to rabbits of an antisense phosphorothioate oligodeoxynucleotide designed to inhibit the production of antisense A_1 receptor has been reported to reduce receptor numbers in the airway smooth muscle and to inhibit adenosine, house-dust mite allergen, and histamine-induced bronchoconstriction (Nyce et al. 1997). Neither control nor oligonucleotide complementary to bradykinin B_2 receptors reduced the density of adenosine A_1 receptors, although the oligonucleotides complementary to bradykin in B_2 receptor mRNA reduced the density of these receptors.

In addition to local and regional effects of antisense oligonucleotides, a growing number of well-controlled studies have demonstrated systemic effects of phosphorothioate oligodeoxynucleotides. Expression of interleukin-1 in mice was inhibited by systemic administration of antisense oligonucleotides (Burch et al. 1991). Oligonucleotides to the NF-kB p65 subunit administered intraperitoneally at 40 mg/kg every 3 days slowed tumor growth in mice transgenic for the human T-cell leukemia viruses (Kitajima et al. 1992). Similar results with other antisense oligonucleotides were shown in another in vivo tumor model after either prolonged subcutaneous infusion or intermittent subcutaneous injection (Higgins et al. 1993).

Several recent reports further extend the studies of phosphorothioate oligonucleotides as antitumor agents in mice. In one study, a phosphorothioate oligonucleotide directed to inhibition of the *bcr-abl* oncogene was administered at a dose of 1 mg/day for 9 days intravenously to immunodeficient mice injected with human leukemic cells. The drug was shown to inhibit the development of leukemic colonies in the mice and to reduce *bcr-abl* RNA levels selectively in peripheral blood lymphocytes, spleen, bone marrow, liver, lungs, and brain (Skorski et al. 1994). However, it is possible that the effects on the RNA levels were secondary to effects on the growth of various cell types. In the second study, a phosphorothioate oligonucleotide antisense to the protooncogene *myb* inhibited the growth of human melanoma in mice. Again, *myb* mRNA levels appeared to be reduced selectively (Hijiya et al. 1994).

Several studies from our laboratories that directly examined target RNA levels, target protein levels, and pharmacological effects using a wide range of control oligonucleotides and examination of the effects on closely related isotypes have been completed. Single and chronic daily administration of a phosphorothioate oligonu-cleotide designed to inhibit mouse protein kinase C-a, (PKC-a) selectively inhibited expression of PKC-a RNA in mouse liver without effects on any other isotype. The effects lasted at least 24 h after a dose, and a clear dose response curve was observed with a dose of 10–15 mg/kg intraperitoneally, which reduced PKC-a RNA levels in liver by 50% 24 h after a dose (Dean et al. 1994).

A phosphorothioate oligonucleotide designed to inhibit human PKC-a expression selectively inhibited expression of PKC-a RNA and PKC-a protein in human tumor cell lines implanted subcutaneously in nude mice after intravenous administration (Dean et al. 1996). In these studies, effects on RNA and protein levels were highly specific and observed at doses lower than 6 mg/kg/day, and antitumor effects were detected at doses as low as 0.6 mg/kg/day. A large number of control oligonucleotides failed to show activity.

In a similar series of studies, Monia et al. demonstrated highly specific loss of human c-*raf* kinase RNA in human tumor xenografts and antitumor activity that correlated with the loss of RNA. Moreover, a series of control oligonucleotides with 1–7 mismatches showed decreasing potency in vitro and precisely the same rank-order potencies in vivo (Monia et al. 1995; Monia et al. 1996).

Finally, a single injection of a phosphorothioate oligonucleotide designed to inhibit c-AMP-dependent protein kinase type 1 was reported to reduce RNA and protein levels selectively in human tumor xenografts and to reduce tumor growth (Nesterova et al. 1995).

Thus, there is a growing body of evidence that phosphorothioate oligonucleotides can induce potent systemic and local effects in vivo. More importantly, there are now several studies with sufficient controls and direct observation of target RNA and protein levels to suggest highly specific effects that are difficult to explain via any mechanism other than antisense. As would be expected, the potency of these effects varies depending on the target, the organ, and the end point measured as well as the route of administration and the time after a dose when the effect is measured.

In conclusion, although it is of obvious importance to interpret in vivo activity data cautiously, and it is clearly necessary to include a range of controls and to evaluate effects on target RNA and protein levels and control RNA and protein levels directly, it is difficult to argue with the conclusion today that some effects have been observed in animals that are most likely primarily due to an antisense mechanism.

Additionally, in studies on patients with cytomegalovirus-induced retinitis, local injections of ISIS 2922 have resulted in impressive efficacy, though it is obviously impossible to prove the mechanism of action is antisense in these studies (Hutcherson et al. 1995). More recently, ISIS 2302, an ICAM-1 inhibitor, was reported to result in statistically significant reductions in steroid doses and prolonged remissions in a small group of steroid-dependent patients with Crohn's disease. Because this study was a randomized, double-blind one and included serial colonoscopies, it may be considered the first study in humans to demonstrate the therapeutic activity of an antisense drug after systemic administration (Yacyshyn et al. 1997).

Toxicological Properties

In Vitro

In our laboratory, we have evaluated the toxicities of scores of phosphorothioate oligodeoxynucleotides in a significant number of cell lines in tissue culture. As a general rule, no significant cytotoxicity is induced at concentrations below 100 μM of oligonucleotide. Additionally, with a few exceptions, no significant effect on macromolecular synthesis is observed at concentrations below 100 μM (Crooke 1993a; Crooke 1993b).

Polynucleotides and other polyanions have been shown to cause release of cytokines (Colby 1971). Also, bacterial DNA species have been reported to be mitogenic for lymphocytes in vitro (Messina et al. 1991). Furthermore, oligodeoxynucleotides (30–45 nucleotides in length) were reported to induce interferons and enhance natural killer cell activity (Kuramoto et al. 1992). In the latter study, the oligonucleotides that displayed natural killer cell (NK)-stimulating activity contained specific palindromic sequences and tended to be guanosine rich. Collectively, these observations indicate that nucleic acids may have broad immunostimulatory activity.

It has been shown that phosphorothioate oligonucleotides stimulate B-lymphocyte proliferation in a mouse splenocyte preparation (analogous to bacterial DNA), and the response may underlie the observations of lymphoid hyperplasia in the spleen and lymph nodes of rodents caused by repeated administration of these compounds (see below) (Pisetsky et al. 1994). We also have evidence of enhanced cytokine release by immunocompetent cells when exposed to phosphorothioates in vitro (Crooke et al. 1996). In this study, both human keratinocytes and an in vitro model of human skin released interleukin 1-a when treated with 250 μM–1 mm of phosphorothioate oligonucleotides. The effects seemed to be dependent on the phosphorothioate backbone and independent of sequence or 2'-modification. In a study in which murine B-lymphocytes were treated with phosphodiester oligonucleotides, B-cell activation was induced by oligonucleotides with unmethylated CpG dinucleotides (Krieg et al. 1995). This has been extrapolated to suggest that the CpG motif may be required for immune stimulation of oligonucleotide analogs such as phosphorothioates. This clearly is not the case with regard to release of IL-1α from keratinocytes (Crooke et al. 1996), nor is it the case with regard to in vivo immune stimulation (see In Vivo section).

Genotoxicity

As with any new chemical class of therapeutic agents, concerns about genotoxicity cannot be dismissed, for little in vitro testing has been performed, and no data from long-term studies of oligonucleotides are available. Clearly, given the limitations in our understanding about the basic mechanisms that might be involved, empirical data must be generated. We have performed mutagenicity studies on two phosphorothioate oligonucleotides, ISIS 2105 and ISIS 2922, and have found them to be nonmutagenic at all concentrations studied (Crooke et al. 1994).

Two mechanisms of genotoxicity that may be unique to oligonucleotides have been considered. One possibility is that an oligonucleotide analog could be integrated into the genome and produce mutagenic events. Although integration of an oligonucleotide into the genome is conceivable, it is likely to be extremely rare. For most viruses, viral DNA integration is itself a rare event and, of course, viruses have evolved specialized enzyme-mediated mechanisms to achieve integration. Moreover, preliminary studies in our laboratory have shown that phosphorothioate oligodeoxynucleotides are generally poor substrates for DNA polymerases, and it is unlikely that enzymes such as integrases, gyrases, and topoisomerases (that have obligate DNA cleavage as intermediate steps in their enzymatic processes) will accept these compounds as substrates. Consequently, it would seem that the risk of genotoxicity due to genomic integration is no greater and is probably less than that of other potential mechanisms, such as alteration of the activity of growth factors, cytokine release, nonspecific effects on membranes that might trigger arachidonic acid release, or inappropriate intracellular signaling. Presumably, new analogs that deviate more significantly from natural DNA would be even less likely to be integrated.

A second concern that has been raised about possible genotoxicity is the risk that oligonucleotides may be degraded to toxic or carcinogenic metabolites. However, metabolism of phosphorothioate oligodeoxynucleotides by base excision would release normal bases, which presumably would be nongenotoxic. Similarly, oxidation of the phosphorothioate backbone to the natural phosphodiester structure would also yield nonmutagenic (and probably nontoxic) metabolites. Finally, it is possible that phosphorothioate bonds can be hydrolyzed slowly, releasing nucleoside phosphorothioates that presumably will be rapidly oxidized to natural (nontoxic) nucleoside phosphates. However, oligonucleotides with modified bases, backbones, or both, may pose different risks.

In Vivo

The acute LD_{50} in mice of all phosphorothioate oligonucleotides tested to date is in excess of 500 mg/kg (Douglas Kornbrust, unpublished observations). In rodents, we have had the opportunity to evaluate the acute and chronic toxicities of multiple phosphorothioate oligonucleotides administered by multiple routes (Henry et al. 1997; Henry et al.). The consistent dose-limiting toxicity was immune-stimulation-manifested by lymphoid hyperplasia, spelnomegaly and a multiorgan monocellular infiltrate. These effects occurred only with chronic dosing at doses greater than 20 mg/kg and were dose dependent. The liver and kidney were the organs most prominently affected by monocellular infiltrates. All of these effects appeared to be reversible, and chronic intradermal administration seemed to be the most toxic route—probably because of high local concentrations of the drugs that resulted in local cytokine release and initiation of a cytokine cascade. There were no obvious effects of sequence. At doses of

100 mg/kg and greater, minor increases in liver enzyme levels and mild thrombocytopenia were also observed.

In monkeys, however, the toxicological profile of phosphorothioate oligonucleotides is quite different. The most prominent dose-limiting side effect is sporadic reductions in blood pressure associated with bradycardia. When these events are observed, they are often associated with activation of C5 complement, and they are dose-related and peak-plasma-concentration related. This finding appears to be connected with the activation of the alternative pathway (Henry et al. 1997c). All phosphorothioate oligonucleotides tested to date appear to induce these effects, although there may be slight variations in potency as a function of sequence, length, or both (Cornish et al. 1993; Galbraith et al. 1994; Henry et al. 1997c).

A second prominent toxicologic effect in the monkey is the prolongation of activated partial thromboplastin time (APTT). At higher doses, evidence of clotting abnormalities is observed. Again, these effects are dose- and peak-plasma-concentration dependent (Galbraith et al. 1994; Henry et al. 1997b). Although no evidence of sequence dependence has been observed, there appears to be a linear correlation between number of phosphorothioate linkages and potency between 18–25 nucleotides (Nicklin, Paul, unpublished observations). The mechanisms responsible for these effects are likely very complex, but preliminary data suggest that direct interactions with thrombin may be at least partially responsible for the effects observed (Henry et al. 1997d).

In man, again, the toxicological profile differs a bit. When ISIS 2922 is administered intravitreally to patients with cytomegalovirus retinitis, the most common adverse event is anterior chamber inflammation that is easily managed with steroids. A relatively rare and dose-related adverse event is morphological changes in the retina associated with loss in peripheral vision (Hutcherson et al. 1995).

The 20-mer phosphorothioate ISIS 2105, which is designed to inhibit the replication of human papilloma viruses that cause genital warts, is administered intradermally at doses as high as 3 mg/wart weekly for 3 weeks. Essentially no toxicities have been observed, and, remarkably, a complete absence of local inflammation has been noted (Grillone, Lisa, unpublished results).

Every-other-day administration of 2-h intravenous infusions of ISIS 2302 at doses as high as 2 mg/kg resulted in no significant toxicities, including no evidence of immune stimulation and no hypotension. A slight subclinical increase in APTT was observed at the 2 mg/kg dose (Glover et al. 1997).

Therapeutic Index

In Figure 2, an attempt to put the toxicities and their dose response relationships in a therapeutic context is shown. This is particularly important because considerable confusion has arisen concerning the potential utility of phosphorothioate oligonucleotides for selected therapeutic purposes deriving from unsophisticated interpretation of toxicological data. As can be readily seen, the immune stimulation induced by these compounds appears to be particularly prominent in rodents and is unlikely to be dose

Plasma Concentration of ISIS 2302

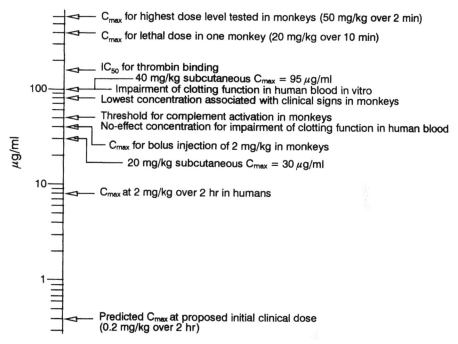

FIG. 2. Plasma concentrations of ISIS 2302 at which various activities are observed. These concentrations are determined by plasma extraction and capillary gel electrophoresis analysis and represent intact ISIS 2302.

limiting in man, nor have we, to date, observed hypotensive events in humans. Thus, this toxicity appears to occur at lower doses in monkeys than man and certainly is not dose-limiting in man.

On the basis of our experience to date, we believe that the dose-limiting toxicity in man will be clotting abnormalities, and this will be associated with peak plasma concentrations well in excess of 10 μg/ml. In animals, pharmacological activities have been observed with intravenous bolus doses from 0.006 mg/kg to 10–15 mg/kg, depending on the target, the end point, the organ studied, and the time after a dose when the effect is measured. Thus, it would appear that phosphorothioate oligonucleotides have a therapeutic index that supports their evaluation for several therapeutic indications.

Conclusions

Phosphorothioate oligonucleotides have perhaps outperformed many expectations. They display attractive parenteral pharmacokinetic properties. They have produced

TABLE 2. *Phosphorothioate oligonucleotides*

Limits
• Pharmacodynamic
– Low affinity per nucleotide unit
– Inhibition of RNase H at high concentrations
• Pharmacokinetic
– Limited bioavailability
– Limited blood–brain barrier penetration
– Dose-dependent pharmacokinetics
– Possible drug–drug interactions
• Toxicologic
– Release of cytokines
– Complement-associated effects on blood pressure?
– Clotting effects

potent systemic effects in several animal models and, in many experiments, the anti-sense mechanism has been directly demonstrated as the hoped-for selectivity. Further, these compounds appear to display satisfactory therapeutic indices for many indications.

Nevertheless, phosphorothioates clearly have significant limits (Table 2). Pharmacodynamically, they have relatively low affinity per nucleotide unit. This means that longer oligonucleotides are required for biological activity and that invasion of many RNA structures may not be possible. At higher concentrations, these compounds inhibit RNase H as well. Thus, the higher end of the pharmacologic dose response curve is lost. Pharmacokinetically, phosphorothioates do not cross the blood–brain barrier, are not orally bioavailable to a significant degree, and may display dose-dependent pharmacokinetics. Toxicologically, the release of cytokines, activation of complement, and interference with clotting will clearly pose dose limits if they are encountered in the clinic.

Because several clinical trials are in progress with phosphorothioates and others will be initiated shortly, we shall soon have more definitive information about the activities, toxicities, and value of this class of antisense drugs in human beings.

THE MEDICINAL CHEMISTRY OF OLIGONUCLEOTIDES

The core of any rational drug discovery program is medicinal chemistry. Although the synthesis of modified nucleic acids has been a subject of interest for some time, the intense focus on the medicinal chemistry of oligonucleotides predates this chapter by perhaps no more than 5 years. Consequently, the scope of medicinal chemistry has recently expanded enormously, but the biological data to support conclusions about synthetic strategies are only beginning to emerge.

Modifications in the base, sugar, and phosphate moieties of oligonucleotides have been reported. The subjects of medicinal chemical programs include approaches to create enhanced affinity and more selective affinity for RNA or duplex structures; the ability to cleave nucleic acid targets; enhanced nuclease stability, cellular uptake, and distribution; and in vivo tissue distribution, metabolism, and clearance.

Heterocycle Modifications

Pyrimidine Modifications

A relatively large number of modified pyrimidines have been synthesized and are now being incorporated into oligonucleotides and evaluated. The principal sites of modification are C-2, C-4, C-5, and C-6. These and other nucleoside analogs have recently been thoroughly reviewed (Sanghvi 1993). Consequently, a very brief summary of the analogs that displayed interesting properties is incorporated here.

Inasmuch as the C-2 position is involved in Watson–Crick hybridization, C-2 modified pyrimidine containing oligonucleotides has shown unattractive hybridization properties. An oligonucleotide containing 2-thiothymidine was found to hybridize well to DNA and, in fact, even better to RNA ΔT_m 1.5 °C modification (Swayze et al., unpublished results).

In contrast, several modifications in the C-4 position that have interesting properties have been reported. Some 4-thiopyrimidines have been incorporated into oligonucleotides with no significant negative effect on hybridization (Nikiforov et al. 1991). A bicyclic and an N4-methoxy analog of cytosine were shown to hybridize with both purine bases in DNA with T_ms approximately equal to natural base pairs (Lin and Brown 1989). Additionally, a fluorescent base has been incorporated into oligonucleotides and shown to enhance DNA–DNA duplex stability (Inoue et al. 1985).

A large number of modifications at the C-5 position have also been reported, including halogenated nucleosides. Although the stability of duplexes may be enhanced by incorporating 5-halogenated nucleosides, the occasional mispairing with G and the potential that the oligonucleotide may degrade and release toxic nucleoside analogs cause concern (Sanghvi 1993).

Furthermore, oligonucleotides containing 5-propynylpyrimidine modifications have been shown to enhance the duplex stability ΔT_m 1.6 °C/modification and support the RNase H activity. The 5-heteroarylpyrimidines were also shown to influence the stability of duplexes (Wagner et al. 1993; Gutierrez et al. 1994). A more dramatic influence was reported for the tricyclic 2′-deoxycytidine analogs, which exhibit an enhancement of 2–5 °C/modification, depending on the positioning of the modified bases (Lin et al. 1995). It is believed that the enhanced binding properties of these analogs is due to extended stacking and increased hydrophobic interactions.

In general, as expected, modifications in the C-6 position of pyrimidines are highly duplex-destabilizing (Sanghvi et al. 1993). Oligonucleotides containing 6-aza pyrimidines have been shown to reduce T_m by 1–2 °C per modification but to enhance the nuclease stability of oligonucleotides and to support RNase H-induced degradation of RNA targets (Sanghvi 1993).

Purine Modifications

Although numerous purine analogs have been synthesized, they usually have resulted in destabilization of duplexes when incorporated into oligonucleotides. However, there are a few exceptions, in which a purine modification had a stabilizing effect. A brief summary of some of these analogs is presented below.

Generally, N1 modifications of purine moiety has resulted in destabilization of the duplex (Manoharan 1993). Similarly, C-2 modifications have usually resulted in destabilization. However, 2-6-diaminopurine has been reported to enhance hybridization by approximately 1 °C per modification when paired with T (Sproat et al. 1991). Of the 3-position substituted bases reported to date, only the 3-deaza adenosine analog has been shown to have no negative effective on hybridization.

Modifications at the C-6 and C-7 positions have likewise resulted in only a few interesting bases from the point of view of hybridization. Inosine has been shown to have little effect on duplex stability, but because it can pair and stack with all four normal DNA bases, it behaves as a universal base and creates an ambiguous position in an oligonucleotide (Martin et al. 1985). Incorporation of 7-deaza inosine into oligonucleotides was destabilizing, and this was considered to be due to its relatively hydrophobic nature (SantaLucia et al. 1991). Although 7-deaza guanine was similarly destabilizing, when 8-aza-7-deaza guanine was incorporated into oligonucleotides, it enhanced hybridizations (Seela et al. 1989). Thus, on occasion, introduction of more than one modification in a nucleobase may compensate for destabilizing effects of some modifications. Interestingly, 7-iodo 7-deazaguanine residue was recently incorporated into oligonucleotides and shown to enhance the binding affinity dramatically ($\Delta T_m 10.0$ °C/modification compared with 7-deazaguanine) (Seela et al. 1995). The increase in T_m value was attributed to (1) the hydrophobic nature of the modification, (2) increased stacking interaction, and (3) favorable (Note: The concentration at which 1/2 of molecule is in acid form) pK_a of the base.

In contrast, some C-8 substituted bases have yielded improved nuclease resistance when incorporated in oligonucleotides but seem to be somewhat destabilizing (Sanghvi 1993).

Oligonucleotide Conjugates

Although conjugation of various functionalities to oligonucleotides has been reported to have achieved several important objectives, the data supporting some of the claims are limited, and generalizations are not possible on the basis of the data presently available.

Nuclease Stability

Numerous 3'-modifications have been reported to enhance the stability of oligonucleotides in serum (Manoharan 1993). Both neutral and charged substituents have been reported to stabilize oligonucleotides in serum and, as a general rule, the stability of a conjugated oligonucleotide tends to be greater as bulkier substituents are added. Inasmuch as the principal nuclease in serum is a 3'-exonuclease, it is not surprising that 5'-modifications have resulted in significantly less stabilization. Internal modifications of base, sugar, and backbone have also been reported to enhance nuclease stability at or near the modified nucleoside (Manoharan 1993). In a recent study, thiono-triester-(adamantyl, cholesteryl, and others) modified oligonucleotides have shown improved nuclease stability, cellular association, and binding affinity (Zhang et al. 1995).

The demonstration that modifications may induce nuclease stability sufficient to enhance activity in cells in tissue culture and in animals has proven to be much more complicated because of the presence of 5'-exonucleases and endonucleases. In our laboratory, 3'-modifications and internal point modifications have not provided sufficient nuclease stability to demonstrate pharmacological activity in cells (Hoke et al. 1991). In fact, even a 5-nucleotide-long phosphodiester gap in the middle of a phosphorothioate oligonucleotide resulted in sufficient loss of nuclease resistance to cause complete loss of pharmacological activity (Monia et al. 1992).

In mice, neither a 5'-cholesterol nor 5'-C-18 amine conjugate altered the metabolic rate of a phosphorothioate oligodeoxynucleotide in liver, kidney, or plasma (Crooke et al. 1996). Furthermore, blocking the 3'- and 5'-termini of a phosphodiester oligonucleotide did not markedly enhance the nuclease stability of the parent compound in mice (Sands et al. 1995). However, 3'-modification of a phosphorothioate oligonucleotide was reported to enhance its stability in mice relative to the parent phosphorothioate (Temsamani et al. 1993). Moreover, a phosphorothioate oligonucleotide with a 3'-hairpin loop was reported to be more stable in rats than its parent (Zhang et al. 1995). Thus, 3'-modifications may enhance the stability of the relatively stable phosphorothioates sufficiently to be of value.

Enhanced Cellular Uptake

Although oligonucleotides have been shown to be taken up by several cell lines in tissue culture (perhaps the most compelling data relate to phosphorothioate oligonucleotides), a clear objective has been to improve cellular uptake of oligonucleotides (Crooke 1991; Crooke et al. 1994). Inasmuch as the mechanisms of cellular uptake of oligonucleotides are still very poorly understood, the medicinal chemistry approaches have been largely empirical and based on many unproven assumptions.

Because phosphodiester and phosphorothioate oligonucleotides are water soluble, the conjugation of lipophilic substituents to enhance membrane permeability has been a subject of considerable interest. Unfortunately, studies in this area have not been systematic and, at present, there is precious little information about the changes in physicochemical properties of oligonucleotides actually affected by specific lipid conjugates. Phospholipids, cholesterol and cholesterol derivatives, cholic acid, and simple alkyl chains have been conjugated to oligonucleotides at various sites in the oligonucleotide. The effects of these modifications on cellular uptake have been assessed using fluorescent, or radiolabeled, oligonucleotides or by measuring pharmacological activities. From the perspective of medicinal chemistry, very few systematic studies have been performed. The activities of short alkyl chains, adamantine, daunomycin, fluorescein, cholesterol, and porphyrin-conjugated oligonucleotides were compared in one study. A cholesterol modification was reported to be more effective at enhancing uptake than the other substituents. It also seems likely that the effects of various conjugates on cellular uptake may be affected by the cell type and target studied. For example, we have studied cholic acid conjugates of phosphorothioate deoxyoligonucleotides or phosphorothioate 2'-methoxy oligonucleotides and observed enhanced activity against HIV and no effect on the activity of ICAM-directed oligonucleotides.

Additionally, polycationic substitutions and various groups designed to bind to cellular carrier systems have been synthesized. Although many compounds have been synthesized, the data reported to date are insufficient to draw firm conclusions about the value of such approaches or structure activity relationships (Manoharan 1993).

RNA Cleaving Groups

Oligonucleotide conjugates were recently reported to act as artificial ribonucleases, albeit in low efficiencies (De Mesmaeker et al. 1995). Conjugation of chemically reactive groups such as alkylating agents, photoinduced azides, prophine, and psoralene has been utilized extensively to effect a cross-linking of oligonucleotide and the target RNA. In principle, this treatment may lead to translation arrest. In addition, lanthanides and complexes thereof have been reported to cleave RNA via a hydrolytic pathway. Recently, a novel europium complex was covalently linked to an oligonucleotide and shown to cleave 88% of the complementary RNA at physiological pH (Hall et al. 1994).

In Vivo Effects

To date, relatively few studies have been reported in vivo. The properties of a 5'-cholesterol and 5'-C-18 amine conjugates of a 20-mer phosphorothioate oligodeoxynucleotide have been determined in mice. Both compounds increased the fraction of an intravenous bolus dose found in the liver. The cholesterol conjugate, in fact, resulted in more than 80% of the dose accumulating in the liver. Neither conjugate enhanced stability in plasma, liver, or kidney (Crooke et al. 1996). Interestingly, the only significant change in the toxicity profile was a slight increase in effects on serum transaminases and histopathological changes indicative of slight liver toxicity associated with the cholesterol conjugate (Henry et al. 1997d). A 5'-cholesterol phosphorothioate conjugate was also recently reported to have a longer elimination half-life, to be more potent, and to induce greater liver toxicity in rats (Desjardins et al. 1995).

Sugar Modifications

The focus of second-generation oligonucleotide modifications has centered on the sugar moiety. In oligonucleotides, the pentafuranose sugar ring occupies a central connecting manifold that also positions the nucleobases for effective stacking. Recently, a symposium series has been published on the carbohydrate modifications in antisense research that covers this topic in great detail (Sanghvi et al. 1994). Therefore, the content of the following discussion is restricted to a summary of the main events in this area.

A growing number of oligonucleotides in which the pentafuranose ring is modified or replaced have been reported (Breslauer et al. 1986). Uniform modifications at the 2'-position have been shown to enhance hybridization to RNA, and in some cases, to enhance nuclease resistance (Breslauer et al. 1986). Chimeric oligonucleotides containing 2'-deoxyoligonucleotide gaps with 2'-modified wings have been shown to be more potent than parent molecules (Monia et al. 1993).

Other sugar modifications include α-oligonucleotides, carbocyclic oligonucleotides, and hexapyranosyl oligonucleotides (Breslauer et al. 1986). Of these, α-oligonucleotides have been most extensively studied. They hybridize in parallel fashion to single-stranded DNA and RNA and are nuclease resistant. However, they have been reported to be oligonucleotides designed to inhibit Ha-*ras* expression. All these oligonucleotides support RNase H and, as can be seen, a direct correlation between affinity and potency exists.

A growing number of oligonucleotides in which the C-2'-position of the sugar ring is modified have been reported (Manoharan 1993; De Mesmaeker et al. 1995). These modifications include lipophilic alkyl groups, intercalators, amphipathic amino-alkyl tethers, positively charged polyamines, highly electronegative fluoro or fluoro-alkyl moities, and sterically bulky methylthio derivatives. The beneficial effects of a C-2'-substitution on the antisense oligonucleotide cellular uptake, nuclease resistance, and binding affinity have been well-documented in the literature. In addition, excellent review articles have appeared in the last few years on the synthesis and properties of C-2'-modified oligonucleotides (Lamond et al. 1993; Sproat et al. 1993; Parmentier et al. 1994; De Mesmaeker et al. 1995).

Other modifications of the sugar moiety have also been studied, including other sites as well as more substantial modifications. However, much less is known about the antisense effects of these modifications (Crooke 1995a).

Recently, 2'-methoxy-substituted phosphorothioate oligonucleotides have been reported to be more stable in mice than their parent compounds and to display enhanced oral bioavailability (Agrawal et al. 1995; Zhang et al. 1995). The analogs displayed tissue distribution similar to that of the parent phosphorothioate.

Similarly, we have compared the pharmacokinetics of 2'-propoxy-modified phosphodiester and phosphorothioate deoxynucleotides (Crooke et al. 1996). As expected, the 2'-propoxy modification increased lipophilicity and nuclease resistance. In fact, in mice the 2'-propoxy phosphorothioate was too stable in liver or kidney to permit measurement of an elimination half-life.

Interestingly, the 2'-propoxy phosphodiester was much less stable than the parent phosphorothioate in all organs except the kidney, in which the 2'-propoxy phosphodiester was remarkably stable. The 2'-propoxy phosphodiester did not bind to albumin significantly, whereas the affinity of the phosphorothioate for albumin was enhanced. The only difference in toxicity between the analogs was a slight increase in renal toxicity associated with the 2'-propoxy phosphodiester analog (Henry et al. 1997d).

Incorporation of the 2'-methoxyethyoxy group into oligonucleotides increased the T_m by 1.1 °C/modification when hybridized to the complement RNA. In a similar manner, several other 2'-O-alkoxy modifications have been reported to enhance the affinity (Martin 1995). The increase in affinity with these modifications was attributed to (1) the favorable gauche effect of the side chain and (2) additional solvation of the alkoxy substituent in water.

More substantial carbohydrate modifications have also been studied. Hexose-containing oligonucleotides were created and found to have very low affinity for RNA (Pitsch et al. 1995). Also, the 4'-oxygen has been replaced with sulfur. Although a single substitution of a 4'-thio-modified nucleoside resulted in destabilization of a duplex, incorporation of two 4'-thio-modified nucleosides increased the affinity of the duplex (Bellon et al. 1994). Finally, bicyclic sugars have been synthesized with the hope that preorganization into more rigid structures would enhance hybridization. Several of these modifications have been reported to enhance hybridization (Sanghvi et al. 1994).

Backbone Modifications

Substantial progress in creating new backbones for oligonucleotides that replace the phosphate or the sugar-phosphate unit has been made. The objectives of these programs are to improve hybridization by removing the negative charge, enhancing stability, and potentially improving pharmacokinetics.

For a review of the backbone modifications reported to date, please see Sanghvi et al. (1994) and Crooke (1995a). Suffice it to say that numerous modifications have been made that replace phosphate, retain hybridization, alter charge, and enhance stability. Because these modifications are now being evaluated in vitro and in vivo, a preliminary assessment should be possible shortly.

Replacement of the entire sugar-phosphate unit has also been accomplished, and the oligonucleotides produced have displayed very interesting characteristics. The PNA oligonucleotides have been shown to bind to single-stranded DNA and RNA with extraordinary affinity and high sequence specificity. They have been shown to be capable of invading some double-stranded nucleic acid structures. The PNA oligonucleotides can form triple-stranded structures with DNA or RNA.

The PNA oligonucleotides were shown to be able to act as antisense and transcriptional inhibitors when microinjected in cells (Hanvey et al. 1992) and appear to be quite stable to nucleases and peptidases as well.

In summary, then, in the past 5 years, enormous advances in the medicinal chemistry of oligonucleotides have been reported. Modifications at nearly every position in oligonucleotides have been attempted, and numerous potentially interesting analogs have been identified. Although it is far too early to determine which of the modifications may be most useful for particular purposes, it is clear that a wealth of new chemicals is available for systematic evaluation and that these studies should provide important insights into the SAR of oligonucleotide analogs.

CONCLUSIONS

Although many more questions about antisense remain to be answered than have been answered, progress has continued to be gratifying. Clearly, as more is learned, we will be in the position to perform progressively more sophisticated studies and to understand more of the factors that determine whether an oligonucleotide actually works via an antisense mechanism. We should also have the opportunity to learn a great deal more about this class of drugs as additional studies are completed in humans.

ABBREVIATIONS

PNA: peptide nucleic acid; mRNA: messenger RNA; Ha-*ras*: Harvey *ras*; IL: interleukin; TAR: transactivator response element; ICAM: intercellular adhesion molecule; PKC: protein kinase C; HIV: human immune deficiency virus; NK: natural killer; CMV: cytomegalovirus; HPV: human papillomavirus; HSV: herpes simplex virus; PTHrP: parathyroid hormone-related peptide.

ACKNOWLEDGMENTS

The author wishes to thank Colleen Matzinger for excellent typographic and administrative assistance.

REFERENCES

Abe, J., Zhou, W., Taguchi, J., Takuwa, N., Miki, K., Okazaki, H., Kurokawa, K., Kumada, M., and Takuwa, Y. 1994. Suppression of neointimal smooth muscle cell accumulation *in vivo* by antisense CDC2 and CDK2 oligonucleotides in rat carotid artery. *Biochem. Biophys. Res. Commun.* 198:16–24.

Adams, J. U., Chen, X. H., deRiel, J. K., Adler, M. W., and Liu-Chen, L.-Y. 1994. *In vivo* treatment with antisense oligodeoxynucleotide to kappa-opioid receptors inhibited kappa-agonist-induced analgesia in rats. *Regul. Pept.* 54:1–2.

Agrawal, S., Goodchild, J., Civeira, M. P., Thornton, A. H., Sarin, P. S., and Zamecnik, P. C. 1988. Oligodeoxynucleoside phosphoramidates and phosphorothioates as inhibitors of human immunodeficiency virus. *Proc. Natl. Acad. Sci. U.S.A.* 85:7079–7083.

Agrawal, S., Temsamani, J., and Tang, J. Y. 1991. Pharmacokinetics, biodistribution, and stability of oligodeoxynucleotide phosphorothioates in mice. *Proc. Natl. Acad. Sci. U.S.A.* 88:7595–7599.

Agrawal, S., Zhang, X., Lu, Z., Zhao, H., Tamburin, J. M., Yan, J., Cai, H., Diasio, R. B., Habus, I., Jiang, Z., Iyer, R. P., Yu, D., and Zhang, R. 1995. Absorption, tissue distribution and *in vivo* stability in rats of a hybrid antisense oligonucleotide following oral administration. *Biochem. Pharmacol.* 50(4):571–576.

Akabayashi, A., Wahlestedt, C., Alexander, J. T., and Leibowitz, S. F. 1994. Specific inhibition of endogenous neuropeptide Y synthesis in arcuate nucleus by antisense oligonucleotides suppresses feeding behavior and insulin secretion. *Mol. Brain Res.* 21:55–61.

Akino, K., Ohtsuru, A., Yano, H., Ozeki, S., Namba, H., Nakashima, M., Ito, M., Matsumoto, T., and Yamashita, S. 1996. Antisense inhibition of parathyroid hormone-related peptide gene expression reduces malignant pituitary tumor progression and metastases in the rat. *Cancer Res.* 56(1):77–86.

Amaratunga, A., Morin, P. J., Kosik, K. S., and Fine, R. E. 1993. Inhibition of kinesin synthesis and rapid anterograde axonal transport in vivo by an antisense oligonucleotide. *J. Biol. Chem.* 268(23):17427–17430.

Ambuhl, P., Gyurko, R., and Phillips, M. I. 1995. A decrease in angiotensin receptor binding in rat brain nuclei by antisense oligonucleotides to the angiotensin AT1 receptor. *Regul. Pept.* 59(2):171–182.

Azad, R. F., Driver, V. B., Tanaka, K., Crooke, R. M., and Anderson, K. P. 1993. Antiviral activity of a phosphorothioate oligonucleotide complementary to RNA of the human cytomegalovirus major immediate-early region. *Antimicrob. Agents Chemother.* 37(9):1945–1954.

Barton, C. M., and Lemoine, N. R. 1995. Antisense oligonucleotides directed against p53 have antiproliferative effects unrelated to effects on p53 expression. *Br. J. Cancer* 71:429–437.

Bayever, E., Iversen, P. L., Bishop, M. R., Sharp, J. G., Tewary, H. K., Arneson, M. A., Pirruccello, S. J., Ruddon, R. W., Kessinger, A., and Zon, G. 1993. Systemic administration of a phosphorothioate oligonucleotide with a sequence complementary to p53 for acute myelogenous leukemia and myelodysplastic syndrome: Initial results of a phase I trial. *Antisense Res. Dev.* 3(4):383–390.

Bellon, L., Leydier, C., and Barascut, J.-L. 1994. 4-Thio RNA: A novel class of sugar-modified B-RNA. In *Carbohydrate modifications in antisense research*, eds. Y. S. Sanghvi and P. D. Cook, 68–79. Washington, DC: American Chemical Society.

Bennett, C. F., Chiang, M.-Y., Chan, H., and Grimm, S. 1993. Use of cationic lipids to enhance the biological activity of antisense oligonucleotides. *J. Liposome Res.* 3:85–102.

Bennett, C. F., Chiang, M.-Y., Chan, H., Shoemaker, J. E. E., and Mirabelli, C. K. 1992. Cationic lipids enhance cellular uptake and activity of phosphorothioate antisense oligonucleotides. *Mol. Pharmacol.* 41:1023–1033.

Bennett, C. F., and Crooke, S. T. 1996. Oligonucleotide-based inhibitors of cytokine expression and function. In *Therapeutic modulation of cytokines*, eds. B. Henderson and M. W. Bodmer, 171–193. Boca Raton, Florida: CRC Press.

Bennett, C. F., Kornbrust, D., Henry, S., Stecker, K., Howard, R., Cooper, S., Dutson, S., Hall, W., and Jacoby, H. I. 1997. An ICAM-1 antisense oligonucleotide prevents and reverses dextran sulfate sodium-induced colitis in mice. *J. Pharmacol. Exp. Ther.* 280(2):988–1000.

Breslauer, K. J., Frank, R., Blocker, H., and Marky, L. A. 1986. Predicting DNA duplex stability from base sequence. *Proc. Natl. Acad. Sci. U.S.A.* 83:3746–3750.

Burch, R. M., and Mahan, L. C. 1991. Oligonucleotides antisense to the interleukin 1 receptor mRNA block the effects of interleukin 1 in cultured murine and human fibroblasts and in mice. *J. Clin. Invest.* 88:1190–1196.

Burgess, T. L., Fisher, E. F., Ross, S. L., Bready, J. V., Qian, Y.-X., Bayewitch, L. A., Cohen, A. M., Herrera, C. J., Hu, S. S.-F., Kramer, T. B., Lott, F. D., Martin, F. H., Pierce, G. F., Simonet, L., and Farrell, C. L. 1995. The antiproliferative activity of c-*myb* and c-*myc* antisense oligonucleotides in smooth muscle cells is caused by a nonantisense mechanism. *Proc. Natl. Acad. Sci. U.S.A.* 92:4051–4055.

Campbell, J. M., Bacon, T. A., and Wickstrom, E. 1990. Oligodeoxynucleoside phosphorothioate stability in subcellular extracts, culture media, sera and cerebrospinal fluid. *J. Biochem. Biophys. Methods* 20:259–267.

Castro-Alamancos, M. A., and Torres-Aleman, I. 1994. Learning of the conditioned eye-blink response is impaired by an antisense insulin-like growth factor I oligonucleotide. *Proc. Natl. Acad. Sci. U.S.A.* 91:10203–10207.

Cha, X. Y., Xu, H., Ni, Q., Partilla, J. S., Rice, K. C., Matecka, D., Calderon, S. N., Porreca, F., Lai, J., and Rothman, R. B. 1995. Opioid peptide receptor studies. 4. Antisense oligodeoxynucleotide to the delta opioid receptor delineates opioid receptor subtypes. *Regul. Pept.* 59(2):247–253.

Phosphorothioate oligodeoxynucleotide distribution in kidney

Saline   20 mg/kg P=S ODN

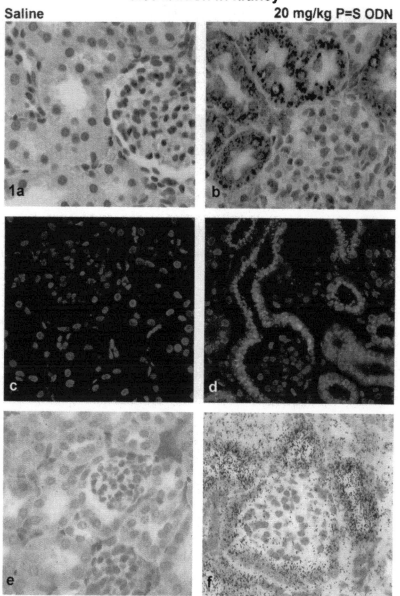

FIG. 1. Phosphorothioate oligodeoxynucleotide distribution in kidney.

Cheng, Y., Gao, W., and Han, F. 1991. Phosphorothioate oligonucleotides as potential antiviral compounds against human immunodeficiency virus and herpes viruses. *Nucleosides Nucleotides* 10:155–166.

Chiang, M.-Y., Chan, H., Zounes, M. A., Freier, S. M., Lima, W. F., and Bennett, C. F. 1991. Antisense oligonucleotides inhibit intercellular adhesion molecule 1 expression by two distinct mechanisms. *J. Biol. Chem.* 266:18162–18171.

Chiasson, B. J., Hooper, M. L., Murphy, P. R., and Robertson, H. A. 1992. Antisense oligonucleotide eliminates in vivo expression of c-*fos* in mammalian brain. *Eur. J. Pharmacol.* 227:451–453.

Chrisey, L. A., Walz, S. E., Pazirandeh, M., and Campbell, J. R. 1993. Internalization of oligodeoxyribonucleotides by *Vibrio parahemolyticus. Antisense Res. Dev.* 3:367–381.

Cohen, J. S. 1993. Phosphorothioate oligodeoxynucleotides. In *Antisense research and applications*, eds. S. T. Crooke and B. Lebleu, 205–222. Boca Raton, Florida: CRC Press.

Colby, C. J. 1971. The induction of interferon by natural and synthetic polynucleotides. *Prog. Nucleic Acid Res. Mol. Biol.* 11:1–32.

Cook, P. D. (1993). Medicinal chemistry strategies for antisense research. In *Antisense research and applications*, eds. S. T. Crooke and B. Lebleu, 149–187. Boca Raton, Florida: CRC Press.

Cornish, K. G., Iversen, P., Smith, L., Arneson, M., and Bayever, E. 1993. Cardiovascular effects of a phosphorothioate oligonucleotide to p53 in the conscious rhesus monkey. *Pharmacol. Comm.* 3:239–247.

Cossum, P. A., Sasmor, H., Dellinger, D., Truong, L., Cummins, L., Owens, S. R., Markham, P. M., Shea, J. P., and Crooke, S. 1993. Disposition of the ^{14}C-labeled phosphorothioate oligonucleotide ISIS 2105 after intravenous administration to rats. *J. Pharmacol. Exp. Ther.* 267:1181–1190.

Cossum, P. A., Truong, L., Owens, S. R., Markham, P. M., Shea, J. P., and Crooke, S. T. 1994. Pharmacokinetics of a ^{14}C-labeled phosphorothioate oligonucleotide, ISIS 2105, after intradermal administration to rats. *J. Pharm. Exp. Ther.* 269:89–94.

Cowsert, L. M. 1993. Antiviral activities of antisense oligonucleotides. In *Antisense research and applications*, eds. S. T. Crooke and B. Lebleu., 521–533. Boca Raton, Florida: CRC Press.

Cowsert, L. M., Fox, M. C., Zon, G., and Mirabelli, C. K. 1993. In vitro evaluation of phosphorothioate oligonucleotides targeted to the E2 mRNA of papillomavirus: Potential treatment of genital warts. *Antimicrob. Agents Chemother.* 37:171–177.

Crooke, R. M. 1991. In vitro toxicology and pharmacokinetics of antisense oligonucleotides. *Anti-Cancer Drug Design* 6:609–646.

Crooke, R. M. 1993a. Cellular uptake, distribution and metabolism of phosphorothioate, phosphodiester, and methylphosphonate oligonucleotides. In *Antisense research and applications*, eds. S. T. Crooke and B. Lebleu, 427–449. Boca Raton, Florida: CRC Press.

Crooke, R. M. 1993b. In vitro and in vivo toxicology of first generation analogs. In *Antisense research and applications*, eds. S. T. Crooke and B. Lebleu, 471–492. Boca Raton, Florida: CRC Press.

Crooke, R. M., Crooke, S. T., Graham, M. J., and Cooke, M. E. 1996. Effect of antisense oligonucleotides on cytokine release from human keratinocytes in an *in vitro* model of skin. *Toxicol. Appl. Pharmacol.* 140:85–93.

Crooke, R. M., Graham, M. J., Cooke, M. E., and Crooke, S. T. 1995. In vitro pharmacokinetics of phosphorothioate antisense oligonucleotides. *J. Pharmacol. Exp. Ther.* 275(1):462–473.

Crooke, S. T. 1992. Therapeutic applications of oligonucleotides. *Ann. Rev. Pharmacol. Toxicol.* 32:329–376.

Crooke, S. T. 1993. Progress toward oligonucleotide therapeutics: Pharmacodynamic properties. *FASEB J.* 7:533–539.

Crooke, S. T. 1995a. Oligonucleotide therapeutics. In *Burger's medicinal chemistry and drug discovery*, Vol. 1, ed. M. E. Wolff, 863–900. New York: John Wiley & Sons, Inc.

Crooke, S. T. 1995b. *Therapeutic applications of oligonucleotides*. Austin, Texas: R. G. Landes Company.

Crooke, S. T., Graham, M. J., Zuckerman, J. E., Brooks, D., Conklin, B. S., Cummins, L. L., Greig, M. J., Guinosso, C. J., Kornbrust, D., Manoharan, M., Sasmor, H. M., Schleich, T., Tivel, K. L., and Griffey, R. H. 1996. Pharmacokinetic properties of several novel oligonucleotide analogs in mice. *J. Pharmacol. Exp. Ther.* 277(2):923–937.

Crooke, S. T., Grillone, L. R., Tendolkar, A., Garrett, A., Fratkin, M. J., Leeds, J., and Barr, W. H. (1994). A pharmacokinetic evaluation of ^{14}C-labeled afovirsen sodium in patients with genital warts. *Clin. Pharma. Ther.* 56:641–646.

Crooke, S. T., and Lebleu, B. 1993. Antisense Research and Applications. Boca Raton, Florida: CRC Press.

Crooke, S. T., Lemonidis, K. M., Nielson, L., Griffey, R., and Monia, B. P. 1995. Kinetic characteristics of *E. coli* RNase H1: Cleavage of various antisense oligonucleotides-RNA duplexes. *Biochem. J.* 312(2):599–608.

De Clercq, E., Eckstein, F., and Merigan, T. C. 1969. Interferon induction increased through chemical modification of synthetic polyribonucleotide. *Science* 165:1137–1140.

De Mesmaeker, A., Haener, R., Martin, P., and Moser, H. E. 1995. Antisense oligonucleotides. *Acc. Chem. Res.* 28(9):366–74.

Dean, N. M., and McKay, R. 1994. Inhibition of protein kinase C-alpha expression in mice after systemic administration of phosphorothioate antisense oligodeoxynucleotides. *Proc. Natl. Acad. Sci. U.S.A.* 91:11762–11766.

Dean, N. M., McKay, R., Miraglia, L., Howard, R., Cooper, S., Giddings, J., Nicklin, P., Meister, L., Zeil, R., Geiger, T., Muller, M., and Fabbro, D. 1996. Inhibition of growth of human tumor cell lines in nude mice by an antisense oligonucleotide inhibitor of PKC-alpha expression. *Cancer Res.* 56(15):3499–3507.

Desjardins, J., Mata, J., Brown, T., Graham, D., Zon, G., and Iversen, P. 1995. Cholesteryl-conjugated phosphorothioate oligodeoxynucleotides modulate CYP2B1 expression in vivo. *J. Drug Targeting* 2:477–485.

Ecker, D. J. 1993. Strategies for invasion of RNA secondary structure. In *Antisense research and applications*, eds. S. T. Crooke and R. Lebleu, 387–400. Boca Raton, Florida: CRC Press.

Ecker, D. J., Vickers, T. A., Bruice, T. W., Freier, S. M., Jenison, R. D., Manoharan, M., and Zounes, M. 1992. Pseudo—half-knot formation with RNA. *Science* 257:958–961.

Flanagan, L. M., McCarthy, M. M., Brooks, P. J., Pfaff, D. W., and McEwen, B. S. 1993. Arginine vasopressin levels after daily infusions of antisense oligonucleotides into the supraoptic nucleus. *Ann. N. Y. Acad. Sci.* 689:520–521.

Freier, S. M. 1993. Hybridization considerations affecting antisense drugs. In *Antisense research and applications*, eds. S. T. Crooke and B. Lebleu, 67–82. Boca Raton, Florida: CRC Press.

Galbraith, W. M., Hobson, W. C., Giclas, P. C., Schechter, P. J., and Agrawal, S. 1994. Complement activation and hemodynamic changes following intravenous administration of phosphorothioate oligonucleotides in the monkey. *Antisense Res. Dev.* 4(3):201–206.

Gao, W.-Y., Han, F.-S., Storm, C., Egan, W., and Cheng, Y.-C. 1992. Phosphorothioate oligonucleotides are inhibitors of human DNA polymerases and RNase H: Implications for antisense technology. *Mol. Pharmacol.* 41:223–229.

Giles, R. V., Spiller, D. G., and Tidd, D. M. 1995. Detection of ribonuclease H-generated mRNA fragments in human leukemia cells following reversible membrane permeabilization in the presence of antisense oligodeoxynucleotides. *Antisense Res. Dev.* 5:23–31.

Gillardon, F., Beck, H., Uhlmann, E., Herdegen, T., Sandkühler, J., Peyman, A., and Zimmermann, M. 1994. Inhibition of c-*fos* protein expression in rat spinal cord by antisense oligodeoxynucleotide superfusion. *Eur. J. Neurosci.* 6:880–884.

Glover, J. M., Leeds, J. M., Mant, T. G. K., Amin, D., Kisner, D., Zuckerman, J. Geary, R. S., Levin, A., and Shanahan, J. W. R. 1997. Phase 1 safety and pharmacokinetic profile of an intercellular adhesion molecule-1 antisense oligodeoxynucleotide (ISIS 2302). *J. Pharmacol. Exp. Ther.* 282:1173–1180, 1997. 282:1173–1180.

Graham, M. J., Freier, S. M., Crooke, R. M., Ecker, D. J., Maslova, R. N., and Lesnik, E. A. 1993. Tritium labeling of antisense oligonucleotides by exchange with tritiated water. *Nucleic Acids Res.* 21:3737–3743.

Gutierrez, A. J., Terhorst, T. J., Matteucci, M. D., and Froehler, B. C. 1994. 5-heteroaryl-2'-deoxyuridine analogs. Synthesis and incorporation into high-affinity oligonucleotides. *J. Am. Chem. Soc.* 116:5540–5544.

Gyurko, R., Wielbo, D., and Phillips, M. I. 1993. Antisense inhibition of AT_1 receptor mRNA and angiotensinogen mRNA in the brain of spontaneously hypertensive rats reduces hypertension of neurogenic origin. *Regul. Pept.* 49:167–174.

Hall, J., Hüsken, D., Pieles, U., Moser, H. E., and Haner, R. 1994. Efficient sequence-specific cleavage of RNA using novel europium complexes conjugated to oligonucleotides. *Chem. Biol.* 1(3):185–90.

Hanvey, J. C., Peffer, N. C., Bisi, J. E., Thomson, S. A., Cadilla, R., Josey, J. A., Ricca, D. J., Hassman, C. F., Bonham, M. A., Au, K. G., Carter, S. G., Bruckenstein, D. A., Boyd, A. L., Noble, S. A., and Babiss, L. E. 1992. Antisense and antigene properties of peptide nucleic acids. *Science* 258:1481–1485.

Hartmann, G., Krug, A., Waller-Fontaine, K., and Endres, S. 1996. Oligodeoxynucleotides enhance lipopolysaccharide-stimulated synthesis of tumor necrosis factor: Dependence on phosphorothioate modification and reversal by heparin. *Mol. Med.* 2(4):429–438.

Hawley, P., and Gibson, I. 1996. Interaction of oligodeoxynucleotides with mammilian cells. *Antisense and Nucleic Drug Dev.* 6:185–195.

Heilig, M., Engel, J. A., and Soderpalm, B. 1993. C-*fos* antisense in the nucleus accumbens blocks the locomotor stimulant action of cocaine. *Eur. J. Pharmacol.* 236:339–340.

Henry, S. P., Giclas, P. C., Leeds, J., Pangburn, M., Auletta, C., Levin, A. A., and Kornbrust, D. J. 1997a. Activation of the alternative pathway of complement by a phosphorothioate oligonucleotide: Potential mechanism of action. *J. Pharmacol. Exp. Ther.* 281:810–816.

Henry, S. P., Grillone, L. R., Orr, J. L., Brunner, R. H., and Kornbrust, D. J. 1997b. Comparison of the toxicity profiles of ISIS 1082 and ISIS 2105, phosphorothioate oligonucleotides, following subacute intradermal administration in Sprague–Dawley rats. *Toxicology* 116(1–3):77–88.

Henry, S. P., Novotny, W., Leeds, J., Auletta, C., and Kornbrust, D. J. 1997c. Inhibition of coagulation by a phosphorothioate oligonucleotide. *Antisense Nuc. Acid Drug Dev.* 7(5):503–510.

Henry, S. P., Taylor, J., Midgley, L., Levin, A. A., and Kornbrust, D. J. 1997d. Evaluation of the toxicity profile of ISIS 2302, a phosphorothioate oligonucleotide in a 4-week study in CD-1 mice. *Antisense Nucleic Acid Drug Dev.* 7(5):473–481.

Henry, S. P., Zuckerman, J. E., Rojko, J., Hall, W. C., Harman, R. J., Kitchen, D., and Crooke, S. T. 1997e. Toxicologic properties of several novel oligonucleotide analogs in mice. *Anti-Cancer Drug Des.* 12(1):1–14.

Hertl, M., Neckers, L. M., and Katz, S. I. 1995. Inhibition of interferon-gamma-induced intercellular adhesion molecule-1 expression on human keratinocytes by phosphorothioate antisense oligodeoxynucleotides is the consequence of antisense-specific and antisense-non-specific effects. *J. Invest. Dermatol.* 104:813–818.

Higgins, K. A., Perez, J. R., Coleman, T. A., Dorshkind, K., McComas, W. A., Sarmiento, U. M., Rosen, C. A., and Narayanan, R. 1993. Antisense inhibition of the p65 subunit of NF-kappaB blocks tumorigenicity and causes tumor regression. *Proc. Natl. Acad. Sci. U.S.A.* 90:9901–9905.

Hijiya, N., Zhang, J., Ratajczak, M. Z., Kant, J. A., DeRiel, K., Herlyn, M., Zon, G., and Gewirtz, A. M. 1994. Biologic and therapeutic significance of *MYB* expression in human melanoma. *Proc. Natl. Acad. Sci. U.S.A.* 91:4499–4503.

Hodges, D., and Crooke, S. T. 1995. Inhibition of splicing of wild-type and mutated luciferase-adenovirus pre-mRNA by antisense oligonucleotides. *Mol. Pharmacol.* 48:905–918.

Hoke, G. D., Draper, K., Freier, S. M., Gonzalez, C., Driver, V. B., Zounes, M. C., and Ecker, D. J. 1991. Effects of phosphorothioate capping on antisense oligonucleotide stability, hybridization and antiviral efficacy versus herpes simplex virus infection. *Nucleic Acids Res.* 19:5743–5748.

Hooper, M. L., Chiasson, B. J., and Robertson, H. A. 1994. Infusion into the brain of an antisense oligonucleotide to the immediate-early gene *c-fos* suppresses production of Fos and produces a behavioral effect. *Neurosci.* 63:917–924.

Hughes, J. A., Avrutskaya, A. V., Brouwer, K. L. R., Wickstrom, E., and Juliano, R. L. 1995. Radiolabeling of methylphosphonate and phosphorothioate oligonucleotides and evaluation of their transport in everted rat jejunum sacs. *Pharm. Res.* 12:817–824.

Hutcherson, S. L., Palestine, A. G., Cantrill, H. L., Lieberman, R. M., Holland, G. N., and Anderson, K. P. 1995. Antisense oligonucleotide safety and efficacy for CMV retinitis in AIDS patients. *35th ICAAC*: 204.

Inoue, H., Imura, A., and Ohtsuka, E. 1985. Synthesis and hybridization of dodecadeoxyribonucleotides containing a fluorescent pyridopyrimidine deoxynucleoside. *Nucleic Acids Res.* 13(19):7119–7128.

Iversen, P. 1991. In vivo studies with phosphorothioate oligonucleotides: Pharmacokinetics prologue. *Anticancer Drug Des.* 6(6):531–538.

Joos, R. W., and Hall, W. H. 1969. Determination of binding constants of serum albumin for penicillin. *J. Pharmacol. Exp. Ther.* 166:113.

Katz, S. M., Browne, B., Pham, T., Wang, M. E., Bennett, C. F., Stepkowski, S. M., and Kahan, B. D. 1995. Efficacy of ICAM-1 antisense oligonucleotide in pancreatic islet transplanation. *Transplant. Proc.* 27(6):3214.

Kindy, M. S. 1994. NMDA receptor inhibition using antisense oligonucleotides prevents delayed neuronal death in gerbil hippocampus following cerebral ischemia. *Neurosci. Res. Commun.* 14:175–183.

Kitajima, I., Shinohara, T., Bilakovics, J., Brown, D. A., Xiao, X., and Nerenberg, M. 1992. Ablation of transplanted HTLV-1 tax-transformed tumors in mice by antisense inhibition of NF-KB. *Science* 258:1792–1795.

Konradi, C., Cole, R. L., Heckers, S., and Hyman, S. E. 1994. Amphetamine regulates gene expression in rat striatum via transcription factor CREB. *J. Neurosci.* 14:5623–5634.

Krieg, A. M., Yi, A.-K., Matson, S., Waldschmidt, T. J., Bishop, G. A., Teasdale, R., Koretzky, G. A., and Klinman, D. M. 1995. CpG motifs in bacterial DNA trigger direct B-cell activation. *Nature* 374: 546–549.

Kulka, M., Smith, C. C., Aurelian, L., Fishelevich, R., Meade, K., Miller, P., and Ts'o, P. O. P. 1989. Site specificity of the inhibitory effects of oligo(nucleoside methylphosphonate)s complementary to the acceptor splice junction of herpes simplex virus type 1 immediate early mRNA 4. *Proc. Natl. Acad. Sci. U.S.A.* 86:6868–6872.

Kumasaka, T., Quinlan, W. M., Doyle, N. A., Condon, T. P., Sligh, J., Takei, F., Beaudet, A. L., Bennett, C. F., and Doerschuk, C. M. 1996. The role of the intercellular adhesion molecule-1 (ICAM-1) in endotoxin-induced pneumonia evaluated using ICAM-1 antisense oligonucleotides, anti-ICAM-1 monoclonal antibodies, and ICAM-1 mutant mice. *J. Clin. Invest.* 97(10):2362–2369.

Kuramoto, E., Yano, O., Kimura, Y., Baba, M., Makino, T., Yamamoto, S., Yamamoto, T., Kataoka, T., and Tokunaga, T. 1992. Oligonucleotide sequences required for natural killer cell activation. *Jpn. J. Cancer Res.* 83(11):1128–1131.

Lai, J., Bilsky, E. J., Rothman, R. B., and Porreca, F. 1994. Treatment with antisense oligodeoxynucleotide to the opioid d receptor selectively inhibits d_2-agonist antinociception. *Neuroreport.* 5:1049–1052.

Lamond, A. I., and Sproat, B. S. 1993. Antisense oligonucleotides made of 2'-O-alkylRNA: Their properties and applications in RNA biochemistry. *FEBS Lett.* 325:123–127.

Liebsch, G., Landgraf, R., Gerstberger, R., Probst, J. C., Wotjak, C. T., Engelmann, M., Holsboer, F., and Montkowski, A. 1995. Chronic infusion of a CRH1 receptor antisense oligodeoxynucleotide into the central nucleus of the amygdala anxiety-related behavior in socially defeated rats. *Regul. Pept.* 59(2):229–239.

Lima, W. F., Monia, B. P., Ecker, D. J., and Freier, S. M. 1992. Implication of RNA structure on antisense oligonucleotide hybridization kinetics. *Biochem.* 31:12055–12061.

Lin, K.-Y., Jones, R. J., and Matteucci, M. 1995. Tricyclic-2'-deoxycytidine analogs: Synthesis and incorporation into oligodeoxynucleotides which have enhanced binding to complementary RNA. *J. Am. Chem. Soc.* 117:3873–3874.

Lin, P. K. T., and Brown, D. M. 1989. Synthesis and duplex stability of oligonucleotides containing cytosine-thymine analogues. *Nucleic Acids Res.* 17:10373–10383.

Loke, S. L., Stein, C. A., Zhang, X. H., Mori, K., Nakanishi, M., Subasinghe, C., Cohen, J. S., and Neckers, L. M. 1989. Characterization of oligonucleotide transport into living cells. *Proc. Natl. Acad. Sci. U.S.A.* 86:3474–3478.

Majumdar, C., Stein, C. A., Cohen, J. S., Broder, S., and Wilson, S. H. 1989. Stepwise mechanism of HIV reverse transcriptase: Primer function of phosphorothioate oligodeoxynucleotide. *Biochem.* 28:1340–1346.

Mani, S. K., Blaustein, J. D., Allen, J. M. C., Law, S. W., O'Malley, B. W., and Clark, J. H. 1994. Inhibition of rat sexual behavior by antisense oligonucleotides to the progesterone receptor. *Endocrinol.* 135:1409–1414.

Manoharan, M. 1993. Designer antisense oligonucleotides: Conjugation chemistry and functionality placement. In *Antisense research and applications*, eds. S. T. Crooke and B. Lebleu, 303–349. Boca Raton, Florida: CRC Press.

Martin, F. H., Castro, M. M., Aboul-ela, F., and Tinoco, I. J. 1985. Base pairing involving deoxyinosine: Implications for probe. *Nucleic Acids Res.* 13:8927–8938.

Martin, P. 1995. Ein neuer zugang zu 2'-O-alkylribonucleosiden und eigenschaften deren oligonucleotide. *Helv. Chim. Acta.* 78:486–489.

McCarthy, M. M., Masters, D. B., Rimvall, K., Schwartz-Giblin, S., and Pfaff, D. W. 1994. Intracerebral administration of antisense oligodeoxynucleotides to GAD_{65} and GAD_{67} mRNAs modulate reproductive behavior in the female rat. *Brain Res.* 636:209–220.

McCarthy, M. M., Nielsen, D. A., and Goldman, D. 1995. Antisense oligonucleotide inhibition of tryptophan hydroxylase activity in mouse brain. *Regul Pept.* 59(2):163–170.

Messina, J. P., Gilkeson, G. S., and Pisetsky, D. S. 1991. Stimulation of in vitro murine lymphocyte proliferation by bacterial DNA. *J. Immunol.* 147(6):1759–1764.

Mirabelli, C. K., and Crooke, S. T. 1993. Antisense oligonucleotides in the context of modern molecular drug discovery and development. In *Antisense research and applications*, eds. S. T. Crooke and B. Lebleu, 7–35. Boca Raton, Florida: CRC Press.

Miyao, T., Takakura, Y., Akiyama, T., Yoneda, F., Sezaki, H., and Hashida, M. 1995. Stability and pharmacokinetic characteristics of oligonucleotides modified at terminal linkages in mice. *Antisense Res. Dev.* 5(2):115–121.

Mizoguchi, H., Narita, M., Nagase, H., and Tseng, L. F. 1995. Antisense oligodeoxynucleotide to a delta-opioid receptor blocks the antinociception induced by cold water swimming. *Regul. Pept.* 59(2):255–259.

Monia, B., Johnston, J. F., Sasmor, H., and Cummins, L. L. 1996. Nuclease resistance and antisense activity of modified oligonucleotides targeted to Ha-*ras. J. Biol. Chem.* 24(14):14533–14540.

Monia, B. P., Johnston, J. F., Ecker, D. J., Zounes, M., Lima, W. F., and Freier, S. M. 1992. Selective inhibition of mutant Ha-*ras* mRNA expression by antisense oligonucleotides. *J. Biol. Chem.* 267:19954–19962.

Monia, B. P., Johnston, J. F., Geiger, T., Muller, M., and Fabbro, D. 1995. Antitumor activity of a phosphorothioate oligodeoxynucleotide targeted against C-*raf* kinase. *Nature Med.* 2(6):668–675.

Monia, B. P., Lesnik, E. A., Gonzalez, C., Lima, W. F., McGee, D., Guinosso, C. J., Kawasaki, A. M., Cook, P. D., and Freier, S. M. 1993. Evaluation of 2′ modified oligonucleotides containing deoxy gaps as antisense inhibitors of gene expression. *J. Biol. Chem.* 268:14514–14522.

Morishita, R., Gibbons, G. H., Ellison, K. E., Nakajima, M., Zhang, L., Kaneda, Y., Ogihara, T., and Dzau, V. J. 1993. Single intraluminal delivery of antisense cdc2 kinase and proliferating-cell nuclear antigen oligonucleotides results in chronic inhibition of neointimal hyperplasia. *Proc. Natl. Acad. Sci. U.S.A.* 90:8474–8478.

Morishita, R., Gibbons, G. H., Kaneda, Y., Ogihara, T., and Dzau, V. J. 1994. Pharmacokinetics of antisense oligodeoxyribonucleotides (cyclin B_1 and CDC 2 kinase) in the vessel wall in vivo: Enhanced therapeutic utility for restenosis by HVJ-liposome delivery. *Gene* 149:13–19.

Morris, M., Li, P., Barrett, C., and Callahan, M. F. 1995. Oxytocin antisense reduces salt intake in the baroreceptor-denervated rat. *Regul. Pept.* 59(2):261–266.

Nagel, K. M., Holstad, S. G., and Isenberg, K. E. 1993. Oligonucleotide pharmacotherapy: An antigene strategy. *Pharmacother.* 13(3):177–188.

Neckers, L. M. 1993. Cellular internalization of oligodeoxynucleotides. In *Antisense research and applications*, eds. S. T. Crooke and B. Lebleu, 451–460. Boca Raton, Florida: CRC Press.

Nesterova, M., and Cho-Chung, Y. S. 1995. A single-injection protein kinase A-directed antisense treatment to inhibit tumor growth. *Nature Med.* 1:528–533.

Neumann, I., Porter, D. W. F., Landgraf, R., and Pittman, Q. J. 1994. Rapid effect on suckling of an oxytocin antisense oligonucleotide administered into rat supraoptic nucleus. *Am. J. Physiol. Regul. Integr. Comp. Physiol.* 267:R852–R858.

Nikiforov, T. T., and Connolly, B. A. 1991. The synthesis of oligodeoxynucleotides containing 4-thiothymidine residues. *Tetrahedron Lett.* 32(31):3851–4.

Nyce, J. W., and Metzger, W. J. 1997. DNA antisense therapy for asthma in an animal model. *Nature* (London) 385(6618):721–725.

Offensperger, W.-B., Offensperger, S., Walter, E., Teubner, K., Igloi, G., Blum, H. E., and Gerok, W. 1993. *In vivo* inhibition of duck hepatitis B virus replication and gene expression by phosphorothioate modified antisense oligodeoxynucleotides. *EMBO J.* 12:1257–1262.

Ogo, H., Hirai, Y., Miki, S., Nishio, H., Akiyama, M., and Nakata, Y. 1994. Modulation of substance P/neurokinin-1 receptor in human astrocytoma cells by antisense oligodeoxynucleotides. *Gen. Pharmacol.* 25:1131–1135.

Osen-Sand, A., Catsicas, M., Staple, J. K., Jones, K. A., Ayala, G., Knowles, J., Grenningloh, G., and Catsicas, S. 1993. Inhibition of axonal growth by SNAP-25 antisense oligonucleotides *in vitro* and *in vivo*. *Nature* (London) 364:445–448.

Osen-Sand, A., Catsicast, M., Staple, J. K., Jones, K. A., Ayala, G., Knowles, J., Grenningloh, G., and Catsicas, S. 1993. Inhibition of axonla growth by SNAP-25 antisense oligonucleotides in vitro and in vivo. *Nature* (London) 364:445–448.

Parmentier, G., Schmitt, G., Dolle, F., and Luu, B. 1994. A convergent synthesis of 2′-O-methyl uridine. *Tetrahedron Lett.* 50(18):5361–5368.

Perlaky, L., Saijo, Y., Busch, R. K., Bennett, C. F., Mirabelli, C. K., Crooke, S. T., and Busch, H. 1993. Growth inhibition of human tumor cell lines by antisense oligonucleotides designed to inhibit p120 expression. *Anti-Cancer Drug Design* 8:3–14.

Phillips, M. I., Wielbo, D., and Gyurko, R. 1994. Antisense inhibition of hypertension: A new strategy for renin–angiotensin candidate genes. *Kidney Int.* 46:1554–1556.

Pisetsky, D. S., and Reich, C. F. 1994. Stimulation of murine lymphocyte proliferation by a phosphorothioate oligonucleotide with antisense activity for herpes simplex virus. *Life Sci.* 54:101–107.

Pitsch, S., Krishnamurthy, R., Bolli, M., Wendeborn, S., Holzner, A., Minton, M., Lesueur, C., Schloenvogt, I., Jaun, B. 1995. Pyranosyl-RNA ('p-RNA'): Base-pairing selectivity and potential to replicate. *Helv. Chim. Acta* 78(7):1621–1635.

Pollio, G., Xue, P., Zanisi, M., Nicolin, A., and Maggi, A. 1993. Antisense oligonucleotide blocks progesterone-induced lordosis behavior in ovariectomized rats. *Mol. Brain Res.* 19:135–139.

Qin, Z. H., Zhou, L. W., Zhang, S. P., Wang, Y., and Weiss, B. 1995. D2 dopamine receptor antisense oligodeoxynucleotide inhibits the synthesis of a functional pool of D2 dopamine receptors: Mol. Pharmacol. *48* 4(730–737).

Quattrone, A., Papucci, L., Schiavone, N., Mini, E., and Capaccioli, S. 1994. Intracellular enhancement of intact antisense oligonucleotide steady-state levels by cationic lipids. *Anti-Cancer Drug Design* 9:549–553.

Rappaport, J., Hanss, B., Kopp, J. B., Copeland, T. D., Bruggeman, L. A., Coffman, T. M., and Klotman, P. E. 1995. Transport of phosphorothioate oligonucleotides in kidney: Implications for molecular therapy. *Kidney Int.* 47:1462–1469.

Ratajczak, M. Z., Kant, J. A., Luger, S. M., Huiya, N., Zhang, J., Zon, G., and Gewirtz, A. M. 1992. In vivo treatment of human leukemia in a scid mouse model with c-*myb* antisense oligodeoxynucleotides. *Proc. Natl. Acad. Sci. U.S.A.* 89:11823–11827.

Sakai, R. R., Ma, L. Y., He, P. F., and Fluharty, S. J. 1995. Intracerebroventricular administration of angiotensin type 1 (AT1) receptor antisense oligonucleotides attenuate thirst in the rat. *Regul. Pept.* 59(2):183–192.

Sands, H., Gorey-Feret, L. J., Cocuzza, A. J., Hobbs, F. W., Chidester, D., and Trainor, G. L. 1994. Biodistribution and metabolism of internally ^{3}H-labeled oligonucleotides. I. Comparison of a phosphodiester and a phosphorothioate. *Mol. Pharmacol.* 45:932–943.

Sands, H., Gorey-Feret, L. J., Ho, S. P., Bao, Y., Cocuzza, A. J., Chidester, D., and Hobbs, F. W. 1995. Biodistribution and metabolism of internally ^{3}H-labeled oligonucleotides. II. 3′,5—blocked oligonucleotides. *Mol. Pharmacol.* 47:636–646.

Sanghvi, Y. S. 1993. Heterocyclic base modifications in nucleic acids and their applications in antisense oligonucleotides. In *Antisense research and applications*, eds. S. T. Crooke and B. Lebleu, 273–288. Boca Raton, Florida: CRC Press.

Sanghvi, Y. S., and Cook, P. D. 1994. *Carbohydrate modifications in antisense research*, ACS Symposium Series No. 580. Washington, D.C.: American Chemical Society.

Sanghvi, Y. S., Hoke, G. D., Freier, S. M., Zounes, M. C., Gonzalez, C., Cummins, L., Sasmor, H., and Cook, P. D. 1993. Antisense oligodeoxynucleotides: Synthesis, biophysical and biological evaluation of oligodeoxynucleotides containing modified pyrimidines. *Nucleic Acids Res.* 21:3197–3203.

SantaLucia, J., Jr., Kierzek, R., and Turner, D. H. 1991. Functional group substitutions as probes of hydrogen bonding between GA mismatches in RNA internal loops. *J. Am. Chem. Soc.* 113:4313–4322.

Schwab, G., Chavany, C., Duroux, I., Goubin, G., Lebeau, J., Hélène, C., and Saison-Behmoaras, T. 1994. Antisense oligonucleotides adsorbed to polyalkylcyanoacrylate nanoparticles specifically inhibit mutated Ha-*ras*-mediated cell proliferation and tumorigenicity in nude mice. *Proc. Natl. Acad. Sci. U.S.A.* 91:10460–10464.

Seela, F., Kaiser, K., and Bindig, U. 1989. 2′-deoxy-.beta.-D-ribofuranosides of N6-methylated 7-deazaadenine and 8-aza-7-deazaadenine: Solid-phase synthesis of oligodeoxyribonucleotides and properties of self-complementary duplexes. *Helv. Chim. Acta* 72(5):868–81.

Seela, F., Ramzaeva, N., and Chen, Y. 1995. Oligonucleotide duplex stability controlled by the 7-substituents of 7-deazaguanine bases. *Bioorg. Med. Chem. Lett.* 5(24):3049–3052.

Simons, M., Edelman, E. R., DeKeyser, J.-L., Langer, R., and Rosenberg, R. D. 1992. Antisense c-*myb* oligonucleotides inhibit arterial smooth muscle cell accumulation in vivo. *Nature* (London) 359:67–70.

Simons, M., Edelman, E. R., and Rosenberg, R. D. 1994. Antisense proliferating cell nuclear antigen oligonucleotides inhibit intimal hyperplasia in a rat carotid artery injury model. *J. Clin. Invest.* 93:2351–2356.

Skorski, T., Nieborowska-Skorska, M., Nicolaides, N. C., Szczylik, C., Iversen, P., Iozzo, R. V., Zon, G., and Calabretta, B. 1994. Suppression of Philadelphia leukemia cell growth in mice by *BCR-ABL* antisense oligodeoxynucleotide. *Proc. Natl. Acad. Sci. U.S.A.* 91:4504–4508.

Skutella, T., Probst, J. C., Jirikowski, G. F., Holsboer, F., and Spanagel, R. 1994. Ventral tegmental area (VTA) injections of tyrosine hydroxylase phosphorothioate antisense oligonucleotide suppress operant behavior in rats. *Neurosci. Lett.* 167:55–58.

Sproat, B. S., Iribarren, A. M., Garcia, R. G., and Beijer, B. 1991. New synthetic routes to synthons suitable for 2′-O-allyloligoribonucleotide assembly. *Nucleic Acids Res.* 19(4):733–738.

Sproat, B. S., and Lamond, A. I. 1993. 2′-O-alkyloligoribonucleotides. In *Antisense research and applications*, eds. S. T. Crooke and B. Lebleu, 351–362. Boca Raton, Florida: CRC Press.

Srinivasan, S. K., Tewary, H. K., and Iversen, P. L. 1995. Characterization of binding sites, extent of binding, and drug interactions of oligonucleotides with albumin. *Antisense Res. Dev.* 5(2):131–139.

Stein, C. A., and Cheng, Y.-C. 1993. Antisense oligonucleotides as therapeutic agents—Is the bullet really magical? *Science* 261:1004–1012.

Stein, C. A., Neckers, M., Nair, B. C., Mumbauer, S., Hoke, G., and Pal, R. 1991. Phosphorothioate oligodeoxycytidine interferes with binding of HIV-1 gp120 to CD4. *J. Acquired Immune Deficiency Synd.* 4:686–693.

Stepkowski, S. M., Tu, Y., Condon, T. P., and Bennett, C. F. 1994. Blocking of heart allograft rejection by intercellular adhesion molecule-1 antisense oligonucleotides alone or in combination with other immunosuppressive modalities. *J. Immunol.* 153:5336–5346.

Stepkowski, S. M., Tu, Y., Condon, T. P. and Bennett, C. F. 1995. Induction of transplantation tolerance by treatment with ICAM-1 antisense oligonucleotides and anti-LFA-1 monoclonal antibodies. *Transplant. Proc.* 27:113.

Suzuki, S., Pilowsky, P., Minson, J., Arnolda, L., Llewellyn-Smith, I. J., and Chalmers, J. 1994. c-*fos* antisense in rostral ventral medulla reduces arterial blood pressure. *Am. J. Physiol.* 266(4 Pt. 2):R1418–R1422.

Takakura, Y., Mahato, R. I., Yoshida, M., Kanamaru, T., and Hashida, M. 1996. Uptake characteristics of oligonucleotides in the isolated rat liver perfusion system. *Antisense Nucleic Acid Drug Del.* 6:177–183.

Tao, L. F., Marx, K. A., Wongwit, W., Jiang, Z., Agrawal, S., and Coleman, R. M. 1995. Uptake, intracellular distribution, and stability of oligodeoxynucleotide phosphorothioate by *Schistosoma mansoni*. *Antisense Res. Dev.* 5(2):123–129.

Temsamani, J., Tang, J., Padmapriya, A., Kubert, M., and Agrawal, S. 1993. Pharmacokinetics, biodistribution, and stability of capped oligodeoxynucleotide phosphorothioates in mice. *Antisense Res. Dev.* 3:277–284.

Tischmeyer, W., Grimm, R., Schicknick, H., Brysch, W., and Schlingensiepen, K.-H. 1994. Sequence-specific impairment of learning by c-*jun* antisense oligonucleotides. *Neuroreport* 5:1501–1504.

Vickers, T., Baker, B. F., Cook, P. D., Zounes, M., Buckheit, R. W., Jr., Germany, J., and Ecker, D. J. 1991. Inhibition of HIV-LTR gene expression by oligonucleotides targeted to the TAR element. *Nucleic Acids Res.* 19:3359–3368.

Vlassov, V. V. 1989. Inhibition of tick-borne viral encephalitis expression using covalently linked oligonucleotide analogs. Proceedings of Nucleic Acid Therapeutics Conference, Washington, D.C.

Wagner, R. W., Matteucci, M. D., Lewis, J. G., Gutierrez, A. J., Moulds, C., and Froehler, B. C. 1993. Antisense gene inhibition by oligonucleotides containing C-5 propyne pyrimidines. *Science* 260:1510–1513.

Wahlestedt, C., Pich, E. M., Koob, G. F., Yee, F. and Heilig, M. 1993. Modulation of anxiety and neuropeptide Y-Y1 receptors by antisense oligodeoxynucleotides. *Science* 259:528–531.

Wang, S., Lee, R. J., Cauchon, G., Gorenstein, D. G., and Low, P. S. 1995. Delivery of antisense oligodeoxyribonucleotides against the human epidermal growth factor receptor into cultured KB cells with liposomes conjugated to folate via polyethylene glycol. *Proc. Natl. Acad. Sci.* 92:3318–3322.

Weiss, B., Zhou, L.-W., Zhang, S.-P., and Qin, Z.-H. 1993. Antisense oligodeoxynucleotide inhibits D_2 dopamine receptor-mediated behavior and D_2 messenger RNA. *Neurosc.* 55:607–612.

Whitesell, L., Rosolen, A., and Neckers, L. M. 1991. In vivo modulation of N-*myc* expression by continuous perfusion with an antisense oligonucleotide. *Antisense Res. Dev.* 1:343–350.

Wickstrom, E. 1986. Oligodeoxynucleotide stability in subcellular extracts and culture media. *J. Biochem. Biophys. Methods* 13:97–102.

Woodburn, V. L., Hunter, J. C., Durieux, C., Poat, J. A., and Hughes, J. 1994. The effect of C-FOS antisense in the formalin-paw test. *Regul. Pept.* 54(1 Pt. 2):327–328.

Wyatt, J. R., Vickers, T. A., Roberson, J. L., Buckheit, R. W., Jr., Klimkait, T., DeBaets, E., Davis, P. W., Rayner, B., Imbach, J. L., and Ecker, D. J. 1994. Combinatorially selected guanosine-quartet structure is a potent inhibitor of human immunodeficiency virus envelope-mediated cell fusion. *Proc. Natl. Acad. Sci. U.S.A.* 91:1356–1360.

Yacyshyn, B., Woloschuk, B., Yacyshyn, M. B., Martini, D., Tami, J., Bennett, F., Kisner, D., and Shanahan, W. 1997. Efficacy and safety of ISIS 2302 (ICAM-1 antisense oligonucleotide) treatment of steroid-dependent Crohn's disease. Paper presented at Annual Meeting of the American Gastroenterological Association and American Association for the Study of Liver Diseases, Washington, D.C.

Yazaki, T., Ahmad, S., Chahlavi, A., Zylber-Katz, E., Dean, N. M., Rabkin, S. D., Martuza, R. L., and Glazer, R. I. 1996. Treatment of glioblastoma U-87 by systemic administration of an antisense protein kinase C-a phosphorothioate oligodeoxynucleotide. *Mol. Pharmacol.* 50(2):236–242.

Zhang, M., and Creese, I. 1993. Antisense oligodeoxynucleotide reduces brain dopamine D_2 receptors: Behavioral correlates. *Neurosci. Lett.* 161:223–226.

Zhang, R., Lu, Z., Zhang, X., Zhao, H., Diasio, R. B., Liu, T., Jiang, Z., and Agrawal, S. 1995. In vivo stability and disposition of a self-stabilized oligodeoxynucleotide phosphorothioate in rats. *Clin. Chem.* 41(6, Pt. 1):836–843.

Zhang, R., Yan, J., Shahinian, H., Amin, G., Lu, Z., Liu, T., Saag, M. S., Jiang, Z., Temsamani, J., Martin, R. R., Schechter, P. J., Agrawal, S., and Diasio, R. B. 1995. Pharmacokinetics of an anti-human immunodeficiency virus antisense oligodeoxynucleotide phosphorothioate (GEM 91) in HIV-infected subjects. *Clin. Pharmacol. Ther.* 58:44–53.

Zhang, S.-P., Zhou, L. W., and Weiss, B. 1994. Oligodeoxynucleotide antisense to the D_1 dopamine receptor mRNA inhibits D_1 dopamine receptor-mediated behaviors in normal mice and in mice lesioned with 6-hydroxydopamine. *J. Pharmacol. Exp. Ther.* 271:1462–1470.

Zhou, L.-W., Zhang, S.-P., Qin, Z.-H., and Weiss, B. 1994. *In vivo* administration of an oligodeoxynucleotide antisense to the D_2 dopamine receptor messenger RNA inhibits D_2 dopamine receptor-mediated behavior and the expression of D_2 dopamine receptors in mouse striatum. *J. Pharmacol. Exp. Ther.* 268:1015–1023.

Biotechnology and Safety Assessment, 2nd ed.
Edited by John A. Thomas
Copyright © 1998 Taylor & Francis

3

Safety Evaluation for New Varieties of Food Crops Developed Through Biotechnology

Bruce G. Hammond and Roy L. Fuchs

Monsanto Company, Saint Louis, Missouri

Throughout history, plant breeders have sought to genetically modify food crops to improve yield and increase resistance to disease and plant pests. Initially these improvements were achieved by selecting seed from superior plants and reproducing these with continual selection and breeding. Traditional breeding methods have increased corn and wheat yields by approximately 100% over the last half century. However, traditional plant breeding methods are slow and unpredictable. To introduce a desired gene or set of genes by conventional breeding methods requires a sexual cross between parental lines followed by repeated backcrossing between the hybrid offspring and one of the parents until progeny with the desired characteristics are obtained. Genes are only accessible from plants that can be sexually crossed, and many genes besides the desired gene(s) will be transferred (Watson et al. 1996).

Biotechnology provides an opportunity to overcome some of the limitations of traditional breeding by enabling plant geneticists to identify and clone specific genes encoding desirable traits, such as protection against insect pests, and to introduce these genes selectively into already useful varieties of plants. Sexual compatibility is no longer a limiting factor for transfer of desired traits. The transformation process is faster and more efficient because successfully transformed plants can readily be identified. Numerous traits are being assessed for their potential to yield products with the ability to (1) protect plants against various insect, fungal, and viral pests and plant pathogens; (2) provide selectivity to preferred herbicides; (3) improve agronomic performance such as crop yields; (4) increase nutritional value of food for humans and farm animals; (5) reduce naturally occurring toxicants, antinutrients, or allergens; (6) modify the ripening process to improve the flavor of fruits and vegetables; (7) use plants as factories to make environmentally friendly biodegradable polymers for packaging materials; and (8) use plants to produce pharmaceutical products more cost effectively, and so on.

Since the initial reports of the first genetically modified plants 15 years ago, nearly all agronomically important crops have been genetically modified. By the end of

1995, more than 70 different crop species had been transformed. At least 56 different crops have been planted in at least 34 countries and grown in more than 15,000 individual field sites (James and Krattiger 1996). According to the U.S. Animal and Plant Health Inspection Service (APHIS), the most frequently field-tested traits are insect protection, virus protection, fungal resistance, herbicide tolerance, and food quality enhancements. The number of transformed crops and field trials will continue to expand as more new crop varieties are developed through biotechnology for eventual entry to the marketplace.

As shown in Table 1, at least 34 genetically modified plants have successfully completed regulatory review by the appropriate regulatory agencies. Genetically modified plants were grown commercially in 1996 on approximately 7 million acres

TABLE 1. *Examples of plant biotechnology products that have successfully completed regulatory review in at least one country*

Company	Genetic Trait
AgrEvo Canada, Inc.	Glufosinate-tolerant canola
	Glufosinate-tolerant corn
	Glufosinate-tolerant soybean
Agritope, Inc.	Modified fruit-ripening tomato
Asgrow Seed Co.	Virus-resistant squash I
	Virus-resistant squash II
Bejo–Baden	Male sterility/glufosinate-tolerant chicory
Calgene, Inc.	Flavr Savr™ tomato
	Bromoxynil-tolerant cotton
	Laurate canola
China	Virus-resistant tomato
Ciba Seeds	Insect-protected corn
Cornell U./U. of Hawaii	Virus-resistant papaya
DeKalb Genetics Corp.	Glufosinate-tolerant corn
	Insect-protected corn
DNA Plant Technology	Improved ripening tomato
DuPont	Sulfonylurea-tolerant cotton
	High-oleic-acid soybean
Florigene	Carnations with increased vase life
	Carnations with modified flower color
Monsanto	Glyphosate-tolerant soybean
	Improved ripening tomato
	Insect-protected potato
	Insect-protected cotton
	Glyphosate-tolerant cotton
	Glyphosate-tolerant canola
	Insect-protected corn
	Glyphosate-tolerant corn
Mycogen	Insect-protected corn
Northrup King	Insect-protected corn
Plant Genetic Systems	Male sterile oilseed rape
	Male sterility/glufosinate-tolerant corn
University of Saskatchewan	Sulfonylurea-tolerant flax
Zeneca/Petoseed	Improved ripening tomato

in various world areas with approximately 30 million acres planted in 1997 (James 1997).

Biotechnology provides plant breeders the opportunity to develop new varieties of food crops more efficiently and with greater potential benefit than has been possible with conventional breeding practices. As with any technological innovation, there must be assurance that the technology will deliver food "as safe as" that developed through traditional breeding programs. This chapter addresses the safety assessment strategies for new varieties of food products developed through biotechnology. The approaches are consistent with the guidance developed by various international organizations such as OECD (1993, 1996, 1997), FAO/WHO (1996), and WHO (1991, 1995). An example of the application of these strategies to assess the safety of a genetically modified food plant, YieldGard™ corn, is provided in Chapter 10 of this book.

To understand how safety assessment approaches have evolved, it is instructive to review briefly how new traits are introduced into food crops through conventional breeding compared with biotechnology.

INTRODUCTION OF GENES INTO FOOD CROPS: CONVENTIONAL BREEDING COMPARED WITH BIOTECHNOLOGY

Conventional Breeding

Genetic modification of plants has been practiced for hundreds of years with considerable success by plant breeders. Plant breeding has become a very sophisticated branch of applied genetics. Breeders have developed elegant procedures for crossing plants to introduce and maintain desirable traits such as increased yield and resistance to disease. With conventional breeding, cultivars highly adapted to cultivation can be improved by crossing them with closely related highly adapted cultivars to combine the desired features of the parents. If the parents are highly adapted to cultivation, their progeny tend to be highly adapted also. However, if a desirable trait is not available in highly adapted cultivars, the breeder will cross-adapt cultivars with cultivars from different geographic areas, more primitive varieties, or wild species. The greater the diversity in genetic material of the parents, the greater the chance that undesirable characteristics (genes) will be introduced into the progeny. The progeny may be an "offtype," which means it has less desirable agronomic properties, such as stunted growth or poor yield, than parent cultivars. Eliminating plants with undesirable traits while retaining plants with desired features may require many backcrosses carried out over several generations. The greater the difference in genetic content of the parents, the more difficult sexual crossing can be. In vitro procedures such as embryo culture and protoplast fusion have made it possible to cross parents that may not be sexually compatible (IFBC 1990).

Biotechnology

Biotechnological methods do not replace conventional breeding practices but can facilitate the introduction of desirable traits into the plant genome more efficiently and with greater precision. Unlike traditional breeding in which thousands of genes may be introduced into progeny from their parents, only one or a few genes are typically introduced using biotechnological techniques. The source of the desired trait the breeder wishes to introduce into new crop varieties is no longer limited to those from sexually compatible species. This greatly expands the opportunity to introduce new traits to improve crop varieties.

The techniques used to introduce new traits into food crops vary depending on the kind of plant being transformed. Seed plants have been divided into two subclasses: monocotyledonous plants (monocots), whose seeds have a single cotyledon (meaning "seed leaf"), and dicotyledonous plants (dicots) or those with two cotyledons. Dicots (broadleaf plants like tomato, cotton, and soybean) were the first seed plants to have genes introduced via biotechnology. The first genetically modified plants were produced in 1982 using *Agrobacterium tumefaciens*, a bacterium that can transfer a portion of its own DNA (T-DNA) into the genome of plants. The disease-causing sequences from the *Agrobacterium* T-DNA have been deleted followed by insertion of DNA that encodes for a desired trait such as insect protection. Regulatory signals are added to enable the gene to function optimally in the plant. Border sequences in the plasmid delineate the desired genes that will typically be inserted in the plant genome. The T-DNA is incorporated into a plasmid derived from *Escherichia coli,* which is transferred to *A. tumefaciens* via a conjugation process. *A. tumefaciens* containing the engineered plasmid is incubated with selected tissue from the host plant, and the desired genes are stably inserted into the plant chromosome. The inserted genes can then be transferred to new plant varieties using traditional breeding methods (Watson et al. 1996).

Monocots, which are represented by agronomically important cereal crops such as wheat, rice, and corn, were not initially amenable to transformation with *A. tumefaciens*. Recently, more aggressive strains of *A. tumefaciens* with broader host ranges have been developed that have been used to introduce genes into some cereal crops. In addition, techniques such as protoplast transformation and particle bombardment have been used to transfer DNA directly into cells where it is stably inserted into the plant genome. These techniques have been particularly valuable in monocot plants for which the *Agrobacterium* transformation method was not effective or efficient. In particle bombardment, very small metal beads (1-μ diameter) are coated with DNA and shot from a "gun" into the target monocot cell. Some of the plant cells will incorporate the desired gene(s) into their genome.

Whether *Agrobacterium* transformation or particle bombardment is used to introduce genes into plant cells, only a small percentage of eligible plant cells will be successfully transformed. To identify the transformed cells in culture, genes for selectable markers have been included with the genetic information inserted into plant cells. The marker genes provide resistance to antibiotics, herbicides, and other substances added to the cell culture to inhibit the growth of nontransformed cells. Plant cells that survive have been successfully transformed and can therefore be identified and regenerated into whole plants.

As with conventional breeding, offtypes are typically discarded. Progenies with normal agronomic properties that express the desired phenotype or trait are backcrossed with commercial crop varieties to generate seed bearing the new trait. The chances of success in identifying a progeny with the desired trait are much higher with recombinant DNA techniques because the breeder knows that the trait was successfully inserted into the plant genome. With conventional breeding, many generations of backcrossing may be required to identify a progeny with the desired trait.

HISTORY OF SAFETY ASSESSMENT FOR NEW CROP VARIETIES DEVELOPED BY CONVENTIONAL BREEDING

During this century, our food supply has steadily improved in quality, variety, nutritional value, safety, and economy through the use of conventional breeding techniques to improve food crops. The history of safe use of new varieties of food crops developed by classical breeding techniques is based on several factors: (1) confidence and experience with the procedures used to generate new crop varieties; (2) knowledge of the composition of the food crop, including important nutrients and toxicants, if present; and (3) observation of the agronomic properties of new crop varieties to eliminate those with undesirable properties. Of the thousands of new crop varieties that have been developed during this century via traditional breeding, only a very limited number of new varieties have presented safety concerns. The more well-known examples are (1) increased psoralens in certain varieties of celery that caused photodermatitis in food handlers; (2) increased glycoalkaloid content in the Lenape variety of potatoes (glycoalkaloids can cause gastrointestinal discomfort); and (3) increased cucurbatin levels in vegetable squashes that leave a bitter taste (IFBC 1990). In these three examples, the food crops contained endogenous toxicants affording protection against plant pests. The levels of these endogenous toxicants were inadvertently increased in these new varieties. These examples have caused plant breeders to monitor new varieties of food crops that naturally contain potentially harmful toxicants or antinutrients more carefully to be certain that levels of these substances are within acceptable limits. For example, the United States and Canada have set acceptable limits for glycoalkaloid levels in potatoes that new potato varieties must meet.

Regulatory Oversight of New Crop Varieties

Conventional Breeding

In the United States, there are no premarket regulatory requirements governing the introduction of new crop varieties developed through conventional breeding (with the exception that the variety must not exceed standards set for the level of certain natural toxicants such as glycoalkaloids in potatoes). Food safety is assessed through postmarketing measures under the Food, Drug, and Cosmetic Act [21 CFR 402(a)1]. This practice is based on the long history of safe introduction of new crop varieties developed through conventional breeding. As discussed earlier, plant breeders monitor the quality of new plant varieties before they are introduced into commerce. In Europe and for some crops in Canada, new varieties of food crops must be registered with the government. This is not done for safety reasons but more as an assurance to the farmer that the new variety will perform at least as well as commercial varieties in the field. In the United States, the market place determines the performance acceptability of new crop varieties.

Biotechnology

Because biotechnology is relatively new, there is considerably more regulatory oversight for the introduction of new crop varieties developed through this approach. In the United States, the regulatory authority to ensure the safety of food and feed products derived from plant biotechnology resides within the Food and Drug Administration (FDA). The U.S. Department of Agriculture (USDA) has the authority to ensure that genetically modified plants will not become plant pests. The Environmental Protection Agency (EPA) has the authority to evaluate the safety of plants that have been genetically modified for protection against plant pests such as insects, fungi, bacteria, and viruses. The EPA also regulates herbicides by establishing herbicide tolerances for plants genetically modified to be herbicide tolerant.

The movement and release of genetically modified plants is regulated by USDA under the Federal Plant Pest Act and the Plant Quarantine Act. Permits or notifications must be filed with USDA to obtain approval for field testing of new varieties of genetically modified plants under development. A determination that the genetically modified plant is not a plant pest (e.g., that the plants do not pose a risk to the environment or production agriculture) must be obtained before market introduction.

The FDA has the regulatory authority to ensure the safety and wholesomeness of food and feed products, including those derived from genetically modified plants. The FDA has adopted a decision-tree approach to ensure safety of products derived from new varieties of food and feed developed by both traditional and newer genetic modification methods, including biotechnology (Kessler et al. 1992). Under the Food Drug and Cosmetic Act [21 CFR 402(a)1] the FDA has the authority to take regulatory action against a new variety of food if the genetic modifications would render the food "ordinarily injurious" to human health (IFBC 1990). Therefore, the FDA uses the same postmarket food adulteration approach and regulations for these products as are used for food and feed products derived from traditionally bred plant varieties. It is recommended that developers of genetically modified food consult with the FDA before commercialization, and this has been done with all genetically modified food products that are in the marketplace. The FDA has completed consultations on at least 29 different genetically modified crop plants to date.

The Federal Insecticide, Fungicide and Rodenticide Act (FIFRA) provides the EPA with the authority to regulate plants with bioengineered pesticidal traits. The EPA has approved at least seven different genetically modified plants with introduced pesticidal traits. These products are also reviewed by the FDA and USDA.

In regard to regulation of genetically modified plants in other world areas, Health Canada regulates food safety, and Agri-Food Canada regulates feed and environmental safety as well as registration of specific plant varieties for certain crops. In Japan, the Ministry of Heath and Welfare regulates food safety, whereas the Ministry of Agriculture, Food, and Fisheries regulates feed and environmental safety. In the European Union, environmental assessments are conducted before placing a product on the market, as outlined in the 90/220 EEC regulations. The recently authorized Novel Foods Regulation governs food safety in the European Union and replaces individual country food regulations that were in place in the United Kingdom, Denmark, and the Netherlands. A Novel Feed Regulation is under consideration for overseeing feed safety under the current 90/220 EEC process. Many other countries either have, or are in the process of developing, regulations for plant biotechnology products. The number of genetically modified plants that have successfully completed regulatory review in various countries include 23 in Canada, 5 in the European Union, 15 in Japan, 3 in Mexico, 2 in Argentina, 1 each in Australia and Brazil.

Safety Assessment Strategies for New Crop Varieties Developed Through Biotechnology

The safety issues that have been raised for genetically engineered plant products are similar to those for new varieties of plants derived from conventional breeding. For example, progenies derived from conventional breeding as well as biotechnology may have progenies with altered agronomic properties when compared with the parents. Varieties with altered agronomic properties are usually readily identified in field trials. It has been suggested that biotechnology may inadvertently cause the production of new toxicants through insertional mutagenesis events that activate dormant biosynthetic pathways. This scenario seems very unlikely—particularly for those crops where there has been considerable experience with traditional breeding. It is more likely that an insertional event would result in the increase or decrease in the expression of already recognized toxicants, which has very infrequently been observed in conventional breeding, as discussed above.

In conventional breeding, the genes bearing the desired trait have often not been identified. The same applies to the protein expression products of these introduced genes. Introduction of genes through biotechnology procedures is much more precise, for the genes have been defined before their introduction. However, unlike conventional breeding, introduced genes can be obtained from almost any source, not just sexually compatible relatives of the food crop. Therefore, as part of the safety evaluation, regulatory agencies have required a molecular characterization of any gene introduced into food crops. In addition, the protein expression product of the cloned gene must be characterized as to its function, specificity of action, and safety. There are no particular concerns about the safety of the genetic material itself. Genetic material from living organisms is made from the same four nucleotide building blocks; the only difference between genes is the nucleotide composition. The human gut at any one time has been estimated to contain hundreds of milligrams of DNA from ingested food and mucosal cells sloughed into the gastrointestinal tract (Karenlampi 1996). These DNA molecules are efficiently degraded by nucleotidases during digestion. The contribution of genetic material from genes cloned into food is trivial compared with the other sources of DNA in the gut. It was concluded that the introduced genes (DNA) in food products pose no more health risk to consumers than the rest of the DNA ingested from food sources (WHO 1991; FDA 1992; WHO 1993; Karenlampi 1996).

Molecular Characterization

For new plant varieties developed through biotechnology, the source of the gene introduced into the plant must be identified. The transformation system used to insert the gene into the plant genome must be defined as well as the number of copies of inserted genes and the integrity and stability of the genetic insert. For genes coding for pesticidal proteins, the EPA requires the following information: (1) description of the vectors, (2) identity of organisms used for the cloning of the vectors, (3) description of the methodologies used to clone the vectors, (4) vector description (size in kilobases), (5) restriction endonuclease sites, (6) location and function of all relevant gene segments, (7) the final delivery system, (8) description of gene segments transferred to the plant, (9) description of whether the inserted genes are expressed constitutively or inducibly, (10) localization and expression of the pesticidal substance in plant parts, and (11) estimation of the gene copies inserted, and so on (Kough 1996).

Substantial Equivalence

A general consensus has existed among regulatory agencies in major world areas regarding the use of "substantial equivalence" as an approach to assess the safety and acceptability of genetically modified crops (WHO 1995; OECD 1993; FAO/WHO 1996). This concept has been adapted in part from procedures that plant breeders have followed to monitor the acceptability of new varieties of food crops developed through conventional breeding.

According to OECD, "the concept of substantial equivalence embodies the idea that existing organisms used as food or food sources can serve as a basis for comparison when assessing the safety of human consumption of a food or food component that has been modified or is new. If a new food or food component is found to be substantially equivalent to an existing food or food component, it can be treated in the same manner with respect to safety, keeping in mind the that establishment of substantial equivalence is not a safety or nutritional assessment in itself, but an approach to compare a potential new food with its conventional counterpart" (FAO/WHO 1996).

The use of substantial equivalence is considered a practical approach to assess the safety of new varieties of food. Safety is defined as a reasonable certainty that no harm will result from intended uses under the anticipated conditions of consumption.

The following are the three possible outcomes relative to assessing the substantial equivalence of a genetically modified food or food component to a conventional counterpart (FAO/WHO 1996):

1. The genetically modified food or food ingredient is substantially equivalent to the conventional counterpart;
2. The genetically modified food or food ingredient is substantially equivalent to the conventional counterpart with the exception of certain well-defined differences; or
3. The genetically modified food or food ingredient is not substantially equivalent to the conventional counterpart.

Examples of new varieties of food crops that could be considered substantially equivalent to conventional counterparts are virus-resistant plants produced by introduction of viral coat protein (viral protein is already present in plant tissue of conventional infected plants). Alternatively, if the food is processed removing any added traits and the composition has not changed (e.g., processed canola oil), the oil would be considered substantially equivalent to its conventional counterpart.

The second outcome listed above will apply to most genetically modified food crops. For example, many genetically modified crops will be substantially equivalent to conventional counterparts with the exception of a new trait that imparts a desired characteristic such as pest resistance. When this is the case, further safety assessment should focus on the new trait or well-defined differences. For some genetically modified crops or food components derived therefrom, it may not be possible to demonstrate substantial equivalence to a conventional counterpart, either because differences are not sufficiently well defined or because there is no appropriate counterpart with which to make a comparison. The absence of substantial equivalence does not imply that the genetically modified crop is any less safe. The safety assessment should focus on the nature of the changes (FAO/WHO 1996).

Key Parameters for Assessment of Substantial Equivalence

An assessment of substantial equivalence should focus on a comparison of agronomic characteristics combined with compositional analysis that includes key nutrients as well as toxicants and antinutrients that have potential health significance.

1. Agronomic traits are a good starting point for evaluating substantial equivalence of genetically modified plants to their conventional counterpart. These traits will normally be examined as an integral part of the development program leading to a new food plant variety. Agronomic traits are those characteristics measured by plant breeders for a given crop. For example, in the case of potatoes these may include yield, tuber size and distribution, dry matter content, and disease

resistance. Agronomic properties may vary depending on local environmental conditions. Therefore, the genetically modified variety should be grown in the same geographical regions in which it will be grown commercially.

2. Key nutrients are those components in a particular food product that may have a substantial health impact in the overall diet. These may be major constituents (fats, proteins, carbohydrates) or minor components (essential minerals, vitamins). Critical nutrients to be assessed may be determined, in part, by knowledge of the function and expression product of the inserted gene (e.g., if an inserted gene expresses an enzyme that is involved in amino acid biosynthesis, the amino acid profile should be determined). Introduction of an invertase into potatoes could influence the carbohydrate metabolism; therefore, starch should be investigated. Alternatively, introduction of a storage protein high in methionine into the soybean could affect the amino acid content; therefore, the amino acid profile should be measured. Examples of the analysis of key nutrients in different genetically modified plants have been published (Padgette et al. 1996; Lavrik et al. 1995; Noteborn et al. 1995; Redenbaugh et al. 1992).

Given differences among consumption patterns and practices in various cultures and societies, the key nutrients to be examined may differ in various countries. The critical nutrients to be addressed should be determined using consumption data for the target region. For example, in Denmark, potatoes provide an important source of vitamin C in the diet (35%), not because of their high vitamin C content (20 mg/100 gm) but because of high consumption of potatoes (140 gm/day) (Mejborn 1997). In the United States, potatoes are not as important a source of vitamin C owing to lower potato consumption and the availability of other dietary sources of this vitamin.

3. Critical toxicants and antinutrients are those compounds known to be inherently present in a crop variety whose potency could have an impact on health if their levels were increased significantly (e.g., solanine glycoalkaloids in potatoes, trypsin inhibitors in soybeans). Knowledge of the biologic function of the protein expression product of the inserted gene could influence the decision about which toxicants or antinutrients to examine.

The analysis of important nutrients and toxicants or antinutrients should use validated or standard methods where available (e.g., methods from the Association of Official Analytical Chemists [AOAC] or other recognized bodies).

When the genetically modified line is compared with the parental line or conventional varieties, the varieties should be grown under similar environmental and agronomic conditions. If there are no statistically significant differences in measured parameters between the genetically modified line and the parent or conventional lines, then the genetically modified variety can be considered substantially equivalent to the parent or conventional variety. If statistically significant differences are observed in measured parameters, then a comparison can be made with values available in literature or other sources. If the parameter for the genetically modified line is within the normal range for conventional varieties, further evaluation is not warranted. If the parameter for the genetically modified line is outside the normal range, further evaluation may be needed. A toxicological or nutritional evaluation may be needed to assess whether the difference between conventional and genetically modified lines for a given parameter is biologically meaningful and requires further investigation.

In the future, genetically modified lines for sexually compatible crops will be crossed using traditional breeding techniques. If substantial equivalence has been demonstrated for (1) corn with a gene producing insect protection and (2) corn with a gene for herbicide resistance, then crossing (1) and (2) will produce a new variety that is likely to be substantially equivalent to the parents. If there are no expected interactions between the two traits, then no additional evaluation will be needed. Potential genetic interactions will need to be considered on a case-by-case basis. For example, if two independent modifications to the same metabolic pathway are combined by traditional breeding, further analysis of the products of the metabolic pathway may be warranted.

Safety Assessment for the Introduced Gene Expression Product

Many genetically modified food crops have been shown to be substantially equivalent to conventional crops with the exception of the introduced trait(s) that may impart one or more characteristics such as pest resistance, selectivity to preferred herbicides, modification of the ripening process, and so forth. For these examples, the safety assessment should focus on the introduced trait, the protein expression product of the cloned gene (Hammond 1997). The developer should do the following:

1. Define the biological function, specificity, and mode of action of the protein. If the protein is an enzyme, assess the potential effects of the enzyme on metabolic pathways and levels of endogenous metabolites based on its mode of action and specificity.
2. Compare the amino acid sequence of the protein to known sequences in protein databases (GenPept, SwissProt, PIR) to determine if the protein has sequence homology to food proteins, toxins, or allergens.
3. Assess the inherent digestibility of the protein in vitro with simulated gastric and intestinal protease preparations.
4. Determine the level of expression of the protein in the food. This effort should focus on the raw agricultural product or a specific processed food component (e.g., oil), as appropriate.

Criteria for Concluding the Introduced Protein Is "As-Safe-As" Proteins Already Present in Foods

The following criteria are important in assessing the safety of a protein introduced into food or food components:

1. The protein has a history of safe consumption in other food crops.
2. The protein is functionally and structurally related to proteins with a history of safe consumption in food.

3. The biological function and specificity or mode of action of the protein raises no safety concerns.
4. The amino acid sequence of the protein is not similar to known protein allergens.
5. The protein is not derived from a food source with a history of allergy.
6. The amino acid sequence of the protein is not similar to known protein toxins or antinutrients.
7. The protein is susceptible to degradation by digestive enzymes.

For certain insect control proteins such as the *Bacillus thuringiensis (B.t.)* family of proteins, regulatory agencies such as the EPA require administration of the protein as a single oral high dose to mice. The scientific rationale for an acute test in mice is that known protein toxins generally manifest their toxicity via acute mechanisms (Pariza and Foster 1983; Sjoblad et al. 1992; Jones and Maryanski 1991). The *B.t.* proteins have a long history of safe use (EPA 1988); as new forms of *B.t.* proteins are introduced, the EPA has stated that an acute dosing test in mice will provide assurance that the new *B.t.* protein varieties are safe (Sjoblad et al. 1992). Such studies have been conducted for the Cry3A protein expressed in potatoes, the Cry1Ab protein expressed in corn, and the CryIAc protein expressed in cotton and the tomato. Enzymes that provide selectivity against herbicides or function as selective markers have also been tested in mice (Harrison et al. 1996; Reed et al. 1996; Fuchs et al. 1993). Acute testing may be recommended if the protein is derived from plants, or microorganisms that have no history of consumption. Acute toxicity testing of proteins is not normally needed if the protein (1) is not present in the food (e.g., removed during processing, as in vegetable oils), (2) has a history of consumption in food, and (3) is closely related functionally and structurally to proteins with a history of consumption in food. If safety testing of proteins is undertaken, it may be possible to isolate and purify the protein from the plant. However, this may not be practical if the protein is expressed in small amounts in plant tissue. Alternatively, the protein could be produced in an appropriate fermentation system generating a protein that is biochemically equivalent to the protein produced in the plant.

Proteins that meet the safety criteria listed and are nontoxic if administered to mice are considered "as-safe-as" other proteins naturally present in foods. No further safety or nutritional assessment of the new variety of food crop is necessary (Hammond 1997).

Criteria for Concluding that Further Safety Evaluation Is Required for the Introduced Protein

Some introduced proteins may require further safety and nutritional evaluation if they meet any of the following criteria (Hammond 1997):

1. The protein is functionally and structurally related to proteins that are known toxins or antinutrients.
2. The protein is derived from a food with a history of allergy.

3. The protein is structurally similar to known protein allergens.
4. The protein is not degraded by digestive enzymes.
5. The biological function of the protein has not been characterized and there are no structurally related proteins identified in protein databases.

If the introduced proteins meet one or more of the criteria listed above, additional testing may be warranted on the basis of a case-by-case assessment. Toxicology or nutritional end points that may need to be developed are summarized in the next section.

Safety Testing Recommendations

Testing for Toxicity and Antinutritional Effects

Certain antinutritional proteins such as lectins or protease inhibitors are naturally present in plants and provide protection against insect pests. There has been an interest in inserting these proteins in food crops to provide an alternative to insecticides for control of insect pests (Pusztai and Bardocz 1995), although the safety implications of such transformations have raised concerns (Franck-Oberaspach and Keller 1997). Some plant lectins exert antinutrient effects when fed to animals by binding to the brush-border epithelium of gut cells and thus disrupting nutrient absorption. If a lectin were to be introduced into a food crop to enhance protection against insect pests, it should be fed to animals to assess whether it acts as an antinutrient. Assessing its potential for binding to brush-border epithelium would be part of the safety evaluation. If the introduced lectin had antinutrient properties, a "no-effect level" for antinutrient effects would have to be determined in animal feeding studies and compared with potential human exposures to determine if an adequate safety margin existed. Human nutritional studies may also be appropriate. Because some lectins are known food allergens, assessment of potential allergenicity will also be necessary, which will be discussed in the next section. A similar testing scheme for protease inhibitors can be envisioned. Any general safety questions regarding the introduced protein could be addressed by animal feeding studies designed to assess toxicity and nutritional endpoints. For proteins containing common amino acids, no additional testing for mutagenic, carcinogenic, or teratogenic potential is indicated because there is no evidence that proteins are genotoxic (Dean 1997; FDA 1997) or that feeding proteins such as food enzymes have ever directly produced carcinogenic or teratogenic effects in laboratory animals (Pariza and Foster 1983; Ashby et al. 1987; Hjortkjaer et al. 1993; Kondo et al. 1994; Stavnsberg et al. 1986; Greenough et al. 1991).

Testing for Allergenicity

Further testing of an introduced protein for potential allergy would be indicated if (1) the protein were derived from a food source with a history of allergy or (2) the amino acid sequence of the protein matched (at least eight contiguous identical amino acids) that of a known protein allergen (Metcalfe et al. 1996). The immunogenic potential of the introduced protein should be tested in one of the various solid-phase immunoassays such as the radio allergosorbent test (RAST) or RAST inhibition assay or the enzyme-linked immunosorbent assay (ELISA). Solid-phase immunoassays use IgE fractions of sera from individuals confirmed allergic to the food from which the gene coding for the introduced protein was derived. Where in vitro tests are negative or equivocal, then in vivo skin prick tests could be carried out. If no positive response was detected in the prick test, then double-blind placebo controlled food challenges with patients known to be allergic to the food could be carried out under controlled clinical conditions (Metcalfe et al. 1996). The in vivo studies would only be warranted if the gene were derived from one of the following eight food groups that account for more than 90% of all food allergies: eggs, milk, fish, crustacea, peanuts, soybeans, wheat, and tree nuts (Metcalfe et al. 1996).

The testing scheme to detect allergens summarized above works effectively as demonstrated in the case of the Brazil nut 2S storage protein (Metcalfe et al. 1996). This protein was introduced into soybeans to increase the sulfur amino acid content and thereby improve the bean's nutritional value for use in animal feeds. Because there are a small number of individuals allergic to Brazil nuts, it was decided to test sera from these individuals to see if their sera contained IgE that would cross-react with the Brazil nut 2S storage protein. Sera from eight out of nine Brazil-nut-allergic individuals reacted with this storage protein. Development of this soybean line containing the introduced Brazil nut storage protein was terminated. If the developer had wished to proceed with commercialization of this genetically modified soybean, all foods derived from this soybean would have had to be labeled as containing Brazil nut protein.

For many genetically modified plants, the protein expression product will be derived from a gene source that has no history of inducing allergy. If the amino acid sequence of the introduced protein does not show sequence homology to known allergens (see above), the physicochemical properties of the protein should still be assessed to determine if it shares the profile for allergens. These properties include (1) stability to food processing conditions (high temperature and pH changes that denature food proteins), and (2) resistance to gastric acidity and digestive proteases. The level of expression of the introduced protein in food should also be determined because allergenic proteins often comprise a significant proportion of the total protein in that food (OECD 1997; FAO/WHO 1996).

Regarding digestibility of proteins, it is well established that protein allergens tend to be resistant to digestion and generally constitute a significant percentage of the total protein (1–18%) in the food (Astwood et al. 1996). Allergens are often stable to heat processing. These properties increase the potential that sufficient quantities of the protein will survive both food processing and the hostile environment of the gut to elicit immunologic reactions in the intestinal mucosa. However, there are undoubtedly examples of proteins in food that represent a small percentage of total protein, are not digestible, and do not elicit food allergy. According to allergenicity experts, there is currently no validated in vivo or in vitro model to predict allergenicity potential of proteins (Metcalfe et al. 1996). There is therefore a need to develop a predictive model for allergenicity.

Safety Considerations for Marker Genes

As stated earlier, marker genes are added to the genetic material inserted into plants to help identify successfully transformed plant cells. The use of marker genes and their protein expression products in plant biotechnology has raised the following safety concerns: (1) potential horizontal transfer of the marker gene from plant cells to gut microflora that might reduce the efficacy of therapeutic antibiotics, (2) potential toxicity of the protein expression products of the marker gene, or (3) compromised antibiotic efficacy due to expression of the antibiotic marker gene product in the food. After detailed scientific review, it was concluded that the potential for horizontal gene transfer from plants to gut microflora is vanishingly small (FAO/WHO 1996; WHO 1993; Karenlampi 1996). Introduced genes are stably incorporated into the genome of plants, and there are no known mechanisms for direct transfer of genes from plants to microorganisms. Moreover, the DNA released from the plant during digestion is rapidly degraded. This degradation occurs well before the plant material reaches the lower intestine, cecum, and colon, where gut microflora are found (Karenlampi 1996). The potential for the protein expression product of the antibiotic marker gene to compromise antibiotic efficacy is limited by (1) the digestibility of the expressed protein, (2) low expression of the protein in food, or (3) lack of available cofactors such as adenosine triphosphate (ATP) in the gastrointestinal tract that are required for antibiotic inactivation (FAO/WHO 1996). The safety assessment of the protein expression product of marker genes should focus on the approach outlined earlier for protein expression products of introduced genes (new traits).

Assessment of Safety for Nonsubstantially Equivalent Genetically Modified Plants

To date, there have been few, if any, examples of genetically modified plants that are not considered substantially equivalent to conventional counterparts (with the exception of well-defined differences in some cases). As discussed earlier, with future developments in biotechnology, products may be developed that have no conventional counterpart for which substantial equivalence can be assessed. These products could arise by transfer of genomic regions that have only partially been characterized (FAO/WHO 1996). Genomics is the study of all the genes of an organism and their organization into chromosomes, and it includes researching the links between gene structure and function, or expression, that are the foundations for bioengineering new plant products. Desirable agronomic traits may result from the linkage of several genes functioning together to produce effects such as resistance to drought or alkaline soil conditions. Such a development would allow plants to survive and grow under harsh environmental conditions, and millions of acres of land that cannot be currently used to grow food crops would be available for cultivation of genetically modified food crops.

In the future, there may be novel foods derived via biotechnology that have intended benefits, but further safety and nutritional evaluation will be needed before these foods can be marketed. In other cases, unintended changes in a new food crop may occur that could require further evaluation to understand what was changed and its significance to the safety and nutritional value of the food crop.

Toxicological and nutritional studies may be indicated in some cases to resolve safety and nutritional questions of food crops that are not substantially equivalent to their conventional counterparts. If animal feeding studies are considered necessary, their objectives must be clear and the experimental design carefully planned to avoid nutritional problems that may confound data interpretation (FAO/WHO 1996). There may be a need to modify existing protocols for testing whole food or food components—in particular by providing adequate nutrition with respect to diet. Nutritional imbalance may mask toxic effects. Incorporation of high levels of a food into the diet may also result in nutritional deficiencies that could cause adverse effects unrelated to the food being tested. The usual concept of safety margins may not be applicable because it is often not possible to feed a whole food at a high enough level to obtain anything close to a hundredfold margin of safety. The testing of specific components or extracts of a novel food may present an option for addressing safety and nutritional testing of the whole food. For certain foods, animals are not good models for predicting safety to man because animals exhibit species-specific sensitivity to the food. For example, dogs experience transient paralysis from consumption of macadamia nuts (Hammond et al. 1996) or develop fatal cardiac arrhythmias owing to sensitivity to theobromine in chocolate; male rats of one particular strain (Sprague Dawley) develop cardiomyopathy when fed diets high in vegetable oils (Gunner 1993; FDA 1985). Feeding animals high levels of various foods in the diet has produced anemia (rats and dogs fed onions), enlarged cecums (rats fed potatoes), mucosal lesions in the stomach (rats fed tomatoes or chili powder), pulmonary emphysema (rats fed beans), and reduced breeding performance and survival (rats fed wheat flour) (Hammond 1996). If nutritional quality is the important issue, it may be appropriate to go directly to nutritional studies in human volunteers rather than to use animal models that may have limited relevance to man. The industry currently uses "sip and spit" tests to evaluate the organoleptic properties of new varieties of vegetable crops as part of the quality assessment. Once the initial safety assessment has been completed for a nonsubstantially equivalent food product, sensory or nutritional evaluations, or both, may be recommended to ensure the quality and acceptability of the product for consumers.

CONCLUSION

Biotechnology provides plant breeders the opportunity to develop new varieties of food crops more efficiently and with greater potential benefit than has been possible with conventional breeding practices. To provide assurance that this technology will generate food "as safe as" that produced by traditional breeding programs, safety assessment strategies have been developed for products of plant biotechnology that have generally been accepted by international regulatory groups (FAO/WHO 1996). This strategy includes comparison of the agronomic properties and important nutrient–toxicant composition of genetically modified crops with their conventional counterparts. If the comparisons show no meaningful differences in the aforementioned parameters, the genetically modified food crop is said to be "substantially equivalent" to its conventional counterpart, and the safety and nutritional assessment is completed. Many genetically modified foods will be substantially equivalent to conventional varieties with the exception of one or more introduced traits. The safety assessment will then focus on the introduced trait to determine if the protein expression product meets the criteria for being "as-safe-as" proteins already in the food supply.

If protein expression products are derived from plants with a history of allergenicity; have a structural or functional similarity to known allergens, toxins, or antinutrients; or have functions that may alter the nutritional quality of the plant, then additional safety and nutritional studies may be needed. In the absence of validated assays to predict potential allergenicity, comparison of amino acid sequence homology of the protein with known allergens and assessment of its physicochemical properties are the best tools currently available to predict potential allergenicity.

As biotechnology develops, future products may be developed that have no conventional counterpart for which substantial equivalence can be assessed. These varieties could be designed to have improved agronomic properties or provide important health and nutritional benefits. The safety and nutritional assessment of these crops should be carried out on a case-by-case basis using the strategies outlined above.

During the last few years, over 34 different genetically modified plant products have successfully completed regulatory review in countries around the world. The experiences gained in bringing these products to market have enabled developers and regulatory agencies to define the key safety questions and the science to answer the questions. This progress will help provide guidance to the development of improved and safe plant biotechnology products for the future.

REFERENCES

Ashby, R., Hjortkjaer, R. K., Stavnsberg, M., Gurtler, H., Pedersen, P. B., Bootman, J., Hodson-Walker, G., Tesh, J. M., Willoughby, C. R., West, H., and Finn, J. P. 1987. Safety evaluation of *Streptomyces murinus* glucose isomerase. *Toxicol. Lett.* 36:23–35.

Astwood, J. D., Leach, J. N., and Fuchs, R. L. 1996. Stability of food allergens to digestion in vitro. *Bio Technology* 14:1269–1273.

Dean, S. 1997. Industrial Genotoxicology Group (IGG): The use of historical data in data interpretation and genotoxicity testing of biotechnology products, Royal Society of Medicine, London, UK, December 1995. *Mutagenesis* 12(1):49–50.

EPA (United States Environmental Protection Agency). 1988. *Guidance for the reregistration of pesticide products containing* Bacillus thuringiensis *as the active ingredient.* NTIS PB 89 164198. Washington, D.C.: EPA Office of Pesticide Programs.

FAO/WHO. 1996. Biotechnology and food safety. In *Report of a joint FAO/WHO consultation, Rome Italy, 30 September–4 October 1996.* FAO, Rome: Food and Nutrition Paper 61.

FDA (Food and Drug Administration). 1985. Direct food substances affirmed as generally recognized as safe: Low erucic acid rapeseed oil. *Fed. Reg.* 50(18):3745–3755.

FDA (Food and Drug Administration). 1992. Statement of policy: Foods derived from new plant varieties: Notice. *Fed. Reg.* 57(104):22984–23005.

FDA (Food and Drug Administration). 1997. International conference on harmonization. Draft guideline for the preclinical testing of biotechnology-derived pharmaceuticals. *Fed. Reg.* 62(65):16437–16442.

Franck-Oberaspach, S. L., and Keller, B. 1997. Consequences of classical and biotechnological resistance breeding for food toxicology and allergenicity. *Plant Breeding* 116:1–17.

Fuchs, R. L., Ream, J. E., Hammond, B. G., Naylor, M. W., Leimgruber, R. M., and Berberich, S. A. 1993. Safety assessment of the neomycin phosphotransferase II (NPTII) protein. *Bio/Technol.* 11:1543–1547.

Greenough, R. J., and Everett, D. J. 1991. Safety evaluation of alkaline cellulase. *Food Chem. Toxicol.* 29(11):781–785.

Gunner, S. W. 1993. Low erucic acid rapeseed oil (LEAR Oil). In *Safety evaluation of foods derived by modern biotechnology. Concepts and principles*, 35–39. Paris: Organization for Economic Co-Operation and Development.

Hammond, B., Rogers, S. G., and Fuchs, R. L. 1996. Limitations of whole food feeding studies in food safety assessment. In *Food safety evaluation. OECD workshop on food safety evaluation, Oxford England, 12–15 September 1994*, 85–97. Paris: Organization for Economic Co-Operation and Development.

Hammond, B. G. 1997. Assessment of potential protein toxicity. *Report of the OECD workshop on the toxicological and nutritional testing of novel foods. Aussois, France, 5–8 March 1997.* Paris: Organization for Economic Co-Operation and Development.

Harrison, L. A., Bailey, M. R., Naylor, M. W., Ream, J. E., Hammond, B. G., Nida, D. L., Burnette, B. L., Nickson, T. E., Mitsky, T. A., Taylor, M. L., Fuchs, R. L., and Padgette, S. R. 1996. The expressed protein in glyphosate-tolerant soybean, 5-enolpyruvylshikimate-3-phosphate synthase from *Agrobacterium* sp. strain CP4, is rapidly digested in vitro and is not toxic to acutely gavaged mice. *J. Nutrit.* 126(3):728–740.

Hjortkjaer, R. K., Stavnsberg, M., Pedersen, P. B., Heath, J., Wilson, J. A., Marshall, R. R., and Clements, J. 1993. Safety evaluation of esperase. *Food Chem. Toxicol.* 31(12):999–1011.

IFBC (International Food Biotechnology Council). 1990. Biotechnologies and food. Assuring the safety of foods produced by genetic modification. *Regul. Toxicol. Pharmacol.* 12:S1–S196.

James, C. 1997. Global status of transgenic crops in 1997. *ISAAA Briefs No. 5.* Ithaca, NY: ISAAA.

James, C., and Krattiger, A. F. 1996. Global review of the field testing and commercialization of transgenic plants, 1986–1995: The first decade of crop biotechnology. *ISAAA Briefs No. 1.* Ithaca, NY: ISAAA.

Jones, D. D., and Maryanski, J. H. 1991. Safety considerations in the evaluation of transgenic plants for human foods. In *Risk assessment in genetic engineering*, eds. M. A. Levin and H. S. Strauss, 64–82. New York: McGraw-Hill.

Karenlampi, S. 1996. *Health effects of marker genes in genetically engineered food plants*, 530. Copenhagen: TemaNord.

Kessler, D. A., Taylor, M. R., Maryanski, J. H., Flamm, E. L., and Kahl, L. S. 1992. The safety of foods developed by biotechnology. *Science* 256:1747–1750.

Kondo, M., Ogawa, T., Matsubara, Y., Mizutani, A., Murata, S., and Kitigawa, M. 1994. Safety evaluation of lipase G from *Penicillium camembertii. Food Chem. Toxicol.* 32(8):685–696.

Kough, J. L. 1996. U.S. EPA considerations for mammalian health effects presented by transgenic plant pesticides. In *Food Safety Eval.* Paris: Organization for Economic Development and Co-Operation.

Lavrik, P. B., Bartnicki, D. E., Feldman, J., Hammond, B. G., Keck, P. J., Love, S. V., Naylor, M. W., Rogan, G. J., Sims, S. R., and Fuchs, R. L. 1995. Safety assessment of potatoes resistant to Colorado potato beetle. Chap. 13 in *Genetically modified foods. Safety issues.* ACS Symposium Series 605. 148–157. Washington, D.C.: American Chemical Society.

Mejborn, H. Individual nutrients—When to compare about changes. *Report of the OECD workshop on the toxicological and nutritional testing of novel foods. Aussois, France, 5–8 March 1997.* Paris: Organization for Economic Co-Operation and Development.

Metcalfe, D. D., Astwood, J. D., Townsend, R., Sampson, H. A., Taylor, S. L., and Fuchs, R. L. 1996. Assessment of the allergenic potential of foods derived from genetically engineered crop plants. *Crit. Rev. Food Science Nutri.* 36:S165–186.

Noteborn, P. J. M., Bienenmann-Ploum, M. E., van den Berg, J. H. J., Alink, G. M., Zolla, L., Reynaerts, A., Pensa, M., and Kuiper, H. A. 1995. Safety assessment of the *Bacillus thuringiensis* insecticidal crystal protein CRYIA(b) expressed in transgenic tomatoes. In *Genetically modified foods, Safety aspects*, eds. K. H. Engel, G. R. Takeoka, and R. Teranishi, 134–147. Washington, D.C.: American Chemical Society.

OECD (Organization for Economic Cooperation and Development). 1993. *Safety Evaluation of Foods Produced by Modern Biotechnology: Concepts and Principles.* Paris: Organization for Economic Cooperation and Development.

OECD (Organization for Economic Cooperation and Development). 1996. *Food safety evaluation. OECD workshop on food safety evaluation, Oxford England, 12–15 September 1994.* Paris: Organization for Economic Cooperation and Development.

OECD (Organization for Economic Cooperation and Development). 1997. *Report of the OECD workshop on the toxicological and nutritional testing of novel foods. Aussois, France, 5–8 March 1997.* Paris: Organization for Economic Cooperation and Development.

Padgette, S. R., Taylor, N. B., Nida, D. L., Bailey, M., MacDonald, J., Holden, L. R., and Fuchs, R. L. 1996. The composition of glyphosate tolerant soybeans is equivalent to that of conventional soybeans. *J. Nutri.* 126:702–716.

Pariza, M. W., and Foster, E. M. 1983. Determining the safety of enzymes used in food processing. *J. Food Protection* 46(5):453–467.

Pusztai, A., and Bardocz, S. 1995. *Lectins. Biomedical perspectives.* Washington, D.C.: Taylor and Francis.

Redenbaugh, K., Hiatt, W., Martineau, B., Kramer, M., Sheehy, R., Sanders, R., Houck, C., and Emlay, D. 1992. *Safety assessment of genetically engineered fruits and vegetables. A case study of the FLAVR SAVR™ tomato.* Boca Raton, Florida: CRC Press.

Reed, A. J., Kretzmer, K. A., Naylor, M. W., Finn, R. F., Magin, K. M., Hammond, B. G., Leimgruber, R. M., Rogers, S. G., and Fuchs, R. L. 1996. Safety assessment of 1-aminocyclopropane-1-carboxylic acid deaminase protein expressed in delayed ripening tomatoes. *J. Agric. Food Chem.* 44:388–394.

Sjoblad, R. D., McClintock, J. T., and Engler, R. 1992. Toxicological considerations for protein components of biological pesticide products. *Regul. Toxicol. Pharmacol.* 15:3–9.

Stavnsbjerg, M. H., Hjortkjaer, R. K., Bille-Hansen, V., Jensen, B. F., Greenough, R. J., McConville, M., Holmstroem, M., and Hazelden, K. P. 1986. Toxicological safety evaluation of a *Bacillus acido-pullulyticus* pullanase. *J. Food Protection.* 49(2):146–153.

Watson, J. D., Gilman, M., Witkowski, J., and Zoller, M. 1996. Genetic engineering of plants. In Recombinant DNA, 2d ed. New York: W. H. Freeman and Company.

WHO (World Health Organization). 1991. *Strategies for assessing the safety of foods produced by biotechnology. Report of a joint FAO/WHO consultation.* Geneva: World Health Organization.

WHO (World Health Organization). 1993. *Health aspects of marker genes in genetically modified plants. Report of a WHO workshop.* WHO/FNU/FOS/93.6. Geneva: World Health Organization.

WHO (World Health Organization). 1995. *Application of the principles of substantial equivalence to the safety evaluation of foods or food components from plants derived from modern biotechnology. Report of a WHO workshop.* WHO/FNU/FOS/95.1. Geneva: World Health Organization, Food Safety Unit.

Biotechnology and Safety Assessment, 2nd ed.
Edited by John A. Thomas
Copyright © 1998 Taylor & Francis

4

Therapeutic Manipulation of Cytokines

Robert V. House

IIT Research Institute, Chicago, Illinois

For most of medical history, therapeutic intervention has relied on exogenous factors. These treatments have run the gamut from magical incantations, to animal- and plant-derived substances, and finally to rational drug design by combinatorial chemistry. Success in treatment has therefore depended on placebo effect, fortuitous similarities in ligands and receptors (e.g., poppy-derived opiates and the opioid receptors in the brain), or a grasp of biological mechanisms at the molecular level. Paradoxically, this increased understanding of the biology of health and disease, as well as the tools that were developed to gain this understanding, has led to the possibility of utilizing the body's own endogenous factors to treat disease and restore health.

One potential target of this therapy is the group of factors collectively known as cytokines. In this chapter we will describe cytokines and their diverse use in human therapeutics, and provide an overview of the multitude of experimental and clinical approaches that have been designed to tap into the body's own healing power. Along the way we will also describe some of the known and potential disadvantages of this exciting new therapy. This review is not meant to be a comprehensive listing of the results of individual studies but rather a broad survey of the types of approaches used.

THE BASICS OF CYTOKINE BIOLOGY

The majority of cytokines (also known by their previous name, lymphokines) are glycoproteins secreted by cells, although membrane-associated forms have also been described. Cytokines are generally not produced constitutively but rather in response to cellular activation (Fraser et al. 1993). Cytokine production is highly regulated in both paracrine and autocrine fashions at the transcriptional and posttranscriptional levels (Taniguchi 1988; Dendorfer 1996). In addition, the short half-life of cytokine mRNA suggests selective degradation of the message as a result, evidently, of a particular sequence common to many cytokines and proto-oncogenes (Cosman 1987). Finally, a variety of mechanisms appear to ensure that cytokines remain compartmentalized (Kroemer et al. 1993). These mechanisms all work together to limit the biological activity of cytokines.

Perhaps the most important biological feature of cytokines is that they are highly pleiotropic and redundant in function. That is to say, most of the cytokines described to date have multiple actions, and these actions often overlap with the actions of other cytokines. In addition, cytokines frequently work via cascading mechanisms that allow them to interact with each other both synergistically and antagonistically. These features are critical from a therapeutic consideration because the ultimate consequences of manipulating any single cytokine must be evaluated in the context of this overall network of factors. Other features of cytokines important from a clinical standpoint are that, in most cases, they act primarily at a local level and that they are rapidly cleared from the circulation. A notable exception to this rule is interleukin-6 (IL-6), whose importance in regulation of the acute phase response has led to a variety of "chaperone" mechanisms for regulating its systemic bioavailability.

Cytokines interact with specific receptors grouped into the hemopoietin super-family, the tumor necrosis factor (TNF) family, the immunoglobulin superfamily, and the tyrosine kinase family. These receptors are generally membrane-bound molecules, although soluble receptors have also been described. Many cytokine receptors share several similar characteristics, including a subunit structure and an association with signal-transducing elements within the cell (Williams and Giroir 1995). Moreover, several of the receptor subunits are shared between various cytokine receptors, which possibly contributes to cytokine pleiotropy and redundancy.

A frequent conceptual mistake is the belief that cytokines are an exclusive product of the immune system. In fact, certain cytokines are phylogenetically ancient, genet-ically conserved, and highly pleiotropic molecules (probably the best examples are IL-1 and TNF); forms of these molecules are found in invertebrates lacking a true immune system (Secombes 1991). Furthermore, IL-1, TNF, and related molecules are intimately involved in the process of apoptosis, a fundamental biological process. Put another way, cytokines should be recognized for their role as conveyers of bio-information rather than as simple effector molecules involved in a single process. Table 1 illustrates the role of cytokines in the continuum of biological information.

This brings us to a crucial detail: cytokines should not, indeed cannot, be seen as operating in a biological vacuum. Rather, as suggested by Table 1, cytokines may be thought of as a member of a triad including (at a minimum) neuropeptides and hormones and possibly the so-called peptide growth factors as well (Weigent and Blalock 1995). It is now well established that these molecules form a complex net-work responsible for regulating many physiological processes. Certain cytokines (e.g.,

TABLE 1. *The continuum of bioinformation*

Signal	Example	Route	Source
Nucleic acids	RNA	Local: intracellular	All living cells, viruses
Cytokines	Interleukin-1	Local: intercellular	Diverse cells
Neuropeptides	β-Endorphin	Local and systemic	Specialized cells
Hormones	Insulin	Systemic	Endocrine glands
Pheromones	Androstenone	Interspecies: external	Exocrine glands
Vocalization	Distress call	Inter/Intraspecies: external	Specialized organs

TABLE 2. *The (abbreviated) universe of human cytokines*

Interleukins (IL)	Chemokines	
IL-1α, β	α (C-X-C)	β (C-C)
IL-1RA	IL-8	RANTES
IL-2	gro (α, β, γ)	I-309
IL-4 – IL-7	GCP-2	eotaxin
IL-9 – IL-18	ENA-78	MIP-1 (α, β)
	SDF-1	MCP (1,2,3)
Colony-stimulating factors (CSF)	**Hematopoietins**	
IL-3 (Multi-CSF)	Erythropoietin	
G-CSF	Thrombopoietin	
M-CSF	Stem cell factor	
GM-CSF	flt3 ligand	
Tumor necrosis factors (TNF)	**Interferons (IFN)**	
Tumor necrosis factor (TNF-α)	Type I (IFN-α, IFN-β)	
Lymphotoxin (TNF-β)	Type II (IFN-γ)	
Miscellaneous		
Oncostatin-M		
Leukemia inhibitory factor		
Transforming growth factor beta		

IL-6) mediate autocrine functions in the nervous and endocrine systems. Likewise, cells of the immune system produce hormones (e.g., prolactin) and neuropeptides (e.g., endorphins); in addition, receptors for hormones and neuropeptides are found on lymphocytes.

Thus, given the redundant and pleiotropic nature of cytokines themselves as well as their role as regulatory molecules for various physiological processes, it becomes apparent that developing therapeutics based on modifying the functions of cytokines is far from straightforward. As detailed later in this chapter, toxicity and other unintended consequences are often the result of initial forays into this area. Nevertheless, great advances have been made and continue apace.

Table 2 illustrates a suggested categorization of the various human cytokines. Cytokine nomenclature is fraught with confusion; beginning with interleukin-1, many newly discovered cytokines were named sequentially regardless of their function or genetic relationship to other cytokines. In most current classification schemes, cytokines are grouped according to function or genetic similarity to each other. Table 2 is not meant to be inclusive; on the contrary, many excellent classification schemes based on function have previously been published (e.g., Whiteside 1994).

A brief explanation of the classification scheme used here may be helpful. In general, the term *interleukins* refers to cytokines exerting their primary effect on cells of the immune system; IL-1 and IL-6 are notable exceptions to this broad categorization because they mediate a tremendous number of effects besides immunomodulation. *Chemokines* are small polypeptides that are important not only in immunity and inflammation but in other regulatory processes as well. *Colony stimulating factors*

(CSFs) are important as mediators of hematopoiesis but also exert effects on the immune system. *Hematopoietins* also regulate the function of bone marrow but have fewer effects on the immune response. *Tumor necrosis factors* (TNFs) regulate a panoply of biological effects. *Interferons* were originally named on the basis of their ability to interfere with viral replication, and thus they function as important mediators of host defense; however, they also affect various other processes. Finally, we may lump together **miscellaneous** cytokines that do not fit quite so neatly into the scheme owing to their multiplicity of action. Not included in this scheme are the various *peptide growth factors* such as epidermal growth factor and insulin-like growth factor—whose functions appear to be primarily related to growth, tissue repair, and homeostasis—but that may function as cytokines in certain circumstances. All of these cytokine classes, to varying degrees, represent viable targets for therapeutic manipulation.

CYTOKINES IN HUMAN HEALTH AND DISEASE

Given the ubiquitous role of cytokines in normal human biology, it is only natural that disruption in cytokine levels could be either the cause or effect of human disease. A major advance in understanding the role of cytokines in disease, and particularly in diseases incorporating an immune component in their etiology, was first described by Mosmann et al. (1986). These investigators found that T-helper lymphocytes could be categorized functionally (although not, as yet, phenotypically) into at least two subsets on the basis of their particular pattern of cytokine production. Originally described in vitro using T-cell clones, these subsets were named T-helper-1 and T-helper-2 (subsequent to this, the nomenclature TH1/TH2, T1/T2, and Type 1/Type 2 has also been used; TH1/TH2 will be the preferred usage in this chapter). Building on the data collected from in vitro clones, researchers found that these patterns of cytokine production resulted in vivo in rodent models. More recently, the TH1/TH2 dichotomy has been confirmed in humans (Romagnani 1995).

In general, TH1 cells predominantly produce interferon-gamma (IFN-γ), IL-2, and TNF, whereas TH2 cells primarily make IL-4, IL-5, IL-10 and IL-13; these patterns appear to be essentially the same in humans. Interleukin-2 acts as an autocrine growth factor for TH1 and TH2 cells, although TH1 cells are much more sensitive to this activity. Interferon-gamma acts to suppress the growth of TH2 cells. Conversely, IL-4 acts as the autocrine growth factor for TH2 cells, whereas IL-10 suppresses the growth and function of TH1 cells. Various other cytokines are produced in common by these subsets, albeit at relatively lower levels. It should be noted that these distinctions are somewhat fluid, for clones have been isolated in both human and laboratory animal cells that display intermediate cytokine production profiles. Moreover, CD8 (cytotoxic–suppressor) T cells also produce cytokines, as do macrophages and B cells (Pistoia 1997), thus complicating the elucidation of a "typical" cytokine response. However, the general pattern described above for T-helper cells appears to hold true in most cases. This paradigm is illustrated in Figure 1.

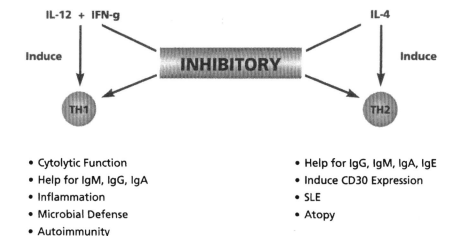

FIG. 1.

The induction of a TH1-type response versus a TH2-type response (i.e., a state characterized by the preponderance of either TH1- or TH2-type cytokine production) depends on a variety of factors, including the type of antigen-presenting cell, the type and amount of antigen, and factors in the cellular milieu such as other cytokines and hormones (Romagnani 1995). This particular pattern of stimulatory–inhibitory–autocrine functions is now recognized to form the basis for differential immune responses as well as certain pathologies. For example, in normal immune responses TH1 cytokines tend to favor cytolytic (cell-mediated) reactions and the activation of macrophages, whereas TH2 cytokines provide T-cell help for the production of various antibodies. On the other hand, TH1 responses are usually associated with autoimmune diseases (e.g., multiple sclerosis and rheumatoid arthritis), whereas TH2 responses include atopic conditions, reduced response to infectious challenge, and even successful pregnancy (Romagnani 1994, 1995). In current and future therapies involving cytokines, it will be of vital importance to understand this interlocking network of cytokines in greater detail to allow possibly for the direct manipulation of immune responses (Powrie and Coffman 1993; Hetzel and Lamb 1994).

Immunomodulation

Ironically, although cytokines are usually associated with the immune system, direct modulation of the immune response has not been the principal therapeutic application of these factors to date. A wide range of options are theoretically available for modulating the immune system, and most of these options are described in this chapter.

Neoplastic Disease

One of the most thoroughly investigated therapeutic uses of cytokines thus far has been in the treatment of neoplastic disease. Owing to the intrinsically complex nature of neoplastic disease, it follows that the use of cytokines for treatment is equally complex. Therapeutic options to date include disruption of autocrine growth, enhancement of natural immune resistance, the use of cytokines in combination with other drugs, and direct tumor cytotoxicity by cytokines.

It is recognized that many tumors, particularly neoplasias of the hematopoietic system (e.g., leukemia) may involve the autocrine growth of cytokine-responsive or cytokine-producing cells. For example, IL-6 is thought to have the potential to induce polyclonal B-cell activation (Whicher and Evans 1990). In such cases, selective reduction in the levels of certain growth factors could theoretically be beneficial (although clinical trials employing extracorporeal removal of cytokine have yet to demonstrate clinical improvements). Conversely, deficiencies in cytokine production may impair the function of cellular immune components required for resistance to tumors, such as cytotoxic T lymphocytes or natural killer cells (Romagnani 1994; Wagstaff et al. 1995). In such cases, the administration of carefully selected cytokines or cytokine combinations may boost a failing immune response.

An intriguing concept that has been explored is the combined administration of cytokines with cytotoxic drugs. Experimental data suggest that administration of cytokines may result in enhanced susceptibility to cytotoxic drugs under certain circumstances (Kreuser et al. 1992; Mitchell 1992). Certain cytokines may also exert a direct cytotoxic effect on tumor cells—either as single agents or in combination (Heaton and Grimm 1993). Finally, a relatively successful use of cytokines in oncology has been the repopulation of bone marrow following transplant subsequent to chemotherapy or radiation.

Transplantation

Another successful therapeutic use of cytokines has been as an invaluable adjunct to transplantation. Ironically, successful treatment may take the form of either cytokine suppression or cytokine addition. In the former situation, cytokines are highly active participants in the rejection process responsible for the failure of organ transplantation and may contribute to engraftment failure by a variety of mechanisms (Soulillou 1993). Successful long-term engraftment has been made possible by new drugs that selectively inhibit cytokine production without completely eliminating host defense (discussed in the section on cytokine inhibition therapy).

Conversely, the availability of recombinant cytokines (both natural and modified) has led to great advances in bone marrow transplantation and, more recently, stem cell transplantation (Stewart 1995). Cytokine therapy may be used in a variety of ways for bone marrow transplantation. For example, administration of colony-stimulating factors and other hematopoietic cytokines (particularly in combination) can be used to enhance repopulation of bone marrow following chemotherapy or radiation treatment and thus lessening potential morbidity from infectious disease or coagulation dysfunction. Alternatively, cytokines may be administered before therapy to enhance the success of autologous transplantation—particularly of stem cells. This expansion of stem cells is usually carried out in vivo (Stewart 1995), although studies are under way to use ex vivo expansion of cell populations for transplant (McAlister et al. 1992).

Infectious Diseases

Resistance to infectious organisms is a primary function of the immune system and therefore requires the precise interplay of most of the cytokine catalog. However, the complex and redundant mechanisms of natural and acquired host resistance are probably not amenable to a simplistic addition or deletion of individual cytokines. Improvements in the control of infectious disease will more likely involve cytokines as a component of the therapeutic regimen (Roilides and Pizzo 1993).

One promising avenue is the use of cytokines as adjuvants. Although not as powerful as some of the best experimental adjuvants, cytokines have been shown to be at least equal, if not superior, to adjuvants currently approved for human use (Heath and Playfair 1992). Much work remains to be done to optimize this approach; some initial attempts have included using combinations of cytokines, increasing exposure time, and physically associating cytokines with antigens (Heath and Playfair 1992).

Another use of cytokines in treatment and prevention of infectious disease is to correct defects in immune function and thus to enhance host resistance to opportunistic infections common in immunocompromised hosts. This approach usually entails the use of colony-stimulating factors to restore depleted bone marrow, although other cytokines such as IL-1, IL-6, and IFN-γ display potential benefits (Roilides and Pizzo 1993). A more ambitious possibility is the manipulation of cytokine levels (along with other variables) to direct the immune response selectively to either a Type 1 or Type 2 response (Heath and Playfair 1992; Golding et al. 1994; Romagnani 1994). At present, this approach remains experimental.

SPECIFIC THERAPEUTIC OPTIONS

Exogenous Cytokine Therapy

"Natural" Cytokines

The discovery of many of the earliest known cytokines took place in human or animal cell cultures, and elucidation of their in vitro and in vivo effects utilized culture supernatants of stimulated primary lymphoid or myeloid cells. These cultures contain a diversity of cytokines that are produced naturally in response to cellular activation. Although these "natural" (contrasted with man-made) cytokines obviously produce desired biological effects, they have not been pursued as intensively as recombinant cytokines for several reasons. First, mammalian cell culture does not routinely produce cytokines as economically as bacterial fermentations, and mammalian cells are technically more difficult to work with than bacteria or other biotechnology workhorses such as insect cells. However, a more important consideration may be the undefined nature of these cells—particularly the primary cell cultures. These culture supernatants often contain a multitude of cytokines and other bioactive molecules. This undefined nature makes it difficult to determine the activity of individual factors, which is an undesirable circumstance for conventional drug discovery efforts.

On the other hand, a teleological assessment would suggest that just such an assortment of cytokines is perhaps the best application of cytokines and that the cellular source of the cytokine mix represents the ultimate best judge of an appropriate "mix." On the basis of this logic, a mix of natural cytokines may represent an extension of the body's own healing process. One such preparation that has demonstrated clinical efficacy is Leukocyte Interleukin Inj., which has the trade name Multikine. Multikine is a serum-free lymphokine mixture (containing, among others, IL-2) prepared from human buffy-coat mononuclear cells (Chirigos et al. 1995). Multikine appears to enhance the function of natural killer cells and cytotoxic T lymphocytes. Injection of this product into humans with head and neck cancer was found to result in tumor regression with infiltration of the tumors by lymphocytes. Products such as Multikine may hold promise as adjuncts to lymphokine-activated killer cell therapy.

Recombinant Natural Sequence Cytokines

At present, the overwhelming majority of preclinical and clinical studies of exogenous cytokine therapy have utilized natural sequences produced recombinantly. The literature on the results of these studies is therefore extensive, and a recounting of their results is well beyond the scope of this chapter. The interested reader is directed to several excellent reviews by Moore (1991), Herrmann (1992), Nemunaitis (1993), Bruton and Koeller (1994), and especially Cardi et al. (1996).

Mutant or Synthetic Cytokines

As described later in this chapter, toxicity is often the limiting factor in protocols involving cytokine administration. Seeking to broaden the therapeutic index of these agents, some investigators have found that rational protein drug design can result in cytokines with enhanced biological activity and reduced "changes in quantitative parameters." This latter phrase was proposed by Brouckaert et al. (1994) as an alternative to "toxic side effects" in describing the sequelae of exogenous TNF administration and implies a belief that the current state of knowledge does not allow for a precise understanding of whether observed affects are truly toxic or simply adaptive changes. This semantic device is undoubtedly cold comfort to patients enduring the side effects of cytokine therapy (see the section on toxicity associated with cytokine therapy).

The majority of published work on cytokine mutant proteins (muteins) has focused on synthetic versions of TNF. Tumor necrosis factor is a potent, pleiotropic cytokine;

because it is involved in a multitude of both normal and pathological processes, it represents an obvious target for cytokine therapy. However, this very diversity of actions has proven to be problematic when TNF is administered systemically, which often results in profound toxicity. This has led to efforts to design TNF molecules that retain beneficial activity but minimize the toxic side effects. Several investigators have reported success in this effort (Brouckaert et al. 1994; Lucas et al. 1997), although clinical data are not yet available.

Another recent example of this approach is the construction of synthetic cytokine (Synthokine) SC-55494, a potent IL-3 receptor antagonist first described by Thomas et al. (1995). By employing oligonucleotide-directed mutagenesis on a synthetic, natural sequence IL-3 DNA followed by iterative combination, these investigators produced several thousand IL-3 mutants that were subsequently screened for biological activity. The results of this screening identified SC-55494 (Synthokine), a molecule with 48 amino acid sequence changes relative to natural IL-3. Synthokine was found to produce a ten- to twentyfold increase in hematopoietic activity, although its potential for induction of inflammatory responses was only about twofold greater than natural sequence IL-3 (Thomas et al. 1995). Subsequent primate studies (MacVittie et al. 1996; Farese et al. 1996) demonstrated that administration of Synthokine, either alone or in combination with G-CSF, significantly enhanced recovery from both neutropenia and thrombocytopenia following radiation-induced bone marrow aplasia. These results suggest that Synthokine and related molecules may hold significant promise in the support of myelosuppressed patients.

Several other cytokine muteins have been constructed and are under evaluation for the treatment of various conditions. Examples include IL-4 muteins (Duschl et al. 1995) and IL-5 mutein (Devos et al. 1995) for the treatment of allergic diseases.

Fusion Proteins

A different form of cytokine is the fusion protein. The concept behind fusion proteins is to connect the functional portion of a cytokine (i.e., the receptor-binding moiety) with another molecule; this second molecule can have biological activity of its own or may be there to assist the function of the cytokine molecule. In the case of fusion proteins consisting of two different cytokines, the resulting molecule is termed a hybrid cytokine or hybrikine. One of the first examples of a hybrikine was PIXY321 (Pixykine), which consists of the active domains of IL-3 and GM-CSF coupled by a flexible amino acid linker sequence. By incorporating elements of both IL-3 and GM-CSF (both of which are potent colony-stimulating factors), PIXY321 was found to stimulate a wide range of progenitor cells. Perhaps more significantly, PIXY321 was found to be more active than either of its constituent cytokines, either alone or in combination (Curtis et al. 1991). Subsequent clinical trials with PIXY321 have shown it to be a useful adjunct in autologous bone marrow transplantation among other indications (Vose et al. 1997). Other cytokine–cytokine fusion proteins are now being described such as CH925, which is a fusion protein comprising rhIL-6 and rhIL-2. This hybrid molecule exhibits erythropoietin-like activity (Zhao et al. 1996).

Cytokines may also be combined with other molecules. For example, construction of cytokine–antibody hybrids produces fusion proteins retaining the properties of both parents (i.e., antigen binding and cellular activation). Such constructs display promise as potent immunostimulants (Penichet et al. 1997) or as treatments for conditions such as septic shock (Abraham et al. 1997). A different but related immunostimulant property can be obtained by fusing cytokines directly to antigens, as demonstrated by Kim et al. (1997).

An interesting example of a novel problem that was overcome by constructing a mutant cytokine was reported by Anderson et al. (1997). These investigators worked with IL-12, a cytokine that has been somewhat problematic to produce recombinantly owing to its heterodimeric structure and because the distinct subunits of the molecule are produced by separate genes on different chromosomes. This problem has been overcome by constructing a single-chain fusion protein (Flexi-12) that retains the biological activity of the natural cytokine molecule. No doubt this approach will find greater utility in the future.

Methods for Delivering Cytokines

As previously mentioned, cytokines are proteins or small polypeptide molecules; therefore, oral administration has been impractical because these molecules cannot withstand the digestive environment. Their administration thus far has been primarily parenteral—mainly by intravenous injection. Although certainly the most convenient method clinically, this approach is associated with many serious disadvantages, including the necessity of bolus administration (or at least short-term infusion), which induces toxicities, as well as the need to administer high doses of cytokines to yield sufficient locally effective concentrations (i.e., lack of targeting).

A variety of advanced methodologies are under development for a more efficient and clinically manageable delivery system for therapeutic cytokines. Although many of these techniques are in the developmental phase, they show exciting promise as future treatment options. Table 3 illustrates some of the methods now under development in this area.

Cytokine Gene Transfer

One type of gene transfer is designed specifically for treating neoplastic disease. In this approach, cytokine genes are inserted directly into tumor cells, which subsequently produce cytokines. Cytokines exhibit several properties that make them good candidates for treatment of neoplasia, including immunostimulatory activity, the ability to enhance tumor antigen presentation or antigenicity, and direct cytotoxic activity (Colombo and Forni 1994; Qin and Blankenstein 1996). However, the toxicity associated with systemic cytokine treatment (discussed later in this chapter) is

TABLE 3. *Methodologies for cytokine delivery*

Delivery system	Example	Reference
Topical administration	Delivery of TGF-β for wound healing	Puolakkainen et al. 1995
Ultrasound-mediated transdermal delivery	Delivery of erythropoietin and interferon-gamma demonstrated	Mitragotri et al. 1995
Biodegradable microparticles	Oculotopical administration of IL-5 and IL-6	Rafferty et al. 1996
Expression plasmid DNA	IL-5 expressed in myofibers and released systemically	Tokui et al. 1997
Cytokine-containing copolymers	Administration of soluble TNF-R	Eliaz et al. 1996
Microencapsulated cytokines	Systemic administration of CSF-1	Maysinger et al. 1996
Vitamin B12-mediated oral delivery	Delivery of orally active EPO and G-CSF	Russell-Jones et al. 1995
Subcutaneous infusion	Stimulation of megakaryocytopoiesis by IL-11	Leonard et al. 1996
Liposomes	Reduced toxicity and improved pharmacokinetics of systemic TNF	Kedar et al. 1997
Polyethylene glycol hybrid	Increased plasma half-life of TNF	Tsutsumi et al. 1995

a major drawback. On the other hand, a more localized administration might lessen this toxicity.

In practice, tumor cells from a patient would be removed, specific cytokine genes would be inserted, and these cells would then be reinfused. The theory behind this technique is that the tumor cells will secrete cytokines in vivo; when these cells are encountered by immune effector cells, the secreted cytokines will enhance the antitumor effect, and the tumor cells will be eliminated. Although some success has been demonstrated in experimental animal systems, some cytokines actually appear to enhance tumor growth rather than suppress it (Qin and Blankenstein 1996). Another potential problem is that gene transfer may induce growth autonomy in the tumor cells and actually exacerbate the problem (Gansbacher et al. 1993). Clearly, many issues remain to be addressed before implementing this technique in human therapy.

Cytokine Inhibition Therapy

Chemical (Nonbiological) Agents Inhibiting Cytokine Production and Action

A variety of nonbiological drugs have been found to suppress cytokine production either specifically or nonspecifically. These drugs act via myriad mechanisms, including alteration in cytokine gene transcription or mRNA translation, inhibition of cytokine release, or inhibition of cytokine processing (Henderson and Blake 1992). To date, the primary clinical use of these agents has been as adjuncts to organ and tissue transplantation, in which they have provided great advances in clinical success. Increasingly, these agents are now being evaluated for treatment of other immune-related diseases. Some of these agents are discussed in the following paragraphs.

Glucocorticoids are well known and powerful immunosuppressive drugs and enjoy wide clinical use for a variety of indications in which inappropriate immune reactions are factors in disease etiology. These drugs act at a variety of molecular targets; germane to the present discussion is their role in modulating cytokine production. Glucocorticoids affect cytokine production via a number of mechanisms including transcriptional repression, posttranscriptional alterations, induction of the suppressive cytokine TGF-β, antagonisms of transcription factors, and cooperation with transcription factors (Almawi et al. 1996). Glucocorticoids also alter the expression of cytokine receptors. Owing to their highly potent anti-inflammatory and immunosuppressive effects, these drugs are associated with numerous side effects and toxicity.

Cyclosporin (SandImmune) is probably the most thoroughly studied of the immunosuppressive drugs that specifically target cytokine production. Cyclosporin is a fungal metabolite with several unique features, including a novel amino acid in its composition. Cyclosporin acts primarily at the level of the T lymphocyte and effectively and reversibly inhibits the action of these cells. The drug exerts this highly specific effect by binding to the cytosolic protein cyclophilin, which is a member of a class of molecules generically known as immunophilins. The cyclosporin–cyclophilin complex targets calcineurin and essentially binds calcineurin to the immunophilin. The calcineurin is unable to interact with downstream transcription factors such as NF-AT, which ultimately prevents the transcription of cytokine genes, including IL-2, IL-3, IL-4, IL-5, and TNF (Morris 1994; Jirapongsananuruk and Leung 1997).

Tacrolimus (FK506) is another immunosuppressive fungal metabolite. Like cyclosporin, FK506 seems to exert its principal effects by altering the expression of cytokine genes—in particular the gene for IL-2. This agent shares little structural similarity with cyclosporin, although it has a similar mechanism of action (Morris 1994). Like cyclosporin, FK506 acts by binding to a cytosolic immunophilin; unlike cyclosporin, for FK506 this immunophilin is termed FK506 binding protein (FKBP). Inhibition of IL-2 gene transcription then occurs in a molecular manner similar to that following treatment with cyclosporin (Thomson et al. 1995).

Rapamycin (Rapamune, Sirolimus) shares structural similarities, including identical binding domains, with FK506; like FK506, it also binds to the cyclophilin FKBP. However, unlike its molecular cousin, rapamycin appears to have only limited (or no) effect on cytokine production but rather appears to affect cell cycle progression in late G1 phase by inhibiting growth factor signal transduction pathways (Sehgal 1995). Although rapamycin does not appear to affect cytokine production, there is some evidence that it may decrease the stability of cytokine RNA within the cell. Work on elucidating the precise mechanism of this drug is under way.

Leflunomide (HWA 486) is a synthetic immunomodulatory drug (an isoxazole derivative) that has been demonstrated to be effective in animal models of autoimmunity and transplantation rejection. Early reports on its mechanism of action were conflicting, although it was recognized at the outset that the drug operates via different mechanisms than the macrolides described above. In particular, early studies provided conflicting data regarding its ability to suppress cytokine production or the expression of cytokine receptors (Bartlett et al. 1994). More recently, Cao et al. (1996) have demonstrated that leflunomide does in fact work via modulation of cytokine function and thus enhances the production of the immunosuppressive cytokine TGF-β. It is clear that more work is required to identify the drug's exact mechanism of action.

Thalidomide has languished for many years owing to its teratogenic effects following its use in pregnant humans. In recent years, however, this drug has demonstrated potential as a useful inhibitor of cytokine production and thus may be enjoying a comeback. For several years thalidomide has been known to be a potent suppressor of TNF production (Zwingenberger and Wnendt 1995) and has recently been demonstrated to suppress production of the immunoregulatory cytokine IL-12 (Moller et al. 1997b). The mechanism of cytokine suppression by thalidomide is currently unknown, although it may act as an immunomodulator (but not necessarily an immunosuppressant) via selective gene regulation. Thalidomide is currently being evaluated for clinical treatment of several conditions such as autoimmune disease and graft-versus-host (GVH) disease (Moller et al. 1997b).

Pentoxifylline is a methylxanthine drug that has reproducibly been shown to inhibit the production of TNF and IL-12 (Moller et al. 1997a) and to affect the production of other cytokines such as IL-1, IL-6, IL-8, and IL-10 variably (D'Hellencourt et al. 1996). Pentoxifylline is showing some promise in maintenance of AIDS patients (Dezube et al. 1995), although several other clinical conditions may be candidates for treatment with this drug.

Pentamidine is an antiprotozoal drug used primarily to treat pneumonia caused by *Pneumocystis carinii* as well as certain inflammatory conditions. In addition to its antiprotozoal activity, pentamidine has been demonstrated to inhibit the production of IL-1 via a posttranslational event—possibly altering the cleavage of the precursor form of the cytokine (Rosenthal et al. 1991). More recently, pentamidine has been shown to inhibit the production of inflammatory chemokines (Van Wauwe et al. 1996).

Tenidap is an antirheumatic drug that combines cyclooxygenase inhibition with suppression of the acute phase response. Tenidap has been demonstrated to suppress the production of IL-1, IL-6, and IFN-γ (McNiff et al. 1995; Bondeson 1996). It appears specifically to suppress cytokine production by altering ionic homeostasis, although the exact mechanism of action remains unknown.

Other drugs have also been shown to inhibit cytokine production. The precise mechanism of action of many of these drugs is currently unknown, and they have not been developed extensively for clinical use as specific cytokine inhibitors. Some of these agents are listed in Table 4. As knowledge of the molecular biology of cellular activation grows, drugs will undoubtedly be formulated to modulate specific mechanisms (Alessandro et al. 1996).

TABLE 4. *Experimental drugs with anticytokine activity*

Drug	Reference
Phosphodiesterase isozyme inhibitors	Yoshimura et al. 1997
Metalloproteinase inhibitors	Gallea-Robache et al. 1997
p38 kinase inhibitors	Badger et al. 1996
SK&F86002	Triplett et al. 1996
MDL 201,449A	Edwards et al. 1996
CGP 47969A	Casini-Raggi et al. 1995

Biological Agents That Inhibit Cytokine Production or Action

Specific Inhibitory Cytokines

To date, only one naturally occurring cytokine antagonist has been identified, namely interleukin-1 receptor antagonist (IL-1RA), which exists in two forms: secreted and intracellular. This antagonist is a member of the IL-1 cytokine family (which includes IL-1α and IL-β) according to amino acid sequence, receptor binding avidity, and gene structure and location (Arend 1995). This agent seems to function as a pure receptor antagonist and does not exhibit any discernible agonist activity (e.g., internalization of receptor complexes, etc). Experimental evidence suggests that IL-1RA plays a role in various disease states such as rheumatoid arthritis (RA), sepsis, diabetes, and other diseases (Arend 1995). Understandably, the clinical application of a natural receptor antagonist represents an ideal clinical tool; clinical trials are under way for treatment of RA (Campion et al. 1996).

Mutant proteins have also been constructed that act as a specific receptor antagonist for both IL-4 and IL-13 activity (Aversa et al. 1993). This early success should lead to the creation of additional specific cytokine antagonists that are therapeutic.

Cytokine Receptors

Although most cytokines appear to exert their biological activity via interaction with specific cell-surface-bound receptors, many cytokine receptors are known to be present in a soluble form in the circulation of normal, healthy individuals. These soluble receptors may bind their cognate antigen with the same affinity as the surface receptors and have been postulated to have various physiological functions. One possible function is to serve as a chaperone molecule protecting the cytokines as they are ferried through the circulation. Another possible function, and one that serves as the basis for their possible use as therapeutics, is as a binding molecule for free cytokines in the circulation, which serves to control their bioavailability (Kroemer et al. 1993). Most of the work in this area has involved the IL-2 receptor (Waldmann et al. 1990; Verheul et al. 1992), although several other soluble receptors have been investigated as well (Arend 1995).

A potential problem (or benefit, depending on the desired outcome) inherent in the use of soluble cytokine receptors is their propensity for actually enhancing the activity of certain cytokines—possibly by extending their presence in the circulation (Arend 1995). Pharmacokinetic studies with recombinant cytokine receptors have demonstrated variability in the clearance between the different receptors as well as with different forms of the same receptor (Jacobs et al. 1993). Given the structural and functional differences between the different cytokine receptor subfamilies, this is perhaps not very surprising. However, this finding does point to the need to evaluate these molecules on a case-by-case basis in their development as human therapeutics.

Anticytokine Antibodies

Like soluble cytokine receptors, naturally occurring anticytokine antibodies have been demonstrated in the circulation of normal healthy individuals. Also, like the soluble receptors, these antibodies are thought to represent a mechanism for controlling the bioactivity of cytokines and to prevent them from inducing inappropriate responses. These molecules have been invaluable research reagents in understanding the role of cytokines in myriad biological processes (Kroemer et al. 1993) and are increasingly being investigated as therapeutics.

Several anticytokine antibodies (particularly anti-IL-4 and anti-IL-5 antibodies) have been evaluated with some success in the treatment of allergic disease (Jirapongsananuruk et al. 1997). A variety of other conditions have been treated with anticytokine antibodies with varying degrees of success (Arend 1995). As with the addition of exogenous cytokines, treatment of disease with single anticytokines antibodies may be confounded by the highly redundant nature of cytokines.

Fusion Toxins, Immunotoxins, and Chimera Toxins

In a slightly different use of anticytokine biotherapeutics, the cytokine-producing cell rather than the cytokine itself is targeted. As discussed elsewhere, the overproduction of cytokines, or at least an overly vigorous response by cytokine receptor-bearing cells (e.g., lymphocytes), can lead to a number of pathological conditions such as autoimmunity or neoplasia. The greatest degree of specificity could be achieved by developing a so-called "magic bullet," a molecule that is both highly specific and toxic. Potential candidates for this role are fusion toxins, which are also termed immunotoxins or chimera toxins.

Numerous highly potent toxins are found in nature, and three of these in particular have been used in the construction of immunotoxins: diphtheria toxin and *Pseudomonas* exotoxin (both bacterial products) and ricin (a plant product). These molecules share several structural features, including a binding domain allowing them to attach to cells, a component that allows the molecule to cross the target cell membrane, and an enzymatically active domain (the toxophore) that poisons the cell (Strom et al. 1991; Kreitman et al. 1992). Unfortunately, the binding domain allows

TABLE 5. *Various cytokine fusion toxins*

Toxin	Cytokine	Reference
Diphtheria toxin	G-CSF	Chadwick et al. 1993a
	IL-6	Chadwick et al. 1993b
	IL-15	van der Spek et al. 1995
	GM-CSF	Bendel et al. 1997
	IL-3	Liger et al. 1997
Ricin	IL-1	Frankel et al. 1996
	GM-CSF	Burbage et al. 1997
Pseudomonas exotoxin	IL-4	Ogata et al. 1989
	IL-6	Siegall et al. 1988

the toxins to attach to many different cell types. However, by removing this binding domain and replacing it with a portion of the cytokine or growth factor of interest the toxicity of the resulting molecule may be tailored to specific targets. The resulting hybrid molecule is now specific only for cells bearing a particular receptor; the molecular components mediating translocation and intoxication remain functional. Two of the earliest successes with this approach were the toxins DAB_{486}-IL-2 and DAB_{389}-IL-2, which contain fragmentary portions of the IL-2 molecule (Strom et al. 1991). Clinical trials and experimental studies with these immunotoxins have found them to be relatively safe and effective for a wide variety of conditions, including inhibition of HIV-1 RNA (Zhang et al. 1992), mycosis fungoides (Kuzel et al. 1993), rheumatoid arthritis (Schrohenloher et al. 1996), and IL-2 receptor-expressing malignancies (Nichols et al. 1997).

Several other fusion toxins have been constructed (Table 5). Most of these novel toxins are still experimental, and thus their actual utility as therapeutics remains to be established.

Antisense Technology

With a more complete understanding of molecular genetics has come the ability to alter fundamental biological processes experimentally and therapeutically. Nowhere is this manifested more than in the emerging field of antisense technology. Antisense technology is based on the construction of synthetic oligonucleotide sequences that will bind specifically with "sense" (i.e., message sequence) nucleic acids and in effect mask them and prevent sequence-specific actions such as translation of mRNA into protein (Stein and Cheng 1993). In theory, antisense therapy could represent a highly specific tool able to interrupt undesired biological processes at their most fundamental level. The advantages relative to cytokine therapy are obvious, for antisense technology allows a highly selective shutdown of specific cytokines or related growth factors to occur while theoretically leaving other factors unaffected. This would make antisense more desirable than even fusion toxins because it involve deactivation—rather than destruction—of a cell.

Although the concept of antisense technology has been demonstrated repeatedly in the laboratory, many important problems remain to be solved before (or if) this technique becomes a routine therapeutic option. These include knowledge of the proper target sequence, target accessibility, specificity, and drug stability and delivery (Maher 1996). One potential anticytokine use that appears to have bypassed this last problem is the administration of antisense oligonucleotides topically as an adjunct to wound healing (Choi et al. 1996). Another and more serious problem with antisense therapy has been unanticipated toxicity—particularly cardiovascular changes (Black et al. 1994). At present, antisense technology appears to be more valuable as a research tool for understanding cytokine biology than as a component of cytokine therapy (Lefebvre d'Hellencourt et al. 1995).

Extracorporeal Removal of Cytokines

In situations where in situ inactivation of cytokines may be impractical or impossible, a more brute force option may be the physical removal of cytokines. One approach that has been investigated is removal of cytokines during continuous hemofiltration. To date, the results of this approach have been equivocal; although removal of certain proinflammatory cytokines from the filtrate has been demonstrated, there has been little evidence of a reduction in systemic cytokine levels. Furthermore, improvements in patient outcome do not appear to be clinically relevant (Heering et al. 1997; Sander et al. 1997). Further clinical trials of this approach are under way.

A more selective approach for physical removal of cytokines from the circulation is by extracorporeal binding to antibodies. This technique was reported by Weber and Falkenhagen (1996) and was termed the Microspheres Based Detoxification System. In essence, polyclonal antibodies specific for IL-1, IL-6, and TNF were covalently linked to microspheres, and human plasma spiked with these cytokines was processed through the system. The authors reported efficient removal of these cytokines using this technique, for the rates of removal were greater than those reported following ultrafiltration. This technique has at least two advantages over removal by hemofiltration. First, the process allows for a highly selective removal of targeted cytokines rather than bulk removal of molecules. Second, antibodies to other molecules can also be attached concomitant with the anticytokine antibodies; this might facilitate the removal of other molecules associated with a particular disease state such as endotoxin during sepsis. The clinical effectiveness of this strategy has yet to be demonstrated.

MONITORING THE RESULTS OF CYTOKINE THERAPY: SOME CONSIDERATIONS

With the capability of altering the balance of cytokines in vivo, the accurate determination of the level of these molecules naturally becomes of great importance (Whiteside and Hank 1997). Cytokines are routinely measured by one method or a combination of three methods: bioassay, immunoassay, and molecular assay. Although each of these assay types has its own spectrum of advantages and disadvantages, an inclusive comparison is beyond the scope of this review. Nonetheless, by far the most commonly employed type is the enzyme-linked immunosorbent assay (ELISA), which is fairly sensitive (and with certain modifications, highly sensitive), highly specific, relatively inexpensive to perform, and technically straightforward. Also, ELISAs are easily adapted to automation, which is an obvious advantage in the clinical laboratory. Given their current and almost certain future acceptance, it is prudent to take a closer look at the potential problems and technical issues associated with the use of ELISAs for quantitating cytokine levels in clinical samples (Cannon et al. 1993; Whiteside 1994). Examples of some of the more critical considerations are as follows:

- Effect of sample processing: Several studies have demonstrated that variables in sample collection, processing, and storage can affect the results of cytokine assays (Thavasu et al. 1992).
- Existence of alternative molecular forms of the antigen: Certain cytokines can exist in variant forms (e.g., IL-6) (Ndubuisi et al. 1998) or as precursor molecules (e.g., IL-1β) (Dinarello 1992). The type of antibody used in the ELISA becomes an issue in these circumstances in that monoclonal antibodies (often used in commercial ELISAs) may not necessarily detect the presence of such alternative forms.
- Presence of interfering substances: A wide variety of substances may be present in clinical samples and can interfere with immunoassays. Some of these include soluble receptors (Corti et al. 1994), natural antagonists such as IL-1RA, chaperone molecules such as macroglobulin (James et al. 1994), and cytokine-specific autoantibodies (Mae et al. 1991).
- Assay precision: Although excellent reagents and kits are increasingly available, assay precision is still an important consideration in cytokine measurement (Heney and Whicher 1995). At present the most reliable method for increasing precision is strict attention to methodological details.
- Reference standards: A variable that is receiving increased attention is the source of the reference material used in the assay. Because experiments are ultimately extrapolated back to the reference preparation, this represents a weak link in the assay. However, until recently there has been limited effort to standardize the reference among commercial suppliers of cytokine ELISAs; thus, results obtained with any given kit may not match those obtained with other kits. It is to be hoped that this essential item of quality control will soon be addressed (Mire-Sluis et al. 1995).

Unlike many other biological entities (e.g., hormones, enzymes, etc.) in which normal human clinical ranges have been established, the expected or "normal" values for cytokines in various biological samples have not yet been published. A comprehensive review of the literature suggests that the vast majority of such evaluations have been made to compare cytokine levels present in humans during particular disease states with normal (i.e., appropriately matched control) individuals. A secondary source of such information is the product inserts for various commercial immunoassays kits that quote the results of internal studies performed by the supplying company. Unfortunately, an exhaustive compilation of all available data is still lacking, although such a future compilation will be of great value.

Even if some day technical assays are perfected and comprehensive baseline values are compiled, a final consideration must be made; namely, Which cytokines should be evaluated and under what circumstances? For example, monitoring the levels of an exogenously administered cytokine for pharmacokinetics would be relatively straightforward—particularly if a novel cytokine were administered (e.g., mutant cytokines with a unique protein structure), and might be useful for maintaining "therapeutic levels" (if in fact such levels have been established).

On the other hand, because baseline cytokine levels are low or undetectable, measuring these factors in the circulation would be essentially useless if one were evaluating potential immunosuppression (either as a disease sequela or clinically induced). In such cases, measurement of ex vivo cytokine production by stimulated cells would be required (Whiteside 1994).

A final consideration concerns the evaluation of cytokine-related biomarkers (Huber and Herold 1989). For example, neopterin, a biologically stable molecule induced by interferon-gamma (Fuchs et al. 1993), might be used to monitor therapy with this molecule. Conversely, certain cytokines such as α_2-macroglobulin are known to interact with cytokines (James et al. 1992). Although these molecules do not appear to affect the function or bioavailability of cytokines significantly, more information is needed to assess their contribution fully.

TOXICITY ASSOCIATED WITH CYTOKINE THERAPY

Preclinical toxicology is an integral and vital component of all pharmaceutical and biotechnological drug development. In toxicological evaluation, the potential of a drug candidate to affect various organ systems adversely at up to superphysiological doses, is estimated to establish a margin of safety before the initiation of clinical trials in humans (Ryffel 1996).

With the advent of genetically engineered proteins in recent decades, an era of more "natural" forms of therapy was anticipated. From the early (and, in retrospect, perhaps a bit naive) assumption that "if a little is good, then more is better" view of cytokines and other biological therapeutics, a realization has emerged that the road to rational biotherapy will be somewhat more difficult. At the risk of fatal oversimplification, the preclinical and clinical evaluation of cytokines comes down to two points: (1) traditional animal toxicology models may not reproducibly predict the ultimate human toxicity of recombinant molecules, particularly cytokines, and (2) the sequelae of in vivo cytokine administration (and perhaps cytokine suppression as well) can not necessarily be predicted on the basis of their in vitro actions. Let us examine these points in greater detail.

The first problem, namely the inability of routine animal models to predict the toxicity of cytokines, can best be understood by considering the basic biology of the cytokines themselves. Three essential features are of particular importance:

1. Certain cytokines (e.g., IL-3, IL-4, IL-12, GM-CSF, and IFN-γ) exhibit a relatively high degree of species specificity. Unlike most nonbiological therapeutics,

cytokines mediate their normal physiological (and apparently at least some of their deleterious) effects via their interaction with highly specific receptor molecules. This receptor interaction is also the basis for the strict species specificity of certain cytokines. Thus, administration of such molecules to an inappropriate species may be expected to result in difficulties in interpretation.

2. Although many recombinant proteins are highly homologous to their endogenous counterparts, minor biochemical discrepancies (e.g., base pair substitution, alternative forms of glycosylation, etc.) may result in altered biological function or, more likely, antigenic stimulation and lead to the induction of neutralizing antibodies. Although not as common when cytokines are evaluated in a homologous species (e.g., mouse–mouse), when human sequence cytokines are injected into laboratory animals, neutralization of the protein's bioactivity will result within a few weeks. Consequently, effects that may not be apparent in humans except after chronic administration will effectively be masked in animal models.

3. Whereas structure and function have been relatively well conserved for most mammalian cytokines, certain of these molecules display divergent function between species (Callard and Turner 1990). In such cases, the clinical sequelae of human cytokine therapy may not be predictable even from appropriately designed animal studies.

In spite of these potential difficulties, progress is being made in preclinical assessment of cytokine toxicity. For example, transgenic animals have been created that have a high constitutive level of cytokine production and thus mimic the effect of superphysiological doses of cytokines in humans (Lübbert et al. 1990). Another approach is the construction of transgenic animals expressing human cytokine receptors.

The second problem, namely that the clinical response to in vivo cytokine administration is not necessarily predictable by in vitro data, is likewise a complex issue. Let us once again examine this in the light of basic cytokine biology:

1. As stressed earlier, cytokines are pleiotropic and redundant in function. An example of this is provided by tumor necrosis factor (TNF), which exhibits potent cytotoxicity for isolated (i.e., in vitro) tumor cells and induces the regression of solid tumors following in situ injection. However, TNF's broad diversity of physiological actions (not the least of which is the mediation of systemic shock) precludes its use as a systemic therapeutic. Cytokine redundancy may account for certain toxicities by inappropriately enhancing (or suppressing) the effects of other endogenous cytokines.

2. Cytokines appear to function principally as carriers of bioinformation at a local level. Although cytokines are now known to mediate the functions of the immune, nervous, and endocrine systems (the neuroimmunoendocrine axis), most of these interactions probably take place at discrete sites. This impression is supported by the demonstration that the systemic (circulating) levels of almost all cytokines are extremely low and that exogenously administered cytokines are rapidly cleared from the circulation.

A common unintended consequence of cytokine treatment—although not a toxicity per se—is the development of anticytokine antibodies. This may seem surprising intuitively, for cytokines are endogenous molecules. However, anticytokine antibodies have been demonstrated in the sera of normal healthy individuals (Antonelli 1994) and are now thought to constitute part of a complex regulatory system for controlling the activities of endogenous cytokines. Thus, it should not be surprising that administration of superphysiological concentrations of these proteins induces the production of specific antibodies. Moreover, as mentioned elsewhere regarding animal studies, even recombinant proteins may not duplicate the natural product, which results in the creation of novel antigens and the subsequent development of antibodies.

Anticytokine antibodies may be either binding (i.e., directed against epitopes without functional consequences) or neutralizing (eliminating the biological activity of the molecule). Naturally, of the two, neutralizing antibodies represent the greatest concern because their presence could effectively negate the therapeutic benefit of treatment. Owing to the number of studies evaluating these molecules, most of the neutralizing antibody information has been gained from clinical trials of interferons (Antonelli 1994) and interleukin-2 (Allegretta et al. 1986). However, these are not isolated incidents, and minimizing the development of anticytokine antibodies will be an ongoing challenge to clinicians.

Aside from the issue of neutralizing antibody formation, administration of almost all exogenous cytokines tested to date has been associated with a range of toxicities of various severity; some representative examples of such toxicity are listed in Table 6. This table lists some of the more common toxicities associated with three different classes of therapeutic cytokines; the list of symptoms is representative of such adverse side effects and is by no means a comprehensive catalog.

One toxic side effect that appears to be common to a number of different cytokines (particularly IL-2 and GM-CSF) is the so-called vascular leak or capillary leak syndrome. This syndrome is characterized by such symptoms as peripheral and pulmonary edema, hypoxia, occasional ascites, perhaps hypotension, and sometimes respiratory failure (Bruton and Koeller 1994; Vial and Descotes 1995). Although the syndrome is suspected to be of immunological origin, the exact mechanism of toxicity remains to be elucidated (Cohen et al. 1992).

TABLE 6. *Representative toxicities associated with therapeutic cytokine administration*

Cytokine(s)	Reported toxicities	Reference
Interleukin-2	Hypotension, weight gain, capillary leak syndrome, renal dysfunction, gastrointestinal disturbances, neurological symptoms	Bruton and Koeller 1994
Hematopoietic growth factors	Bone pain, autoimmune disturbances, vascular leak syndrome, arterial thromboses, cutaneous reactions	Vial and Descotes 1995
Interferons	"Flu-like syndrome," neurological toxicity (fatigue, behavioral changes), hypo/hypertension, myalgia, autoimmunity (rare), tachycardia, hematotoxicity	Vial and Descotes 1994

CONCLUSIONS

The concept of harnessing the body's own natural healing powers has long been considered the ideal form of medicine. The traditional avenues available for accomplishing this have taken the form of herbs (whether crude concoctions or semisynthetic formulations) and spiritualism (ranging from magical incantations to prayer). However, in recent years, the discipline of molecular biology has provided researchers and clinicians with the highly precise tools to make this promise a reality at last. Genetic engineering allows us to produce products in essentially unlimited amounts, thus making it possible to replace missing biochemicals where a deficiency exists. Likewise, technologies such as antisense and hybridoma-produced monoclonal antibodies allow us to deplete "unwanted" biomolecules selectively.

Nevertheless, as pointed out in this chapter, a comprehensive understanding of the design of the vertebrate organism still eludes man. The myriad interconnected and interdependent biological processes work together in health; it may be the height of naivete to assume that the simple addition or deletion of a single molecule will always effect a cure, diseases such as diabetes notwithstanding. As is being reaffirmed in clinical studies, such is the case with cytokines.

Not that success is completely elusive. For selected applications such as bone marrow reconstruction, organ and tissue transplant, and several immunological disorders, cytokines are proving to be an exciting addition to the medical armamentarium. Moreover, certain innovations have resulted in "better than nature" molecules that can be tailored to specific pathologies, and the future promises even more improvements. It is only reasonable to temper our enthusiasm with the knowledge that much remains to be learned.

REFERENCES

Abraham, E., Glauser, M. P., Butler, T., Garbino, J., Gelmont, D., Laterre, P. F., Kudsk, K., Bruining, H. A., Otto, C., Tobin, E., Zwingelstein, C., Lesslauer, W., and Leighton, A. 1997. Tumor necrosis factor receptor fusion protein in the treatment of patients with severe sepsis and septic shock. A randomized controlled multicenter trial. *J.A.M.A.* 277:1531–1538.

Alessandro, R., Spoonster, J., Werstro, R. P., and Kohn, E. C. 1996. Signal transduction as a therapeutic target. *Curr. Top. Microbiol. Immunol.* 213:167–188.

Allegretta, M., Atkins, M. B., Dempsey, R. A., Bradley, E. C., Konrad, M. W., Childs, A., Wolfe, S. N., and Mier, J. W. 1986. The development of anti-interleukin-2 antibodies in patients treated with recombinant human interleukin-2 (IL-2). *J. Clin. Immunol.* 6:481–489.

Almawi, W., Beyhum, H., Rahme, A. A., and Rieder, M. J. 1996. Regulation of cytokine and cytokine receptor expression by glucocorticoids. *J. Leukocyte Biol.* 60:563–572.

Anderson, R., Macdonald, I., Corbett, T., Hacking, G., Lowdell, M. W., and Prentice, H. G. 1997. Construction and biological characterization of an interleukin-12 fusion protein (Flexi-12): Delivery to acute myeloid leukemic blasts using adeno-associated virus. *Hum. Gene Ther.* 8:1125–1135.

Antonelli, G. 1994. Development of neutralizing and binding antibodies to interferon (IFN) in patients undergoing IFN therapy. *Antiviral Res.* 24:235–244.

Arend, W. P. 1995. Inhibiting the effects of cytokines in human diseases. *Adv. Intern. Med.* 40:365–394.

Aversa, G., Punnonen, J., Cocks, B. J., de Waal Malefyt, R., Vega, F., Zurawski, S. M., Zurawski, G., and de Vries, J. E. 1993. An IL-4 mutant protein inhibits both IL-4 and IL-13-induced human IgG4 and IgE synthesis and B cell proliferation: Support for a common component shared by IL-4 and IL-13 receptors. *J. Exp. Med.* 178:2213–2218.

Badger, A. M., Bradbeer, J. N., Votta, B., Lee, J. C., Adams, J. L., and Griswold, D. E. 1996. Pharmacological profile of SB 203580, a selective inhibitor of cytokine suppressive binding protein/p38 kinase, in animal models of arthritis, bone resorption, endotoxin shock and immune function. *J. Pharmacol. Exp. Ther.* 279:1453–1461.

Bartlett, R. R., Campion, G., Musikic, P., and Schleyerbach, R. 1994. Leflunomide: A novel immunomodulating drug. In *Nonsteroidal anti-inflammatory drugs: Mechanisms and clinical uses*, eds. A. J. Lewis and D. E. Furst, 349–366. New York: Marcel Dekker.

Bendel, A. E., Shao, Y., Davies, S. M., Warman, B., Yang, C. H., Waddick, K. G., Uckun, F. M., and Parentesis, J. P. 1997. A recombinant fusion toxin targeted to the granulocyte-macrophage colony-stimulating factor receptor. *Leuk. Lymphoma* 25:257–270.

Black, L. E., Farrelly, J. G., Cavagnaro, J. A., Ahn, C.-H., DeGeorge, J. J., Taylor, A. S., DeFelice, A. F., and Jordan, A. 1994. Regulatory considerations for oligonucleotide drugs: Updated recommendations for pharmacology and toxicology studies. *Antisense Res. Devel.* 4:299–301.

Bondeson, J. 1996. Effects of tenidap on intracellular signal transduction and the induction of proinflammatory cytokines: A review. *Gen. Pharmacol.* 26:943–956.

Brouckaert, P., Ameloot, P., Cauwels, A., Everaerdt, B., Libert, C., Takahashi, N., Molle, W. V., and Fiers, W. 1994. Receptor-selective mutants of tumour necrosis factor in the therapy of cancer: Preclinical studies. *Circulatory Shock* 43:185–190.

Bruton, J. K., and Koeller, J. M. 1994. Recombinant interleukin-2. *Pharmacother.* 14:635–656.

Burbage, C., Tagge, E. P., Harris, B., Hall, P., Fu, T., Willingham, M. C., and Frankel, A. E. 1997. Ricin fusion toxin targeted to the human granulocyte-macrophage colony stimulating factor receptor is selectively toxic to acute myeloid leukemia cells. *Leuk. Res.* 21:681–690.

Callard, R. E., and Turner, M. W. 1990. Cytokines and Ig switching: Evolutionary divergence between mice and humans. *Immunol. Today* 11:200–203.

Campion, G. V., Lebsack, M. E., Lookabaugh, J., Gordon, G., and Catalano, M. 1996. Dose-range and dose-frequency study of recombinant human interleukin-1 receptor antagonist in patients with rheumatoid arthritis. The IL-1Ra Arthritis Study Group. *Arthritis Rheum.* 39:1092–1101.

Cannon, J. G., Neras, J. L., Poutsiaka, D. D., and Dinarello, C. A. 1993. Measuring circulating cytokines. *J. Appl. Physiol.* 75:1897–1902.

Cao, W. W., Kao, P. N., Aoki, Y., Xu, J. C., Shorthouse, R. A., and Morris, R. E. 1996. A novel mechanism of action of the immunomodulatory drug, leflunomide: Augmentation of the immunosuppressive cytokine, TGF-beta 1 and suppression of the immunostimulatory cytokine, IL-2. *Transplant. Proc.* 28:3079–3080.

Cardi, G., Ciardelli, T. L., and Ernstoff, M. S. 1996. Therapeutic applications of cytokines for immunostimulation and immunosuppression: An update. *Prog. Drug Res.* 47:211–250.

Casini-Raggi, V., Monsacchi, L., Vosbeck, K., Nast, C. C., Pizarro, T. T., and Cominelli, F. 1995. Anti-inflammatory effects of CGP 47969A, a novel inhibitor of proinflammatory cytokine synthesis, in rabbit immune colitis. *Gastroenterol.* 109:812–818.

Chadwick, D. E., Williams, D. P., Niho, Y., Murphy, J. R., and Minden, M. D. 1993a. Cytotoxicity of a recombinant diphtheria toxin-granulocyte colony-stimulating factor fusion protein on human leukemic blast cells. *Leuk. Lymphoma* 11:249–262.

Chadwick, D. E., Jean, L. F., Jamal, N., Messner, H. A., Murphy, J. R., and Minden, M. D. 1993b. Differential sensitivity of human myeloma cell lines and normal bone marrow colony forming cells to a recombinant diphtheria toxin-interleukin 6 fusion protein. *Brit. J. Haematol.* 85:25–36.

Chirigos, M. A., Talor, E., Sidwell, R. W., Burger, R. A., and Warrent, R. P. 1995. Leukocyte interleukin inj. (LI) augmentation of natural killer cells and cytolytic T-lymphocytes. *Immunopharmacol. Immunotoxicol.* 17:247–264.

Choi, B. M., Kwak, H. J., Jun, C. D., Park, S. D., Kim, K. Y., Kim, H. R., and Chung, H. T. 1996. Control of scarring in adult wounds using antisense transforming growth factor-beta 1 oligonucleotides. *Immunol. Cell Biol.* 74:144–150.

Cohen, R. B., Siegal, J. P., Puri, R. K., and Pluznik, D. H. 1992. The immunotoxicology of cytokines. In *Clinical immunotoxicology*, eds. D. S. Newcombe, N. R. Rose, and J. C. Bloom, 93–106. New York: Raven Press Ltd.

Colombo, M. P., and Forni, G. 1994. Cytokine gene transfer in tumor inhbition and tumor therapy: Where are we now? *Immunol. Today* 15:48–51.

Corti, A., Poiesi, C., Merli, S., and Cassani, G. 1994. Tumor necrosis factor (TNF) α quantification by ELISA and bioassay: Effects of TNFα-soluble TNF receptors (p55) complex dissociation during assay incubations. *J. Immunol. Meth.* 177:191–198.

Cosman, D. 1987. Control of messenger RNA stability. *Immunol. Today* 8:16–18.

Curtis, B. M., Williams, D. E., Broxmeyer, H. E., Dunn, J., Farrah, T., Jeffery, E., Clevenger, W., deRoos, P., Martin, U., Friend, D., Craig, V., Gayle, R., Price, V., Cosman, D., March, C. J., and Park, L. S. 1991. Enhanced hematopoietic activity of a human granulocyte/macrophage colony-stimulating factors-interleukin-3 fusion protein. *Proc. Natl. Acad. Sci. U.S.A.* 88:5809–5813.

Dendorfer, U. 1996. Molecular biology of cytokines. *Artif. Organs* 20:437–444.

Devos, R., Plaetinck, G., Cornelis, S., Guisez, Y., Van der Heyden, J., and Tavernier, J. 1995. Interleukin-5 and its receptor: A drug target for eosinophilia associated with chronic allergic disease. *J. Leukoc. Biol.* 57:813–819.

Dezube, B. J., Lederman, M. M., Spritzler, J. G., Chapman, B., Korvick, J. A., Flexner, C., Dando, S., Mattiacci, M. R., Alhers, C. M., Zhang, L., Novick, W. J., Kasdan, P., Fahey, J. L., Pardee, A. B., Crumpacker, C. S., and the National Institute of Allergy and Infectious Diseases AIDS Clinical Trials Group. 1995. High-dose pentoxyfilline in patients with AIDS: Inhibition of tumor necrosis factor production. National Institute of Allergy and Infectious Diseases AIDS Clinical Trials Group. *J. Infect. Dis.* 171:1628–1632.

D'Hellencourt, C. L., Diaw, L., Cornillet, P., and Guenounou, M. 1996. Differential regulation of TNF-alpha, IL-1 beta, IL-6, IL-8, TNF beta and IL-10 by pentoxyfilline. *Int. J. Immunopharmacol.* 18:739–748.

Dinarello, C. A. 1992. ELISA kits based on monoclonal antibodies do not measure total IL-1β synthesis. *J. Immunol. Meth.* 148:255–259.

Duschl, A., Muller, T., and Sebald, W. 1995. Antagonistic mutant proteins of interleukin-4. *Behring Inst. Mitt.* 96:87–94.

Edwards, C. K., Zhou, T., Zhang, J., Baker, T. J., De, M., Long, R. E., Borcherding, D. R., Bowlin, T. L., Bluethmann, H., and Mountz, J. D. 1996. Inhibition of superantigen-induced proinflammatory cytokine production and inflammatory arthritis in MRL-lpr/lpr mice by a transcriptional inhibitor of TNF-alpha. *J. Immunol.* 157:1758–1772.

Eliaz, R., Wallach, D., and Kost, J. 1996. Long-term protection against the effects of tumour necrosis factor by controlled delivery of the soluble p55 TNF receptor. *Cytokine* 8:482–487.

Farese, A. M., Herodin, F., McKearn, J. P., Baum, C., Burton, E., and MacVittie, T. J. 1996. Acceleration of hemotopoietic reconstitution with a synthetic cytokine (SC-55494) after radiation-induced bone marrow aplasia. *Blood* 87:581–591.

Frankel, A. E., Burbage, C., Fu, T., Tagge, E., Chandler, J., and Willingham, M. 1996. Characterization of a ricin fusion toxin targeted to the interleukin-2 receptor. *Protein Eng.* 9:913–919.

Fraser, J. D., Straus, D., and Weiss, A. 1993. Signal transduction events leading to T-cell lymphokine gene expression. *Immunol. Today* 14:357–362.

Fuchs, D., Weiss, G., and Wachter, H. 1993. Neopterin, biochemistry and clinical use as a marker for cellular immune reactions. *Int. Arch. Allergy Immunol.* 101:1–6.

Gallea-Robache, S., Morand, V., Millet, S., Bruneau, J. M., Bhatnagar, N., Chouaid, S., and Roman-Roman, S. 1997. A metalloproteinase inhibitor blocks the shedding of soluble cytokine receptors and processing of transmembrane cytokine precursors in human monocytic cells. *Cytokine* 9:340–346.

Gansbacher, B., Rosenthal, F. M., and Zier, K. 1993. Retroviral vector-mediated cytokine-gene transfer into tumor cells. *Cancer Invest.* 11:345–354.

Golding, B., Zaitseva, M., and Golding, H. 1994. The potential for recruiting immune responses toward Type 1 or Type 2 T cell help. *Am. J. Trop. Med. Hyg.* 50:33–40.

Heath, A. W., and Playfair, J. H. L. 1992. Cytokines as immunological adjuvants. *Vaccine* 10:427–434.

Heaton, K. M., and Grimm, E. A. 1993. Cytokine combinations in immunotherapy for solid tumors: A review. *Cancer Immunol. Immunother.* 37:213–219.

Heering, P., Morgera, S., Schmitz, F. J., Schmitz, G., Willers, R., Schultheiss, H. P., Strauer, B. E., and Grabensee, B. 1997. Cytokine removal and cardiovascular hemodynamics in septic patients with continuous venevenous hemofiltration. *Intensive Care Med.* 23:288–296.

Henderson, B., and Blake, S. 1992. Therapeutic potential of cytokine manipulation. *TiPS* 13:145–152.

Heney, D., and Whicher, J. T. 1995. Factors affecting the measurement of cytokines in biological fluids: Implications for their clinical measurement. *Ann. Clin. Biochem.* 32:358–368.

Hetzel, C., and Lamb, J. R. 1994. CD4$^+$ T Cell-targeted immunomodulation and the therapy of allergic disease. *Clin. Immunol. Immunopathol.* 73:1–10.

Huber, C., and Herold, M. 1989. The importance of patient monitoring in clinical cytokine trials: Use of serum markers to define biologically active doses. *Cancer Surv.* 8:809–815.

Jacobs, C. A., Beckmann, M. P., Mohler, K., Maliszewski, C. R., Fanslow, W. C., and Lynch, D. H. 1993. Pharmacokinetic parameters and biodistribution of soluble cytokine receptors. *Int. Rev. Exp. Pathol.* 34B:123–135.

James, K., Milne, I., Cunningham, A., and Elliott, S.-F. (1994). The effect of α_2 macroglobulin in commercial cytokine assays. *J. Immunol. Meth.* 168:33–37.

James, K., van den Haan, J., Lens, D. and Farmer, K. 1992. Preliminary studies on the interaction of TNFα and IFNγ with α_2-macroglobulin. *Immunol. Lett.* 32:49–58.

Jirapongsananuruk, O., and Leung, D. Y. M. 1997. Clinical applications of cytokines: New directions in the therapy of atopic diseases. *Ann. Allergy Asthma Immunol.* 79:5–16.

Kedar, E., Palgi, O., Golod, G., Babai, I., and Barenholz, Y. 1997. Delivery of cytokines by liposomes. III. Liposome-encapsulated GM-CSF and TNF-alpha show improved pharmacokinetics and biological activity and reduced toxicity in mice. *J. Immunother.* 20:180–193.

Kim, T. S., DeKruyff, R. H., Rupper, R., Maecker, H. T., Levy, S., and Umetsu, D. T. 1997. An ovalbumin-IL-12 fusion protein is more effective than ovalbumin plus free recombinant IL-12 in inducing a T helper cell type 1-dominated immune response and inhibiting antigen-specific IgE production. *J. Immunol.* .158:4137–4144.

Kreitman, R. J., FitzGerald, D., and Pastan, I. 1992. Targeting growth factor receptors with fusion toxins. *Int. J. Immunopharmcol.* 14:465–472.

Kreuser, E.-D., Wadler, S., and Thiel, E. 1992. Interactions between cytokines and cytotoxic drugs:putative molecular mechanisms in experimental hematology and oncology. *Sem. Oncol.* 19 (Suppl 4):1–7.

Kroemer, G., de Alborán, I. M., Gonzalo, J. A., and Martínez-A., C. 1993. Immunoregulation by cytokines. *Crit. Rev. Immunol.* 13:163–191.

Kuzel, T. M., Rosen, S. T., Gordon, L. I., Winter, J., Samuelson, E., Kaul, K., Roenigk, H. H., Nylen, P., and Woodworth, T. 1993. Phase I trial of the diphtheria toxin/interleukin-2 fusion protein DAB486IL-2: Efficacy in mycosis fungoides and other non-hodgkin's lymphomas. *Leuk. Lymphoma* 11:369–377.

Lefebvre d'Hellencourt, C., Diaw, L., and Guenounou, M. 1995. Immunomodulation by cytokine antisense oligonucleotides. *Eur. Cytokine Netw.* 6:7–19.

Leonard, J. P., Neben, T. Y., Kozitza, M. K., Quinto, C. M., and Goldman, S. J. 1996. Constant subcutaneous infusion of rhIL-11 in mice: efficient delivery enhances biological activity. *Exp. Hematol.* 24:270–276.

Liger, D., vanderSpek, J. C., Gaillard, C., Cansier, C., Murphy, J. R., Lebouch, P., and Gillet, D. 1997. Characterization and receptor specific toxicity of two diphtheria toxin-related interleukin-3 fusion proteins DAB389-mIL3 and DAB389-(Gly4Ser)2-mIL3. *FEBS Lett.* 406:157–161.

Lübbert, M., Jonas, D., and Herrmann, F. 1990. Animal models for the biological effects of continuous high cytokine levels. *Blut* 61:253–257.

Lucas, R., Echtenacher, B., Sablon, E., Juillard, P., Magez, S., Lou, J., Donati, Y., Bosman, F., Van de Voorde, A., Fransen, L., Männel, D. N., Grau, G. E., and de Baetselier, P. 1997. Generation of a mouse tumor necrosis factor mutant with antiperitonitis and desensitization activities comparable to those of the wild type but with reduced systemic toxicity. *Infect. Immun.* 65:2006–2010.

MacVittie, T. J., Farese, A. M., Herodin, F., Grab, L. B., Baum, C. M., and McKearn, J. P. 1996. Combination therapy for radiation-induced bone marrow aplasia in nonhuman primates using Synthokine SC-55494 and recombinant human granulocyte colony-stimulating factor. *Blood* 87:4129–4135.

Mae, N., Liberato, D. J., Chizzonite, R., and Satoh, H. 1991. Identification of high-affinity anti-IL-1α autoantibodies in normal human serum as an interfering substance in a sensitive enzyme-linked immunosorbent assay for IL-1α. *Lymphokine Cytokine Res.* 10:61–68.

Maher, L. J. 1996. Prospects for the therpeutic use of antigene oligonucleotides. *Cancer Invest.* 14:66–82.

Maysinger, D., Berezovskaya, O., and Fedoroff, S. 1996. The hematopoietic cytokine colony stimulating factor 1 is also a growth factor in the CNS: (II). Microencapsulated CSF-1 and LM-10 cells as delivery systems. *Exp. Neurol.* 14:47–56.

McAlister, I. B., Teepe, M., Gillis, S., and Williams, D. E. 1992. Ex vivo expansion of peripheral blood progenitor cells with recombinant cytokines. *Exp. Hematol.* 20:626–628.

McNiff, P. A., Laliberte, R. E., Svensson, L., Pazoles, C. J., and Gabel, C. A. 1995. Inhibition of cytokine activation processes in vitro by tenidap, a novel anti-inflammatory agent. *Cytokine* 7:196–208.

Mire-Sluis, A. R., Gaines-Das, R., and Thorpe, R. 1995. Immunoassays for detecting cytokines: What are they really measuring? *J. Immunol. Meth.* 186:157–160.

Mitchell, M. S. 1992. Principles of combining biomodulators with cytotoxic agents in vivo. *Sem. Oncol.* 19:51–56.

Mitragotri, S., Blankschtein, D., and Langer, R. 1995. Ultrasound-mediated transdermal delivery of proteins. *Science* 269:850–853.

Moller, D. R., Wysocka, M., Greenlee, B. M., Ma, X., Wahl, L., Flockhart, D. A., Trinchieri, G., and Karp, C. L. 1997a. Inhibition of IL-12 production by thalidomide. *J. Immunol.* 159:5157–5161.

Moller, D. R., Wysocka, M., Greenlee, B. M., Ma, X., Wahl, L., Flockhart, D. A., Trinchieri, G., and Karp, C. L. 1997b. Inhibition of human IL-12 production by pentoxyfilline. *Immunol.* 91:197–203.

Moore, M. A. S. 1991. The clinical use of colony stimulating factors. *Annu. Rev. Immunol.* 9:159–191.

Morris, R. 1994. Modes of action of FK506, cyclosporin A, and rapamycin. *Transplant. Proc.* 26:3272–3275.

Mosmann, T. R., Chervinski, H., Bond, M. W., Giedlin, M. A., and Coffman, R. L. 1986. Two types of murine helper T-cell clones. I. Definition according to profiles of lymphokine activities and secreted proteins. *J. Immunol.* 136:2348–2357.

Ndubuisi, M. I., Patel, K., Rayanade, R. J., Mittelman, A., May, L. T., and Sehgal, P. B. 1998. Distinct classes of chaperoned IL-6 in human blood: Differential immunological and biological activity. *J. Immunol.* 160:494–501.

Nemunaitis, J. 1993. Granulocyte-macrophage-colony stimulating factor: A review from preclinical development to clinical application. *Transfusion* 33:70–83.

Nichols, J., Foss, F., Kuzel, T. M., LeMaistre, C. F., Platanias, L., Ratain, M. J., Rook, A., Saleh, M., and Schwartz, G. 1997. Interleukin-2 fusion protein: An investigational therapy for interleukin-2 receptor expressing malignancies. *Eur. J. Cancer* 33:S34–S36.

Ogata, M., Chaudhary, V. K., Fitzgerald, D. J., and Pastan, I. 1989. Cytotoxic activity of a recombinant fusion protein between interleukin 4 and *Pseudomonas* exotoxin. *Proc. Natl. Acad. Sci. U.S.A.* 86:4215–4219.

Penichet, M. L., Harvill, E. T., and Morrison, S. L. 1997. Antibody-IL-2 fusion proteins: A novel strategy for immune protection. *Hum. Antibodies* 8:106–118.

Pistoia, V. 1997. Production of cytokines by human B cells in health and disease. *Immunol. Today* 18:343–350.

Powrie, F., and Coffman, R. L. 1993. Cytokine regulation of T-cell function: Potential for therapeutic intervention. *Immunol. Today* 14:270–274.

Puolakkainen, P. A., Twardzik, D. R., Ranchalis, J. E., Pankey, S. C., Reed, M. J., and Gombotz, W. R. 1995. The enhancement in wound healing by transforming growth factor-beta 1 (TGF-beta 1) depends on the topical delivery system. *J. Surg. Res.* 58:321–329.

Qin, Z., and Blankenstein, T. 1996. Influence of local cytokines on tumor metastasis: Using cytokine gene-transfected tumor cells as experimental models. *Curr. Top. Microbiol. Immunol.* 213:55–64.

Rafferty, D. E., Elfaki, M. G., and Montgomery, P. C. 1996. Preparation and characterization of a biodegradable microparticle antigen/cytokine delivery system. *Vaccine* 14:532–538.

Roilides, E., and Pizzo, P. A. 1993. Biologicals and hematopoietic cytokines in prevention or treatment of infections in immunocompromised hosts. *Hematol./Oncol. Clin. N. Am.* 7:841–864.

Romagnani, S. 1994. Lymphokine production by human T cells in disease states. *Annu. Rev. Immunol.* 12:227–257.

Romagnani, S. 1995. Biology of human T_H1 and T_H2 cells. *J. Clin. Immunol.* 15:121–129.

Rosenthal, G. J., Corsini, E., Craig, W. A., Comment, C. E., and Luster, M. I. 1991. Pentamidine: An inhibitor of interleukin-1 that acts via a post-translational event. *Toxicol. Appl. Pharmacol.* 107:555–561.

Russell-Jones, G. J., Westwood, S. W., and Habberfield, A. D. 1995. Vitamin B12 mediated oral delivery systems for granulocyte-colony stimulating factor and erythropoietin. *Bioconjug. Chem.* 6:459–465.

Ryffel, B. 1996. Unanticipated human toxicology of recombinant proteins. *Arch. Toxicol. Suppl.* 18:333–341.

Sander, A., Armbruster, W., Sander, B., Daul, A. E., Lange, R., and Peters, J. 1997. Hemofiltration increases IL-6 clearance in early systemic inflammatory response syndrome but does not alter IL-6 and TNF alpha plasma concentrations. *Intensive Care Med.* 23:878–884.

Schrohenloher, R. E., Koopman, W. J., Woodworth, T. G., and Moreland, L. W. 1996. Suppression of in vitro IgM rheumatoid factor production by diphtheria toxin interleukin 2 recombinant fusion protein (DAB 486IL-2) in patients with refractory rheumatoid arthritis. *J. Rheumatol.* 23:1845–1848.

Secombes, C. J. 1991. The phylogeny of cytokines. In *The cytokine handbook*, eds. A. W. Thomson, 387–411. London: Academic Press.

Sehgal, S. N. 1995. Rapamune (Sirolimus, Rapamycin): An overview and mechanism of action. *Ther. Drug Monitor.* 17:660–665.

Siegall, C. B., Chaudhary, V. K., Fitzgerald, D. J., and Pastan, I. 1988. Cytotoxic activity of an interleukin 6-*Pseudomonas* exotoxin fusion protein on myeloma cells. *Proc. Natl. Acad. Sci. U.S.A.* 85:9738–9742.

Soulillou, J. P. 1993. Cytokines and transplantation. *Tranplant. Proc.* 25:106–108.

Stein, C. A., and Cheng, Y.-C. 1993. Antisense oligonucleotides as therapeutic agnets—Is the bullet really magical? *Science* 261:1004–1011.

Stewart, F. M. 1995. Cytokines in stem cell transplantation. *Ann. N.Y. Acad. Sci.* 770:53–69.

Strom, T. B., Anderson, P. L., Rubin-Kelley, V. E., Williams, D. P., Kiyokawa, T., and Murphy, J. R. 1991. Immunotoxins and cytokine fusion proteins. *Ann. N.Y. Acad. Sci.* 636:233–250.

Taniguchi, T. 1988. Regulation of cytokine gene expression. *Ann. Rev. Immunol.* 6:439–464.

Thavasu, P. W., Longhurst, S., Joel, S. P., Slevin, M. L., and Balkwill, F. R. 1992. Measuring cytokine levels in blood: Importance of anticoagulants, processing and storage conditions. *J. Immunol. Meth.* 153:115–124.

Thomas, J. W., Baum, C. M., Hood, W. F., Klein, B., Monahan, J. B., Paik, K., Staten, N., Abrams, M., and McKearn, J. P. 1995. Potent interleukin 3 receptor antagonist with selectively enhanced hematopoietic activity relative to recombinant human interleukin 3. *Proc. Natl. Acad. Sci. U.S.A.* 92:3779–3783.

Thomson, A. W., Bonham, C. A., and Zeevi, A. 1995. Mode of action of Tacrolimus (FK506): Molecular and cellular mechanisms. *Ther. Drug Monitor.* 17:584–591.

Tokui, M., Takei, I., Tashiro, F., Shimada, A., Kasuga, A., Ishii, M., Ishii, T., Takatsu, K., Saruta, T., and Miyazaki, J. 1997. Intramuscular injection of expression plasmid DNA is an effective means of long-term systemic delivery of interleukin-5. *Biochem. Biophys. Res. Commun.* 233:527–531.

Triplett, E. A., Kruse-Elliott, K. T., Hart, A. P., Schram, B. R., MacWilliams, P. S., Cooley, A. J., Clayton, M. K., and Darien, B. J. 1996. SK&F 86002, a dual cytokine and eicosanoid inhibitor, attenuates endotoxin-induced cardiopulmonary dysfunction in the pig. *Shock* 6:357–364.

Tsutsumi, Y., Kihira, T., Tsunoda, S., Kanamori, T., Nakagawa, S., and Mayumi, T. 1995. Molecular design of hybrid tumour necrosis factor alpha with polyethylene glycol increases its anti-tumour potency. *Brit. J. Cancer* 71:963–968.

Van Wauwe, J., Aerts, F., Van Genechten, H., Blcokx, H., Deleersnijder, W., and Walter, H. 1996. The inhibitory effect of pentamidine on the production of chemotactic cytokines by in vitro stimulated human blood cells. *Inflamm. Res.* 45:357–363.

vanderSpek, J. C., Sutherland, J., Sampson, E., and Murphy, J. R. 1995. Genetic construction and characterization of the diphtheria toxin-related interleukin-15 fusion protein DAB389 sIL-15. *Protein Eng.* 8:1317–1321.

Verheul, H. A. M., Verveld, M., and Bos, E. S. 1992. Immunotherapy through the IL-2 receptor. *Immunol. Res.* 11:42–53.

Vial, T., and Descotes, J. 1994. Clinical toxicity of the interferons. *Drug Safety* 10:115–150.

Vial, T., and Descotes, J. 1995. Clinical toxicity of cytokines used as haemopoietic growth factors. *Drug Safety* 13:371–406.

Vose, J. M., Pandite, A. N., Beveridge, R. A., Geller, R. B., Schuster, M. W., Anderson, J. E., LeMaistre, C. F., Ahmed, T., Granena, A., Keating, A., Ranada, J. M. F., Stiff, P. J., Tabbara, I., Longo, W., Copelan, E. A., Nichols, C., Smith, A., Topolsky, D. L., Bierman, P. J., Lebsack, M. E., Lange, M., and Garrison, L. 1997. Granulocyte-macrophage colony-stimulating factor/interleukin-3 fusion protein versus granulocyte-macrophage colony-stimulating factor after autologous bone marrow transplantation for non-Hodgkins' lymphoma: Results of a randomized double-blind trial. *J. Clin. Oncol.* 15:1617–1623.

Wagstaff, J., Baars, J. W., Wolbink, G.-J., Hoekman, K., Eerenberg-Belmer, A. J. M., and Hack, C. E. 1995. Renal cell carcinoma and interleukin-2: A review. *Eur. J. Cancer* 31A:401–408.

Waldmann, T. A., Grant, A., Tendler, C., Greenberg, S., Goldman, C., Bamford, R., Junghans, R. P., and Nelson, D. 1990. Lymphokine receptor-directed therapy: A model of immune intervention. *J. Clin. Immunol.* 10:19S–29S.

Weber, C., and Falkenhagen, D. 1996. Extracorporeal removal of proinflammatory cytokines by specific absorption onto microspheres. *ASAIO J.* 42:M908–911.

Weigent, D. A., and Blalock, J. E. 1995. Associations between the neuroendocrine and immune systems. *J. Leukocyte Biol.* 57:137–150.

Whicher, J. T., and Evans, S. W. 1990. Cytokines in disease. *Clin. Chem.* 36/7:1269–1281.

Whiteside, T. L. 1994. Cytokine measurements and interpretation of cytokine assays in human disease. *J. Clin. Immunol.* 14:327–339.

Whiteside, T. L., and Hank, J. A. (1997). Monitoring of immunologic therapies. In *Manual of clinical laboratory immunology, 5th ed.,* eds. N. R. Rose, E. C. de Macario, J. D. Folds, H. C. Land, and R. M. Nakamura, 1065–1073. Washington, DC: ASM Press.

Williams, G., and Giroir, B. P. 1995. Regulation of cytokine gene expression: Tumor necrosis factor, interleukin-1, and the emerging biology of cytokine receptors. *New Horiz.* 3:276–287.

Yoshimura, T., Kurita, C., Nagao, T., Usami, E., Nakao, T., Watanabe, S., Kobayashi, J., Yamazaki, F., Tanaka, H., and Nagai, H. 1997. Effects of cAMP-phosphodiesterase isozyme inhibitor on cytokine production by lipopolysaccharide-stimulated human peripheral blood mononuclear cells. *Gen. Pharmcol.* 29:633–638.

Zhang, L., Waters, C., Nichols, J., and Crumpacker, C. 1992. Inhibition of HIV-1RNA production by the diphtheria toxin-related IL-2 fusion proteins DAB486IL-2 and DAB389IL-2. *J. Acquir. Immune Defic. Syndr.* 5:1181–1187.

Zhao, C., Tang, P., Mao, N., Zhang, S., Fan, E., Dong, B., Li, Q., and Du, D. 1996. Erythropoietin-like activity in vivo of the fusion protein rhIL-6/IL-2 (CH925). *Exp. Hematol.* 24:54–58.

Zwingenberger, K., and Wnendt, S. 1995. Immunomodulation by thalidomide: Systematic review of the literature and of unpublished observations. *J. Inflamm.* 46:177–211.

Biotechnology and Safety Assessment, 2nd ed.
Edited by John A. Thomas
Copyright © 1998 Taylor & Francis

5

Triplex and Ribozyme Technology in Gene Repression

Yan Lavrovsky, Shuo Chen, and Arun K. Roy

University of Texas Health Science Center, San Antonio, Texas

Specific base-pairing in nucleic acids is fundamental to the storage and retrieval of all genetic information. During the last two decades, concerted efforts have been made to harness this simple and universal process for the control of aberrant gene expression. These efforts have been directed at three different levels: (1) interference with the regulatory regions of specific genes by means of short, complementary single-stranded oligonucleotides that can form a site-specific triple helix and thereby interfere with the binding of regulatory proteins at the gene promoter; (2) inhibition of the messenger ribonucleic acid (mRNA) function by the formation of double-stranded nucleic acids with antisense oligonucleotides; and (3) sequence-specific degradation of the RNA molecules by catalytic RNAs known as ribozymes. Depending on the nature of the pathological condition, each of these approaches may present certain merits and disadvantages. Inhibition of mRNA function by antisense oligonucleotides is also discussed in chapters 2 and 7 of this book. The purposes of this chapter are to extend these discussions and to provide an overview of the current status of ribozyme and triplex deoxyribonucleic acid (DNA) technology. Furthermore, we will attempt to point out the therapeutic potential of the oligonucleotide-based drugs against infectious agents and degenerative disorders.

Nucleic acids are composed of polymeric chains of adenine and guanine (purines), thymine/uracyl or cytosine (pyrimidines), and monophosphate residues covalently linked through phosphodiester bonds. Short strands of nucleic acids (10–40 bases) are generally called oligonucleotides. Two complementary strands of nucleic acids can form a noncovalent duplex as a result of Watson–Crick base pairing, for adenine can form a hydrogen bond with thymine or uracyl, whereas cytosine can bond with guanine. In the case of DNA, where only the 2-deoxy derivative of the ribose is utilized, one strand serves to store the genetic code and is called the sense strand, whereas the other provides the complementary supporting strand and is known as the antisense strand. Ribonucleic acid is generally copied from the antisense strand and has the same sequence as the DNA sense strand except that it uses the unmodified ribose and also the base uracyl in the place of thymine. Additionally, a third strand of nucleic acid can bind to the DNA duplex through a non-Watson–Crick (Hoogsteen) type of hydrogen bonding, resulting in the triple helical or *triplex* DNA structure. The ability of the nucleic acids to form both the duplex and triplex structures is presently being utilized to develop therapeutic agents for inhibiting specific gene function.

INHIBITION OF SPECIFIC GENE EXPRESSION
BY TRIPLEX-FORMING OLIGONUCLEOTIDES

Several investigators have proposed the possibility of triplex DNA formation result-ing from intramolecular rearrangement of the DNA double helix within the natural genome (Agazie et al. 1996). Immunofluorescent staining of fixed metaphase chro-mosomes with the triplex-specific monoclonal antibodies, Jel 318 and Jel 466, pro-vided more direct evidence for the existence of such triplex structures in normal cells (Burkholder et al. 1991). Thomas et al. reported that hydralazine (an antihypertensive drug) elicits antinuclear antibodies owing to its ability to stabilize triplex structures. Serum samples derived from hydrazine-treated patients showed the presence of a high-affinity antibody directed to the triplex DNA. Furthermore, DNA binding ac-tivity of these antibodies was found to be significantly lower in the absence of any triplex-stabilizing agent. These data suggested that a possible mechanism for antinu-clear antibody production in hydralazine-treated patients might involve the induction and stabilization of immunogenic forms of DNA, including higher-order structures such as triplex DNA (Thomas et al. 1995a).

Unlike the intramolecular triplex due to the rearrangement of the duplex DNA, the intermolecular DNA triple helix formation involves the binding of a separate third strand that offers the opportunity to achieve precise sequence recognition and can, therefore, be harnessed for gene therapy (Rich 1993; Kim and Miller 1995). The formation of a specific complex between two strands of polyuridilic acid with one strand of polyadenylic acid in the presence of divalent cations was first reported in 1957 (Felsenfeld et al. 1957). More recent studies have revealed the site-specific formation of short intermolecular triplexes by a wide variety of detection techniques, including footprinting, affinity cleavage, and nuclear magnetic resonance (NMR) (Postel et al. 1991; Han and Dervan 1994; Mayfield et al. 1994; Radhakrishnan and Patel 1994a; Olivas and Maher 1994; Radhakrishnan et al. 1993). Chemical and physical evidence indicates that the nucleotide bases of the third strand occupy the major groove of the Watson–Crick double helix (Lee et al. 1979; Han and Dervan 1994). The definition of a triple helix in nucleic acid results from specific Hoogsteen- or reverse Hoogsteen-type hydrogen-bonding interactions between the bases in a homopurine strand of a Watson–Crick duplex and an additional third strand. Several DNA triplets have been characterized that fall into two distinct classes of the structural motifs. In the case of pyrimidine–purine–pyrimidine (Y*RY,* indicating Hoogsteen bond) triplexes (so-called "pyrimidine motif"), the third pyrimidine-rich strand runs parallel to the purine strand of the target duplex. It results in the formation of T*AT and C^{+}*GC combinations (Postel et al. 1991; Mayfield et al. 1994; Radhakrishnan and Patel 1994b). The requirement for protonation of cytosines in the third strand suggests that these triplexes are stable only at low pH values (<6.0). In the second class R*RY (purine–purine–pyrimidine triplex), the third purine-rich strand runs antiparallel to the duplex DNA and is generally purine rich (Hanvey et al. 1991; Cheng and Van Dyke 1994; Washbrook and Fox 1994; Hobbs and Yoon 1994). The best-characterized triplets within this "purine motif" are A*AT and G*GC (Radhakrishman et al. 1993; Hanvey et al. 1991; Cheng and Van Dyke 1994; Washbrook and Fox 1994; Hobbs and Yoon 1994). The latter types of triplexes are stable at physiologic pH values and at 37°C. Recent studies have also revealed the possible existence of several new triplet combinations: G*TA (Jayasena and Johnson 1993) and T*CG (Yoon et al. 1992; Durland et al. 1994) within Y*RY class of triplexes and a protonated triplet $A^{+*}GC$ (Malkov et al. 1993) within R*RY triplexes. These noncanonical triplets allow the extension of the existing triplex possibilities to all four naturally occurring base pairs in the target duplex.

Binding of these triplex-forming oligonucleotides is sequence specific. Once a triple-helix-forming oligonucleotide is attached to the target sequence, it can alter the affinity of DNA-binding proteins like restriction endonucleases, DNA methylating enzymes, transcription factors, and both DNA and RNA polymerases (Blume et al. 1992; Maher 1992; Spitzner et al. 1995; Duval-Valentin et al. 1992). Wrange et al. have shown that triple helix DNA alters nucleosomal protein–DNA interactions and may act as a nucleosome barrier (Weston et al. 1995). Helene and coauthors have found that triple-helix-forming oligonucleotides inhibited transcriptional elongation (Escude et al. 1996). Therefore, triplex-forming oligonucleotides are considered to offer significant potential for modulating gene expression with novel strategies for gene therapy.

Triple-helix-forming oligonucleotides are already being explored as promising antiviral and anticancer agents. Different sequences within HIV-1 and HIV-2 genomes were successfully targeted to modulate various steps of the infection process, including the initiation of viral transcription, transcription elongation, reverse transcription, and integration (Escude et al. 1996; Mouscadet et al. 1994; Volkmann et al. 1995; Svinarchuk et al. 1995; Bouziane et al. 1996; Ouali et al. 1996). In recent years, several laboratories have attempted to inhibit the expression of oncogenes with triplex-forming oligonucleotides. In an effort to develop specific transcriptional inhibitors of the human Ki-*ras* oncogene, Mayfield et al. designed oligonucleotides targeted to the 22 bp pyrimidine–purine motif in the human Ki-*ras* promoter (−126 to −147). These anthors have shown that oligonucleotide-directed triplex formation inhibits sequence-specific nuclear protein binding to the Ki-*ras* promoter. Similar results were obtained by Miller et al. (Reddoch and Miller 1995; Kim and Miller 1995) and Thomas et al. (Thomas et al. 1995b) when they used sequences from the upstream region of the oncogene c-*myc* as the target. Olivas and Maher were able to suppress transcription of the p53 antioncogene using sequence-specific triple-forming oligonucleotides (Olivas and Maher 1994). Furthermore, a very stable and specific triplex with the promoter of the c-*pim*-1 proto-oncogene was found to block c-*pim*-1 promoter activity in the cell culture system (Svinarchuk et al. 1996). These data suggest that triplex formation by the oligonucleotides may provide a convenient approach to inhibit transcription of certain proto-oncogenes specifically. Additionally, steroid receptors have been targeted by triplex-forming oligonucleotides. Ing et al. have successfully used a 38-base, single-stranded DNA that forms a triple helix on progesterone response elements that block the progesterone receptor interaction with its target gene. The 38-base, single-stranded DNA also inhibited progesterone receptor-dependent transcription in vitro. Furthermore, the cholesterol derivative of the same oligonucleotide specifically inhibited progesterone receptor-dependent transcription in vivo (Ing et al. 1993).

Triplex structures may also be involved in the physiological control of specific gene expression. On the basis of results of single-strand specific nuclease accessibility and DNase I footprinting, we have proposed a dynamic model for transcriptional control of the androgen receptor (AR) gene through intramolecular triple helix at the homopurine–homopyrimidine (pur-pyr) element of the promoter region. The model proposes that the pur-pyr region of the AR promoter exists in a conformational equilibrium between a B double-stranded DNA or "B form" and two "H forms" involving intramolecular triple helices (Fig. 1). In contrast to (C*GC) pyr-pur-pyr, the pur-pur-pyr (G*GC) structure can form a stable Hoogsteen hydrogen bond at the physiological pH and therefore is considered to be the preferred conformation. The two DNA-binding proteins (Sp1 and ssPyrBF) that specifically bind to this element selectively interact with either one of the B or H conformations. When the pur–pyr element is in the double-stranded B conformation, Sp1 can interact and accumulate at the pur–pyr site, resulting in transcriptional activation. However, when the element is in a pur–pur–pyr H form structure, interaction with the single-strand pyrimidine-specific binding factor (ssPyrBF) can stabilize the triple-helical conformation, thereby preventing the access of Sp1 and inhibiting the gene transcription. Varying ratios of ssPyrBF to Sp1 could potentially play a critical role in the spatio–temporal regulation of the AR gene expression (Chen et al. 1997).

By coupling the oligonucleotide to various DNA damaging agents, a covalent modification of the target duplex can be achieved. Such modification is found to occur in a site-specific manner. Several approaches, including DNA cleavage by EDTA-Fe(II) (Radhakrishnan and Patel 1994a), crosslinking of two strands by psoralen (Takasugi et al. 1991; Giovannangeli et al. 1992a; Gasparro et al. 1994) and chlorambucil (Sunters et al. 1992), cleavage by site-specific phenanthroline-Cu chelate (Giovannangeli et al. 1992b) and ellipticine (Perrouault et al. 1990), and site-specific alkylation by oligonucleotide derivatives bearing an alkylating 4-(3-amino)propyl (N-2-chloroethyl-N-methyl)aniline group at 5′ or 3′ terminal phosphate, or both (Brossalina et al. 1993), have been described. Glazer and coauthors have recently reported successful triplex-mediated targeting of psoralen photoadducts in mammalian cells (Wang et al. 1995) and within the murine genome (Gunther et al. 1996).

Reactive derivatives of homopyrimidine oligonucleotides bearing the 5′- or 3′-terminal DNA alkylation of the aromatic 2-chloroethylamino group were used to form covalent triple helices at the c-*fos* promoter region (Lavrovsky et al. 1996).

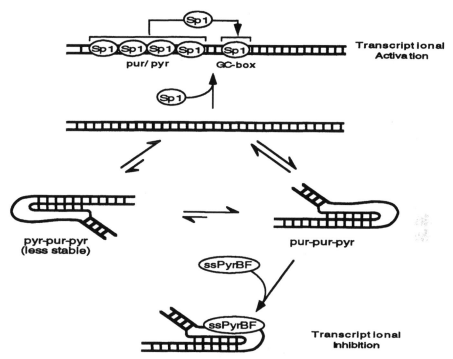

FIG. 1. A model for transcriptional control at the AR homopurine–homopyrimidine element by intramolecular triple helix.

The essence of the model is that the conformational structure of the pur–pyr element can alternate between a B-DNA double-stranded form and two H forms involving intramolecular triple helices. In contrast to (C*GC) pyr–pur–pyr, the pur–pur–pyr (G*GC) structure can form a stable Hoogsteen hydrogen bond (*) at the physiological pH and therefore is the preferred conformation. The two DNA-binding proteins (Sp1 and ssPyrBF) specific to this element can only bind to particular structures. When the pur–pyr element is in double-stranded conformation, Sp1 can interact and accumulate at the pur–pyr site and provide a readily available source for the GC box located immediately downstream. This situation will enhance transcription. When the element is in a pur–pur–pyr H form structure, the binding of the single strand pyrimidine-specific factor (ssPyrBF) can stabilize the triple-helical conformation, thereby preventing the binding of Sp1 and removing the nearby supply source of this transcription factor for the functional GC. Varying ratios of ssPyrBF to Sp1 could box potentially play a role in the differential regulation of the AR gene.

S. Chen et al. 1997. Functional role of a conformationally flexible homopurine–homopyrimidine domain of the androgen receptor gene promoter interacting with SP1 and a pyrimidine single strand DNA binding protein. *Mol. Endocrinol.* 11(1997):3–15. © The Endocrine Society.

Some of these sequences fall within well-characterized transacting elements. Thus, it is possible to interfere with the c-*fos* promoter function by triple helix formation with alkylating derivatives of corresponding oligonucleotides. It may be noted that c-*fos* proto-oncogene is associated with the control of cell differentiation and proliferation and the postsynaptic transduction of neuronal signals (Brown et al. 1996; Bialy and Kaczmarek 1996; Muller et al. 1993; Curran and Morgan 1995).

To obtain a better understanding of the potential of triple-helix-forming oligonucleotide alkylating derivatives as site-specific transcription modulators of transcription, we examined the cellular system designed for measuring in vivo effects of these oligonucleotide reagents on transcription. A reporter gene strategy coupled with the DNA affinity modification technique was used in an effort to facilitate oligonucleotide uptake, optimize triple helix formation, and provide convenient biological assays. Several key features of the system included facilitation of triple helix formation between the third-strand oligonucleotide bearing the reactive group on its terminal phosphate and a promoter region of the reporter *fos*-chloramphenicol acetyltransferase (CAT) plasmid DNA before transfection. Transfection of the complexes formed in vitro followed the measurement of resulting transcriptional effects and demonstrated that four different DNA targets within the c-*fos* promoter region could form triplex structures with synthetic oligonucleotides in a sequence-specific manner. Moreover, modifications of the c-*fos* promoter at position -83 in front of the cAMP-responsive element and FBS3/AP-2-like site at position -431 by triple-helix-forming oligonucleotides cause dramatic suppression of the CAT activity in endothelial cells (Lavrovsky et al. 1996).

Although the formation of intermolecular DNA triple helices offers the potential for designing therapeutic compounds with exquisite sequence-recognition properties, the binding of the third-strand oligonucleotide is certainly less stable than the sense and antisense duplex. One way of improving the interaction is to design compounds that bind to the triplex, but not to the duplex, DNA. There are three different grooves in triple-helical DNA that can be targeted by ligands that bind either covalently or noncovalently to the base pair edges. Because the third strand aligns with the major groove of the Watson–Crick duplex, the unoccupied minor groove has been the principal target for binding ligands that had previously been targeted to the same groove in duplex DNA. In recent years, several such compounds have been described, including 2,6-disubstituted amidoanthraquinones (Fox et al. 1995), the benzopyridoindole derivatives BePI and BgPI (Mergny et al. 1992; Pilch et al. 1993), coralyne (Lee et al. 1993), naphthoquinoline derivatives (Wilson et al. 1993; Chandler et al. 1995; Cassidy et al. 1994), and imidazothioxanthones (Fox et al. 1996).

In recent years, many laboratories have extended the triplex strategy to RNA targets with secondary structures. It was recently reported that RNA strands are excluded from triplexes containing the purine motif (Semerad and Maher 1994). On the other hand, studies with sequences corresponding to the various combinations of DNA and RNA strands show that two types of triplexes, namely DNA:RNA*DNA and RNA:RNA*DNA, are unstable with an RNA purine strand (Roberts and Crothers 1992; Escude et al. 1993). The formation of double-hairpin complexes with an RNA stem-loop target and an antisense oligodeoxynucleotide generates these unfavorable triplexes. Pascolo and Toulme (1996) have demonstrated that the double-hairpin complex strategy allowed formation of a triple-stranded structure with an RNA second strand. This approach offers the possibility of blocking RNA function by selectively targeting the RNA hairpins.

mRNA DEGRADATION BY SEQUENCE-SPECIFIC RIBOZYMES

Regulation of gene expression can occur at several levels, including transcription, mRNA stabilization, and translation. One oligonucleotide-based approach to inhibit expression of specific genes is to target the corresponding mRNA for enzymatic degradation. Once DNA is transcribed into mRNA, a series of mRNA processing events occur (i.e., splicing, polyadenylation, etc.) to produce mature mRNA. Translation of the mRNA can be blocked by either antisense oligonucleotides or by RNA enzymes (ribozymes). Such targeting of the messenger RNA is also a part of the physiological regulatory processes, as indicated by studies with silk moths. Circadian rhythms in the insects are regulated by the Period (Per) gene, which is essential for adult eclosion (Sasson-Corsi 1996; Sauman and Reppert 1996). The Per gene in the silk moth is expressed in an oscillatory manner and functions as a circadian clock. Although the exact structure of the Per gene in the silk moth is not known, an upstream mechanism regulates the circadian clock in the silk moth by an antisense transcript that can selectively hybridize with the sense messenger RNA. When the antisense RNA is paired with its complementary message RNA molecules, it interferes with the ability of the organism's translation machinery to produce the specific protein (Sassone-Corsi 1996; Sauman and Reppert 1996). Through this mechanism, the oscillating levels of the Per protein are regulated.

In many cases, the RNA splicing reactions are catalyzed by ribozymes (Symons 1992). The RNA enzymes include group I and group II introns, the RNA subunit of RNaseP, hairpin ribozyme, hepatitis delta virus ribozymes, ribosomal RNA, and hammerhead ribozymes (Altman 1989; Cech 1989; Symons 1992; Bratty et al. 1993; Gesteland and Alkins 1993). The group I intron was first found in *Tetrahymena thermophia* by Cech and coworkers (Zaug et al. 1986). The natural function of the group I ribozyme is to process ribosomal RNA. Splicing of the rRNA intron involves a two-step mechanism. In the first step, guanine nucleotide is added to the 5' end of the intron as the intron–exon junction is cleaved. In the second step, the freed 3' exon attacks at the 3' intron–exon junction to release the intron and produces a spliced exon (Zaug et al. 1986). Ribonuclease P was initially identified in bacteria by Altman and coworkers (Guerrier-Takada et al. 1983). The ribozyme consists of a small subunit protein and a catalytic RNA. The biological role of the ribozyme is to generate the mature tRNA by endonuclecatalytic cleavage of precursor tRNA (Guerrier-Takada et al. 1983). The hepatitis delta virus is a helper virus, and its RNA was observed by Wu and coworkers (Wu et al. 1989). The hepatitis delta ribozyme has autocatalytic RNA processing activity. The hammerhead ribozyme, discovered in viroids by Symons and coworkers (Forster and Symons 1987), is synthesized in a rolling circle and cleaves the polycisteronic message into individual RNA transcripts (Forster and Symons 1987). The hairpin ribozyme, described by Bruening and coworkers (Feldstein et al. 1989), is found in plant viroids and has a similar biological function as the hammerhead ribozyme (Haseloff and Gerlach 1989; Perreault et al. 1991).

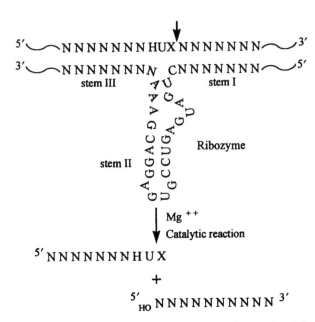

FIG. 2. Reaction scheme for *trans*-hammerhead ribozyme. The hammerhead ribozyme binds to its target RNA substrate through stems I and III of the hammerhead ribozyme, which hybridizes with the substrate RNA (N:N′). Stem II is part of the catalytic core. The arrow represents the cleavage site of the substrate (H can be any nucleotide; X is A, C, or U). Cleavage of a phosphodiester bond occurs at a unique site in this motif, generating 2′ 3′-cyclic phosphate and 5′-hydroxyl terminus.

The hammerhead ribozyme is one of the smallest RNA enzymes. Naturally, it acts "in *cis*" during viral replication by the rolling circle mechanism. The hammerhead ribozyme has been developed by Uhlenbeck (Uhlénbeck 1987) "in *trans*" against other RNA molecules. The *trans*-acting hammerhead ribozyme consists of antisense sections (stems I and III) and a catalytic domain with a flanking stem II as well as a loop section (Fig. 2). The mRNA substrate and the hammerhead ribozyme hybridize through stem I and stem III, orienting the -HUX- nucleotides (H can be any nucleotide; X can be an adenine, or a cytosine or uridine, but not a guanine). The GUC nucleotide triplet yields a higher activity of cleavage when compared with other cleavage triplet sequences such as GUA, CUC (Ruffner et al. 1990). Because of this target specificity, the RNA phosphodiester bond immediately following the nucleotide residue, which has its ribose sugar group held in a south conformation (that is C2–endo-C3′-Exo) is most preferably cleaved by the hammerhead ribozyme (Plavec et al. 1994). Furthermore, compared with other nucleotidyl 3′-ethylphosphates, cytidne 3′-ethylphosphate can most readily assume the south conformation at the ribose moiety, thus explaining the preference for C at the third base of the triplet preceding the cleavage site (Plavec et al. 1994). The guanine is the preferred first base in the triplet. This is deduced from the analysis of Kcat and Km of the cleavage reactions using substrate in which the first base is changed from G to another base (Shimayama et al. 1995). The base preference at the first position of the triplet, despite its distance from the cleavage site, indicates that the entire triplet contributes to the structure of the transition state intermediate formed during the phosphadiester bond cleavage reaction (Hendrix et al. 1996). However, it has been also found that the optimal triplet depends upon the binding arm composition of the stem I and stem III and the nucleotide lengths of the stems as well as secondary structure of the hammerhead ribozyme. Because -UX- dinucleotides occur frequently in RNA (on an average of about once in every six nucleotides) (Thompson et al. 1996), the hammerhead ribozyme theoretically can be designed to cleave any targeted mRNA. Cleavage specificity and efficiency of the hammerhead ribozyme are determined by the binding arms (stems I and III) that hybridize to the complementary sequences flanking the -UX-cleavage site within the targeted mRNA. If stems I and III, which are formed by the hammerhead ribozyme and its substrate, are short enough, the hammerhead ribozyme easily dissociates from the cleavage products, allowing the hammerhead ribozyme to bind to another substrate molecule. By repeating these cycles, one molecule of the hammerhead ribozyme can catalyze several substrate molecules in a multiple turnover reaction. The cleavage reaction catalyzed by the hammerhead ribozyme follows the Michaelis–Menten kinetics (Hampel and Tritz 1989; Fedor and Uhlenbeck 1990). Kinetic studies of numerous cleavage reactions have demonstrated that the binding of the substrate RNA to the hammerhead ribozyme is equivalent to the rate of helix formation of RNA duplexes (Long and Uhlenbeck 1993). It has been known that the length and sequence of stem I and stem III affect the dissociation between the hammerhead ribozyme and the mRNA substrate as well as the catalytic properties of the hammerhead ribozyme. Theoretically, the arm lengths of each section of stems I and III are optimized at 10–20 nucleotides (Herschlag 1991; Jarvis et al. 1996). However, optimization should also be considered with other factors such as the secondary structure of the RNA substrate and avoidance of a U–G wobble pair by creating A-rich sequences (Herschlag 1991; Bertrand et al. 1994). Computer analysis of the RNA secondary structures based on the energy minimization method can predict primary and secondary structures of the targeted RNA substrate for open loops that contain the potential cleavage site for hammerhead ribozyme (Zuker 1989). Additionally, a comprehensive and elegant procedure for the selection of RNA enzymes, referred as systematic evolution of ligand by exponential enrichment, has been developed and successfully used (Joyce 1989; Tuerk and Gold 1990). The principle is to synthesize a large pool of random-sequence RNA molecules using a bacteriophage RNA polymerase. This pool is subjected to a selection procedure in which the molecules with the desired chemical or physical properties are separated from the rest of the pool. Using reverse transcription and polymerase chain reaction (PCR), a DNA pool is then generated from which, in turn, a new RNA pool, enriched with the desired molecules, can be made. This method was successfully used in the selection of the hammerhead ribozyme capable of cleaving the growth hormone RNA substrate (Lieber and Strauss 1995).

The cleavage of the RNA substrate by the hammerhead ribozyme occurs via a transesterification reaction which generates 5' hydroxyl and 2'-3' cyclic phosphate termini. Divalent cations are required for catalysis and enhance the structural stability of fold RNA (Pyle 1993). The hammerhead ribozyme is one of the smallest RNA enzymes known and may have a potential in specific gene therapy. It has been shown that the inhibition of gene function by the hammerhead ribozyme is primarily due to the catalytic action rather than to an antisense effect. However, the antisense effect also contributes to suppressive function of the hammerhead ribozyme (Inokuchi et al. 1994). Initial studies by Sarver et al. showed that the hammerhead ribozyme inhibits the replication of HIV-1 in culture cells (Sarver et al. 1990). Since then, other investigators have reported that the hammerhead and other types of ribozymes inhibit different viruses in cell systems, including the HIV (Sioud and Drlica 1991; Weerasinghe et al. 1991; Zaia et al. 1992; Heidenreich and Eckstein 1992; Ojwang et al. 1992; Chen et al. 1992; Ho and Britton 1992; Homann et al. 1993; Joseph and Burke 1993; Sullenger and Cech 1993; Yu et al. 1993; Tang et al. 1994). Application of this approach to other fields such as inhibition of cancer cell growth (Scanlon et al. 1991; Lange et al. 1993; Ha and Kim 1994; Kobayashi et al. 1994; Ferbeyre et al. 1996; Jarvis et al. 1996) has also been investigated. The hammerhead ribozyme has been extensively studied, not only with respect to its mechanisms of its action but also for its possible application in vivo (Sarver et al. 1990; Lange et al. 1993; Ohkawa et al. 1995). For example, a hammerhead ribozyme designed to cleave mRNA encoding an activated *ras* oncogene in a tumor cell line resulted in a reversion of the malignant phenotype (Kashani-Sabet et al. 1992; Eastham and Ahlering 1996). In this case, the hammerhead ribozyme selectively cleaved the activated *ras* sequence but did not cleave the normal *ras* sequence (Kashani-Sabet et al. 1992; Eastham and Ahlering 1996).

A ribozyme-expressing transgenic mice model has also been generated. Efrat et al. found a 70% reduction in the level of glucokinase activity under the control of rat insulin II gene promoter compared with nonribozyme transgenic mice (Efrat et al. 1994). Similarly, Larsson and coworkers observed a 70, 22, and an 81% reduction in β-microglobulin mRNA in three different transgenic mouse lines expressing the hammerhead ribozyme under the control MMTV-LTR promoter (Larsson et al. 1994). Another group (L'Huillier et al. 1996) reported a 78, 58 and 50% reduction of bovine α-lactoglobulin mRNA in three different transgenic mouse lines expressing a hammerhead ribozyme under the control of the CMV promoter. All of these studies indicate the potential of using the ribozyme transgene in experimental animal models for human gene therapy.

For cell culture studies, two major approaches are generally used to introduce hammerhead ribozyme into the cell system. Each has its own advantages and limitations. The first strategy involves synthesis of the hammerhead ribozyme in vitro and incorporation of the in vitro synthesized ribozyme into the cellular environment. The second approach involves subcloning the hammerhead ribozyme gene into a mammalian expression vector and transfection of the expression vector to produce the hammerhead ribozyme in cellulo. A number of transfection methods are used to introduce hammerhead ribozymes into cells. These include the N-[1-(2,3-dioleyloxy)propyl]-N,N,N,-trimethylammonium chloride (DOTMA) cationic liposome-mediated transfection, electroporation, microinjection, or calcium phosphate coprecipitation. Malone et al. reported that a few micrograms of synthetic RNA are associated with ten million cells by the DOTMA technique (Malone et al. 1989). This corresponds to a delivery of approximately one million molecules per cell. Microinjection is a highly efficient method to deliver the hammerhead ribozyme into cells and can be useful to test the intracellular behavior of hammerhead ribozymes. The natural hammerhead ribozyme is prone to degradation by nucleases present in both the intra- and extracellular environments. A modified hammerhead ribozyme resistant to nucleases can be designed to increase its rate of accumulation in the cell, thereby decreasing the doses required for effective inhibition of the target gene. Recently, different modified ribozymes have been generated (Beigelman et al. 1995). A DNA–RNA chimeric ribozyme in which stems I and III of the hammerhead ribozyme are replaced by DNA increases its stability in vivo (Sawata et al. 1993; Hendry and McCall 1995). Santora and Joyce used an in vitro selection procedure to develop a DNA hammerhead ribozyme that can cleave almost all targeted RNA substrate under physiological condition. The DNA enzyme has a high efficiency on the targeted RNA molecules (Santora and Joyce 1997). The development of a synthetic ribozyme resistance to nuclease degradation is a key consideration for its successful use in pharmaceutical technology. However, for a prolonged inhibition, the use of ribozyme cloned into a mammalian expression vector appears a preferable approach. Two ribozyme-expression strategies have so far been tested with success (Altman 1989; Cotten 1990). First, the hammerhead ribozyme gene is inserted behind a strong promoter for RNA polymerase II, which may be of viral origin, a retroviral long-terminal repeat, or a strong endogenous promoter such as actin gene promoter. A polyadenylation signal is added at the 3' end of the gene to enable transcription termination and the addition of the poly(A) tail which, together with the m7G cap, can increase RNA stability and transport to the cytoplasm if required. An intron can also be inserted to ensure transcription through the normal splicing machinery in the nucleus, which may provide an enhancement of the hammerhead ribozyme effect. Promoters that are transcribed by RNA polymerase II are generally not suitable for production of short RNA molecules such as the hammerhead ribozyme (Sanfacon and Hohn 1990). In this system, at least several hundred nucleotides are needed between the promoter and the terminator for effective transcription. Although extra sequences can be added at both ends of the hammerhead ribozyme, the additional sequences can have undesirable effects on cleavage activity (Chowrira et al. 1994; Cameron and Jennings 1989). Also, a hundred- or thousandfold molar excess of hammerhead ribozyme expression over the target RNA is generally required for successful suppression of gene expression (Cameron and Jennings 1989). However, one of the main advantages of the RNA polymerase II promoter is the availability of a tissue-specific and regulatable promoter function. Such specificity is essential when hammerhead ribozyme expression is desired only in certain tissues or when expression needs to be turned on or off within a given tissue with regulatable switches (Zhao and Pick 1993).

The hammerhead ribozyme transcripts usually accumulate to relatively low levels and only within the cytoplasm. An alternative to RNA polymerase II transcription is to use an RNA polymerase III specific promoter. The RNA polymerase III transcribes a variety of small nuclear and cytoplasmic RNAs that are abundant in all cell types. The RNA polymerase III-specific promoters from genes encoding tRNA, U6 small nuclear RNAs, and "virus-associated" RNAs that are expressed at high levels in adenovirus-infected cells, have been used to drive the expression of hammerhead ribozymes. The hammerhead ribozyme under the control of the RNA polymerase III promoter is expressed at high levels in the cell culture system (Michienzi et al. 1996; Cotten and Birnstiel 1989). Cotten and Birnstiel observed that the ribozyme expression driven by the RNA polymerase III promoter was about tenfold higher compared with those driven by the RNA polymerase II promoter (Cotton and Birnstiel 1989). Using the RNA polymerase III promoter to produce the ribozyme, Yu et al. reported a high-efficiency inhibition of the HIV-1 gene expression in cell culture (Yu et al. 1993). Unlike RNA polymerase II transcript, RNA polymerase III transcript can be designed to localize either within the cytoplasm or the nuclei. However, these promoters cannot be regulated in a cell-specific manner, thus limiting their use in certain situations.

The potential for use of the hammerhead ribozyme against several types of diseases such as cancer, HIV, and other viral infection diseases has been tested in human culture cells without significant negative effects on the normal cell growth. As already mentioned, the hammerhead ribozyme can be delivered as a synthetic RNA or expressed from an expression vector. Thus, the hammerhead ribozyme can be used as a tool for gene therapy against a number of diseases.

FUTURE PROSPECTS

Over the last few years, tremendous progress toward using short DNA molecules as therapeutic agents has been made. Additionally, a substantial transition from laboratory applications to clinical applications is being observed. Clinical trials with oligonucleotides have already been initiated, and currently more than a dozen commercial enterprises selectively focus their attention on the development of short DNA molecules for therapeutic use.

Results of studies on chromatin structure, DNA methylation, and alternative DNA structures are expected to make particular DNA targets more accessible to exogenous oligonucleotides or their analogs. Covalent linking and cooperative interaction between multiple ligands will help to create more stable complexes, and pharmacological studies in animal models will answer questions regarding the pharmacodynamics, pharmacokinetics, toxicity, cellular uptake, and stability of these oligonucleotides.

Thus, oligonucleotide technology is emerging as a powerful tool with a growing number of applications. With improvements in intracellular stability, delivery methods, and oligonucleotide design, and with rapid advances in our knowledge of gene sequences, oligonucleotide technology is certain to become a major player in gene therapy.

ACKNOWLEDGMENTS

Studies in our laboratories are supported by NIH grants RO3 AG14215 (YL), T32 AG00165 (SC), R37 AG10486 (AKR) and R01 DK14744 (AKR). We thank Bandana Chatterjee for her comments and Nyra White for secretarial assistance.

REFERENCES

Agazie, Y. M., Burkholder, G. D., and Lee, J. S. 1996. Triplex DNA in the nucleus: Direct binding of triplex-specific antibodies and their effect on transcription, replication and cell growth. *Biochem. J.* 316:461–466.

Altman, S. 1989. Ribonuclease P: An enzyme with a catalytic RNA subunit. *Adv. Enzymol.* 62:1–36.

Beigelman, L., Karpeisky, A., Matulic-Adamic, J., Haeberli, P., Sweedler, D., and Usman, N. 1995. Synthesis of 2' modified nucleotides and their incorporation into hammerhead ribozymes. *Nucleic Acids Res.* 23(21):4434–4442.

Bertrand, E., Pictet, R., and Grange, T. 1994. Can hammerhead ribozymes be efficient tools to inactivate gene function? *Nucleic Acids Res.* 22(3):293–300.

Bialy, M., and Kaczmarek, L. 1996. c-Fos expression as a tool to search for the neurobiological base of the sexual behaviour of males. *Acta Neurobiologiae Experimentalis (Warszawa)* 56:567–577.

Blume, S. W., Gee, J. E., Shrestha, K., and Miller, D. B. 1992. Triple helix formation by purine-rich oligonucleotides targeted to the human dihydrofolate reductase promoter. *Nucleic Acids Res.* 20:1777–1784.

Bouziane, M., Cherny, D. I., Mouscadet, J. F., and Auclair, C. 1996. Alternate strand DNA triple helix-mediated inhibition of HIV-1 U5 long terminal repeat integration in vitro. *J. Biol. Chem.* 271:10359–10364.

Bratty, J., Chartrand, P., Ferbeyre, G., and Cedergren, R. 1993. The hammerhead RNA domain, a model ribozyme. *Biochim. Biophys. Acta* 1216(3):345–59.

Brossalina, E. B., Demchenko, E. N., and Vlassov, V. V. 1993. Specificity of interaction of pyrimidine oligonucleotides with DNA at acidic pH in the presence of magnesium ions: Affinity modification study. *Antisense Res. Dev.* 3:357–365.

Brown, J. R., Ye, H., Bronson, R. T., Dikkes, P., and Greenberg, M. E. 1996. A defect in nurturing in mice lacking the immediate early gene fosB. *Cell* 86:297–309.

Burkholder, G. D., Latimer, L. J., and Lee, J. S. 1991. Immunofluorescent localization of triplex DNA in polytene chromosomes of *Chironomus* and *Drosophila. Chromosoma* 101:11–18.

Cameron, F. H., and Jennings, P. A. 1989. Specific gene suppression by engineered ribozymes in monkey cells. *Proc. Natl. Acad. Sci. U.S.A.* 86:9139–9143.

Cassidy, S. A., Strekowski, L., Wilson, W. D., and Fox, K. R. 1994. Effect of a triplex-binding ligand on parallel and antiparallel DNA triple helices using short unmodified and acridine-linked oligonucleotides. *Biochem.* 33:15338–15347.

Cech, T. R. 1989. Self-splicing and enzymatic activity of an intervening sequence RNA from *Tetrohymena. A New Chem. Int.* 29:759–768.

Chandler, S. P., Strekowski, L., Wilson, W. D., and Fox, K. R. 1995. Footprinting studies on ligands which stabilize DNA triplexes: Effects on stringency within a parallel triple helix. *Biochem.* 34:7234–7242.

Chen, C. J., Banerjea, A. C., Haglund, K., Harmison, G. G., and Schubert, M. 1992. Inhibition of HIV-1 replication by novel multitarget ribozymes. *Ann. N.Y. Acad. Sci.* 660:271–273.

Chen, S., Supakar, P. C., Vellanoweth, R. L., Song, C. S., Chatterjee, B., and Roy, A. K. 1997. Functional role of a conformationally flexible homopurine/homopyrimidine domain of the androgen receptor gene promoter interacting with SP1 and a pyrimidine single strand DNA binding protein. *Mol. Endocrinol.* 11:3–15.

Cheng, A.-J., and Van Dyke, M. W. 1994. Oligodeoxyribonucleotide length and sequence effects on inter-molecular purine–purine–pyrimidine triple-helix formation. *Nucleic Acids Res.* 22:4742–4747.

Chowrira, B. M., Pavco, P. A., and McSwiggen, J. A. 1994. In vitro and in vivo comparison of hammerhead, hairpin, and hepatitis delta virus self-processing ribozyme cassettes. *J. Biol. Chem.* 269(41):25856–25864.

Cotten, M., and Birnstiel, M. 1989. Ribozyme mediated destruction of RNA in vivo. *EMBO J.* 8(12):3861–3866.

Cotten, M. 1990. The in vivo application of ribozymes. *Trends Biotechnol.* 8(7):174–178.

Curran, T., and Morgan, J. I. 1995. Fos: An immediate–early transcription factor in neurons. *J. Neurobiol.* 26:403–412.

Durland, R. H., Rao, T. S., Revankar, G. R., Tinsley, J. H., Myrick, M. A., Seth, D. M., Rayford, J., Singh, P., and Jayaraman, K. 1994. Binding of T and T analogs to CG base pairs in antiparallel triplexes. *Nucleic Acids Res.* 22:3233–3240.

Duval-Valentin, G., Thuong, N. T., and Helene, C. 1992. Specific inhibition of transcription by triple helix-forming oligonucleotides. *Proc. Natl. Acad. Sci. U.S.A.* 89:504–508.

Eastham, J. A., and Ahlering, T. E. 1996. Use of an anti-*ras* ribozyme to alter the malignant phenotype of a human bladder cancer cell line. *J. Urol.* 156(3):1186–1188.

Efrat, S., Leiser, M., Wu, Y. J., Fusco-DeMane, D., Emran, O. A., Surana, M., Jetton, T. L., Magnuson, M. A., Weir, G., and Fleischer, N. 1994. Ribozyme-mediated attenuation of pancreatic beta-cell glucokinase expression in transgenic mice results in impaired glucose-induced insulin secretion. *Proc. Natl. Acad. Sci. U.S.A.* 91(6):2051–2055.

Escude, C., Francois, J. C., Sun, J. S., Ott, G., Sprinzl, M., Garestier, T., and Helene, C. 1993. Stability of triple helices containing RNA and DNA strands: Experimental and molecular modeling studies. *Nucleic Acids Res.* 21:5547–5553.

Escude, C., Giovannangeli, C., Sun, J. S., Lloyd, D. H., Chen, J. K., Gryaznov, S. M., Garestier, T., and Helene, C. 1996. Stable triple helices formed by oligonucleotide N3'(P5' phosphoramidates inhibit transcription elongation. *Proc. Natl. Acad. Sci. U.S.A.* 93:4365–4369.

Fedor, M. J., and Uhlenbeck, O. C. 1990. Substrate sequence effect on "hammerhead" RNA catalytic efficiency. *Proc. Natl. Acad. Sci. U.S.A.* 87:1668–1672.

Feldstein, P. A., Buzayan, J. M., and Bruening, G. 1989. Two sequences participating in the autolytic processing of satellite tobacco ringspot virus complementary RNA. *Gene* 82(1):53–61.

Felsenfeld, G., Davies, D. R., and Rich, A. 1957. Formation of a three-stranded polynucleotide molecule. *J. Am. Chem. Soc.* 79:2023–2024.

Ferbeyre, G., Bratty, J., Chen, H., and Cedergren, R. 1996. Cell cycle arrest promotes *trans*-hammerhead ribozyme action in yeast. *J. Biol. Chem.* 271(32):19318–19323.

Forster, A. C., and Symons, R. H. 1987. Self-cleavage of plus and minus RNAs of a virusoid and a structural model for the active sites. *Cell* 49(2):211–220.

Fox, K. R., Polucci, P., Jenkins, T. C., and Neidle, S. 1995. A molecular anchor for stabilizing triple-helical DNA. *Proc. Natl. Acad. Sci. U.S.A.* 92:7887–7891.

Fox, K. R., Thurston, D. E., Jenkins, T. C., Varvaresou, A., Tsotinis, A., and Siatra-Papastaikoudi, T. 1996. A novel series of DNA triple helix-binding ligands. *Biochem. Biophys. Res. Commun.* 224:717–720.

Gasparro, F. P., Havre, P. A., Olack, G. A., Gunther, E. J., and Glazer, P. M. 1994. Site-specific targeting of psoralen photoadducts with a triple helix-forming oligonucleotide: Characterization of psoralen monoadduct and crosslink formation. *Nucleic Acids Res.* 22:2845–2852.

Gesteland, R. F., and Alkins, J. F. 1993. *The RNA world,* Monograph 24. Plainview, New York: Cold Spring Harbor Laboratory Press.

Giovannangeli, C., Rougee, M., Garestier, T., Thuong, N. G., and Helene, C. 1992a. Triple-helix formation by oligonucleotides containing the three bases thymine, cytosine, and guanine. *Proc. Natl. Acad. Sci. U.S.A.* 89:8631–8635.

Giovannangeli, C., Thuong, N. T., and Helene, C. 1992b. Oligodeoxynucleotide-directed photo-induced cross-linking of HIV proviral DNA via triple-helix formation. *Nucleic Acids Res.* 20:4275–4281.

Guerrier-Takada, C., Gardiner, K., Marsh, T., Pace, N., and Altman, S. 1983. The RNA moiety of ribonuclease P is the catalytic subunit of the enzyme. *Cell* 35:849–57.

Gunther, E. J., Havre, P. A., Gasparro, F. P., and Glazer, P. M. 1996. Triplex-mediated, in vitro targeting of psoralen photoadducts within the genome of a transgenic mouse. *Photochem. Photobiol.* 63:207–212.

Ha, J., and Kim, K. H. 1994. Inhibition of fatty acid synthesis by expression of an acetyl-CoA carboxylase-specific ribozyme gene. *Proc. Natl. Acad. Sci. U.S.A.* 91(21):9951–9955.

Hampel, A., and Tritz, R. 1989. RNA catalytic properties of the minimum (−) s TRSV sequence. *Biochem.* 28(12):4929–4933.

Han, H., and Dervan, P. B. 1993. Sequence-specific recognition of double helical RNA and RNA.DNA by triple helix formation. *Proc. Natl. Acad. Sci. U.S.A.* 90:3806–3810.

Han, H., and Dervan, P. G. 1994. Different conformational families of pyrimidine.purine.pyrimidine triple helices depending on backbone composition. *Nucleic Acids Res.* 22:2837–2844.

Hanvey, J. C., Williams, E. M., and Besterman, J. M. 1991. DNA triple-helix formation at physiologic pH and temperature. *Antisense Res. Dev.* 1:307–317.

Haseloff, J., and Gerlach, W. L. 1989. Sequences required for self-catalysed cleavage of the satellite RNA of tobacco ringspot virus. *Gene* 82(1):43–52.

Heidenreich, V., and Eckstein, F. 1992. Hammerlead ribozyme-mediated cleavage of the long terminal repeat RNA of human immunodeficiency virus type I. *J. Biol. Chem.* 267:1904–1909.

Hendrix, C., Anne, J., Joris, B., Van-Aerschot, A., and Herdewijn, P. 1996. Selection of hammerhead ribozyme for optimun cleavage of interleukin 6 mRNA. *Biochem. J.* 314:655–661.

Hendry, P., and McCall, M. J. 1995. A comparison of the in vitro activity of DNA-armed and all-RNA hammerhead ribozymes. *Nucleic Acids Res.* 23:3928–36.

Herschlag, D. 1991. Implications of ribozyme kinetics for targeting the cleavage of specific RNA molecules in vivo: More isn't always better. *Proc. Natl. Acad. Sci. U.S.A.* 88(16):6921–6925.

Ho, S. P., and Britton, D. H. 1992. Catalytic RNA molecules and their cleavage of viral mRNA. *Ann. N.Y. Acad. Sci.* 660:265–267.

Hobbs, C. A., and Yoon, K. 1994. Differential regulation of gene expression in vivo by triple helix-forming oligonucleotides as detected by a reporter enzyme. *Antisense Res. Dev.* 4:1–8.

Homann, M., Tzortzakaki, S., Rittner, K., Sczakiel, G., and Tabler, M. 1993. Incorporation of the catalytic domain of a hammerhead ribozyme into antisense RNA enhances its inhibitory effect on the replication of human immunodeficiency virus type 1. *Nucleic Acids Res.* 21(12):2809–2814.

Ing, N. H., Beekman, J. M., Kessler, D. J., Murphy, M., Jayaraman, K., Zendegui, J. G., Hogan, M. E., O'Malley, B. W., and Tsai, M. J. 1993. In vivo transcription of a progesterone-responsive gene is specifically inhibited by a triplex-forming oligonucleotide. *Nucleic Acids Res.* 21:2789–2796.

Inokuchi, Y., Yuyama, N., Hirashima, A., Nishikawa, S. A. U., Ohkawa, J., and Taira, K. 1994. A hammerhead ribozyme inhibits the proliferation of an RNA coliphage SP in *Escherichia coli. J. Biol. Chem.* 269(15):11361–11366.

Jarvis, T. C., Wincott, F. E., Alby, L. T., McSwiggen, G., Beigelman, L., Gustofson, G., Direnzo, A., Levy, K., Arthur, M., Matulik-Adamic, G., Karpiesky, A., Gonzalez, C., Wolf, T., Usman, N., and Stinchcomb, D. T. 1996. Optimizing the cell efficiency of synthetic ribozyme. *J. Biol. Chem.* 268:24515–24518.

Jayasena, S. D., and Johnson, B. H. 1993. Sequence limitations of triple helix formation by alternate-strand recognition. *Biochem.* 32:2800–2807.

Joseph, S., and Burke, J. M. 1993. Optimization of an anti-HIV hairpin ribozyme by in vitro selection. *J. Biol. Chem.* 268(33):24515–24518.

Joyce, G. F. 1989. Amplication, mutation and selection of catalytic RNA. *Gene* 82:83–87.

Kashani-Sabet, M., Funato-Tone, T., Jiao, L., Wang, W., Yoshida, E., Kashfinn, B. I., Shitara,T., Wu, A. M., Moreno, J. G. et al. 1992. Reversal of the malignant phenotype by an anti-*ras* ribozyme. *Antisense Res. Dev.* 2(1):3–15.

Kim, H.-G., and Miller, D. M. 1995. Inhibition of in vitro transcription by a triplex-forming oligonucleotide targeted to human c-*myc* P2 promoter. *Biochem.* 34:8165–8171.

Kobayashi, H., Dorai, T., Holland, J. F., and Ohnuman, T. 1994. Reversal of drug sensitivity of multidrug-resistant tumor cells by an MDR1(RGY1) ribozyme. *Cancer Res.* 54:1271–1275.

Lange, W., Cantin, E. M., Finke, J., and Dolken, G. 1993. In vitro and in vivo effects of synthetic ribozymes targeted against BCR/ABL mRNA. *Leukemia* 7(11):1786–1794.

Larsson, S., Hotchkiss, G., Andang, M., Nyholm, T., Inzunza, J., Jansson, I., and Ahrlund-Richter, L. 1994. Reduced beta 2-microglobulin mRNA levels in transgenic mice expressing a designed hammerhead ribozyme. *Nucleic Acids Res.* 22(12):2242–2248.

Lavrovsky, Y., Stoltz, R. A., Vlassov, V. V., and Abraham, N. G. 1996. C-*fos* protooncogene transcription can be modulated by oligonucleotide-mediated formation of triplex structures in vitro. *Eur. J. Biochem.* 238:582–590.

Lee, J. S., Johnson, D. A., and Morgan, A. R. 1979. Complexes formed by (pyrimidine)n:(purine)n DNAs on lowering the pH are three-stranded. *Nucleic Acids Res.* 6:3073–3091.

Lee, J. S., Latimer, L. J., and Hampel, K. J. 1993. Coralyne binds tightly to both T.A.T.- and C.G.C.(+)-containing DNA triplexes. *Biochem.* 32:5591–5597.

L'Huillier, P. J., Soulier, S., Stinnakre, M. G., Lepourry, L., Davis, S. R., Mercier, J. C., and Vilotte, J. L. 1996. Efficient and specific ribozyme-mediated reduction of bovine alpha-lactoalbumin expression in double transgenic mice. *Proc. Natl. Acad. Sci. U.S.A.* 93:6698–6703.

Lieber, A., and Strauss, M. 1995. Selection of efficient cleavage sites in target RNAs by using a ribozyme expression library. *Mol. Cell. Biol.* 15:540–551.

Long, D. M., and Uhlenbeck, O. C. 1993. Self-cleaving catalytic RNA. *FASEB J.* 7(1):25–30.

Maher, L. J. 1992. DNA triple-helix formation: An approach to artificial gene repressors? *BioEssays* 14:807–815.

Malkov, V. A., Voloshin, O. N., Veselkov, A. G., Rostapshov, V. M., Jansen, I., Soyfer, V. N., and Frank-Kamenetskii, M. D. 1993. Protonated pyrimidine-purine- purine triplex. *Nucleic Acids Res.* 21:105–111.

Malone, R. W., Felgner, D. L., and Verma, I. M. 1989. Cationic liposome-mediated RNA transfection. *Proc. Natl. Acad. Sci. U.S.A.* 86:6077–6081.

Mayfield, C., Squibb, M., and Miller, D. 1994. Inhibition of nuclear protein binding to the human Ki-*ras* promoter by triplex-forming oligonucleotides. *Biochem.* 33:3358–3363.

Mergny, J. L., Duval-Valentin, G., Nguyen, C. H., Perrouault, L., Faucon, B., Rougee, M., Montenay-Garestier, T., Bisagni, E., and Helene, C. 1992. Triple helix-specific ligands. *Science* 256:1681–1684.

Michienzi, A., Prislei, S., and Bozzoni, I. 1996. U1 small nuclear RNA chimeric ribozymes with substrate specificity for the Rev pre-mRNA of human immunodeficiency virus. *Proc. Natl. Acad. Sci. U.S.A.* 93(14):7219–7224.

Mouscadet, J. F., Carteau, S., Goulaouic, H., Subra, F., and Auclair, C. 1994. Triplex-mediated inhibition of HIV DNA integration in vitro. *J. Biol. Chem.* 269:21635–21638.

Muller, R., Mumberg, D., and Lucibello, F. C. 1993. Signals and genes in the control of cell-cycle progression. *Biochim. Biophys. Acta* 1155(2):151–179.

Ohkawa, J., Koguma, T., Kohda, T., and Taira, K. 1995. Ribozymes: From mechanistic studies to applications in vivo. *J. Biochem.* 118(2):251–8.

Ojwang, J. O., Hampel, A., Looney, D. J., Wong-Staal, F., and Rappaport, J. Inhibition of human immunodeficiency virus type 1 expression by a hairpin ribozyme. *Proc. Natl. Acad. Sci. U.S.A.* 89(22):10802–10806.

Olivas, E. M., and Maher, L. J. III. 1994. DNA recognition by alternate strand triple helix formation: affinities of oligonucleotides for a site in the human p53 gene. *Biochem.* 33:983–991.

Ouali, M., Bouziane, M., Ketterle, C., Gabarro-Arpa, J., Auclair, C., and LeBret, M. 1996. A molecular mechanics and dynamics study of alternate triple-helices involving the integrase-binding site of the HIV-1 virus and oligonucleotides having a 3'-3' internucleotide junction. *J. Biomol. Struct. Dynam.* 13:835–853.

Pascolo, E., and Toulme, J. J. 1996. Double hairpin complexes allow accommodation of all four base pairs in triple helices containing both DNA and RNA strands. *J. Biol. Chem.* 271:24187–24192.

Perreault, J. P., Labuda, D., Usman, N., Yang, J. H., and Cedergren, R. 1991. Relationship between 2'-hydroxyls and magnesium binding in the hammerhead RNA domain: a model for ribozyme catalysis. *Biochem.* 30(16):4020–4025.

Perrouault, L., Asseline, U., Rivalle, C., Thuong, N. T., Bisagni, E., Giovannangeli, C., LeDoan, T., and Helene, C. 1990. Sequence-specific artificial photo-induced endonucleases based on triple helix-forming oligonucleotides. *Nature* (London) 344:358–360.

Pilch, D. S., Waring, M. J., Sun, J. S., Rougee, M., Nguyen, C. H., Bisagni, E., Garestier, T., and Helene, C. 1993. Characterization of a triple helix-specific ligand. BePI (3-methoxy-7H-8-methyl-ll-[(3'-amino)propylamino]-benzo[e]pyrido[4,3-b]indole) intercalates into both double-helical and triple-helical DNA. *J. Mol. Biol.* 232:926–946.

Plavec, J., Thibaudeau, C., and Chattopadhyaya, J. 1994. How does the 2'-hydroxy group drive the pseudorotational equilibrium in nucleoside and nucleotide by the tuning of 3' gauche effect. *J. Am. Chem. Soc.* 116:6558–6560.

Postel, E. H., Flint, S. J., Kessler, D. J., and Hogan, M. E. 1991. Evidence that a triplex-forming oligodeoxyribonucleotide binds to the c-myc promoter in HeLa cells, thereby reducing c-*myc* mRNA levels. *Proc. Natl. Acad. Sci. U.S.A.* 88:8227–8231.

Pyle, A. M. 1993. Ribozymes: A distinct class of metalloenzymes. *Science* 261:709–714.

Radhakrishnan, I., de los Santos, C., and Patel, D. J. 1993. Nuclear magnetic resonance structural studies of A:AT base triple alignments in intramolecular purine.purine.pyrimidine DNA triplexes in solution. *J. Mol. Biol.* 234:188–197.

Radhakrishnan, I., and Patel, D. J. 1994a. DNA triplexes: Solution structures, hydration sites, energetics, interactions, and function. *Biochem.* 33:11405–11416.

Radhakrishnan, I., and Patel, D. J. 1994b. Solution structure of a pyrimidine:purine:pyrimidine DNA triplex containing T:AT, C+.GC and G:TA triples. *Struct.* 2:17–32.

Reddoch, J. F., and Miller, D. M. 1995. Inhibition of nuclear protein binding to two sites in the murine c-*myc* promoter by intermolecular triplex formation. *Biochem.* 34:7659–7667.

Rich, A. 1993. DNA comes in many forms. *Gene* 135:99–109.

Roberts, R. W., and Crothers, D. M. 1992. Stability and properties of double and triple helices: Dramatic effects of RNA or DNA backbone composition. *Science* 258:1463–1466.

Rossi, J. J., and Sarver, N. 1990. RNA enzymes (ribozymes) as antiviral therapeutic agents. *Trends Biotechnol.* 8(7):179–183.

Ruffner, D. E., Stormo, G. D., and Uhlenbeck, O. C. 1990. Sequence requirements of the hammerhead RNA self-cleavage reaction. *Biochem.* 29(47):10695–10702.

Sanfacon, H., and Hohn, T. 1990. Proximity to the promoter inhibits recognition of cauliflower mosaic virus polyadenylation signal. *Nature* (London) 346:81–84.

Santoro, S., and Joyce, G. 1997. A general purpose RNA-cleaving DNA enzyme. *Proc. Natl. Acad. Sci. U.S.A.* 94:4262–4266.

Sarver, N., Cantin, E. M., Chang, P. S., Zaia, J. A., Ladne, P. A., Stephens, D. A., and Rossi, J. J. 1990. Ribozymes as potential anti-HIV-1 therapeutic agents. *Science* 247:1222–1225.

Sasson-Corsi, P. 1996. Same clock, different work. *Nature* 384:613–614.

Sauman, I., and Reppert, S. M. 1996. Circadian clock neurons in the silkmoth *Anthreaea pernyi:* Novel mechanisms of period protein regulation. *Neuron* 17:889–900.

Sawata, S., Shimayama, T., Komiyama, M., Kumar, P. K. R., Nishikama, S., and Taira, K. 1993. Enhancement of the cleavage rates of DNA-armed hammerhead ribozyme by various divalent metal ions. *Nucleic Acids Res.* 21:5656–5660.

Scanlon, K. J., Jiao, L., Funato, T., Wang, W., Tone, T., Rossi, J. J., and Kashani- Sabet, M. 1991. Ribozyme-mediated cleavage of c-fos mRNA reduces gene expression of DNA synthesis enzymes and metallothionein. *Proc. Natl. Acad. Sci. U.S.A.* 88(23):10591–10595.

Semerad, C. L., and Maher, L. J. 1994. Exclusion of RNA strands from a purine motif triple helix. *Nucleic Acids Res.* 22:5321–5325.

Shimayama, T., Nishikawa, S., and Taira, K. 1995. Generality of the UX rule: Kinetic analysis of the results of systematic mutations in trinucleotide at the cleavage site of hammerhead ribozyme. *Biochem.* 34:3649–3654.

Sioud, M., and Drlica, K. 1991. Prevention of human immunodificiency virus type I intergrase expression in *Escherichia coli* by a ribozyme. *Proc. Natl. Acad. Sci. U.S.A.* 88:7303–7707.

Spitzner, J. R., Chung, I. K., and Muller, M. T. 1995. Determination of 5' and 3' DNA triplex interference boundaries reveals the core DNA binding sequence for topoisomerase II. *J. Biol. Chem.* 270:5932–5943.

Sullenger, B. A., and Cech, T. R. 1993. Tethering ribozymes to a retroviral packaging signal for destruction of viral RNA. *Science* 262:1566–1569.

Sunters, A., Springer, C. J., Bagshawe, K. D., Souhami, R. L., and Hartley, J. A. 1992. The cytotoxicity, DNA crosslinking ability and DNA sequence selectivity of the aniline mustards melphalan, chlorambucil and 4-[bis(2-chloroethyl)amino] benzoic acid. *Biochem. Pharmacol.* 44:59–64.

Svinarchuk, F., Debin, A., Bertrand, J. R., and Malvy, C. 1996. Investigation of the intracellular stability and formation of a triple helix formed with a short purine oligonucleotide targeted to the murine c-*pim*-1 proto-oncogene promotor. *Nucleic Acids Res.* 24:295–302.

Svinarchuk, F., Monnot, M., Merle, A., Malvy, C., and Fermandjian, S. 1995. The high stability of the triple helices formed between short purine oligonucleotides and SIV/HIV-2 vpx genes is determined by the targeted DNA structure. *Nucleic Acids Res.* 23:3831–3836.

Symons, R. H. 1992. Small catalytic RNAs. *Annu. Rev. Biochem.* 61:641–671.

Takasugi, M., Guendouz, A., Chassignol, M., Decout, J. L., L'homme, J., Thuong, N. T., and Helene, C. 1991. Sequence-specific photo-induced cross-linking of the two strands of double-helical DNA by a psoralen covalently linked to a triple helix forming oligonucleotide. *Proc. Natl. Acad. Sci. U.S.A.* 88:5602–5606.

Tang, X. B., Hobom, G., and Luo, D. 1994. Ribozyme mediated destruction of influenza A virus in vitro and in vivo. *J. Med. Virol.* 42(4):385–395.

Thomas, T. J., Faaland, C. A., Gallo, M. A., and Thomas, T. 1995a. Suppression of c-myc oncogene expression by a polyamine-complexed triplex forming oligonucleotide in MCF-7 breast cancer cells. *Nucleic Acids Res.* 23:3594–3599.

Thomas, T. J., Seibold, J. R., Adams, L. E., and Hess, E. V. 1995b. Triplex–DNA stabilization by hydralazine and the presence of anti-(triplex DNA) antibodies in patients treated with hydralazine. *Biochem. J.* 311:183–188.

Thompson, J. D., Macejak, D., Couture, L., and Seinchcomb, D. T. 1996. Ribozyme in gene therapy. *Nature Med.* 1:277–278.

Tuerk, C., and Gold, L. 1990. Systematic evolution of ligand by exponential enrichment: RNA ligand to bacteriophage T4 DNA polymerase. *Science* 249:505–510.

Uhlenbeck, O. C. 1987. A small catalytic oligonucleotide. *Nature* (London) 328:596–600.

Volkmann, S., Jendis, J., Frauendorf, A., and Moelling, K. 1995. Inhibition of HIV-1 reverse transcription by triple-helix forming oligonucleotides with viral RNA. *Nucleic Acids Res.* 23:1204–1212.

Wang, G., Levy, D. D., Seidman, M. M., and Glazer, P. M. 1995. Targeted mutagenesis in mammalian cells mediated by intracellular triple helix formation. *Mol. Cell. Biol.* 15:1759–1768.

Washbrook, E., and Fox, K. R. 1994. Comparison of antiparallel A:AT and T:AT triplets within an alternate strand DNA triple helix. *Nucleic Acids Res.* 22:3977–3982.

Weerasinghe, M., Liem, S. E., Asad, S., Read, S. E., and Joshi, S. 1991. Resistance to human immuno-deficiency virus type 1 (HIV-1) infection in human CD4+ lymphocyte-derived cell lines conferred by using retroviral vectors expressing an HIV-1 RNA-specific ribozyme. *J. Virol.* 65(10):5531–5534.

Weston, L., Blomquist, P., Milligan, J. F., and Wrange, O. 1995. Triple helix DNA alters nucleosomal histone-DNA interactions and acts as a nucleosome barrier. *Nucleic Acids Res.* 23:2184–2191.

Wilson, W. D., Tanious, F. A., Mizan, S., Yao, S., Kiselyov, A. S., Zon, G., and Strekowski, L. 1993. DNA triple-helix specific intercalators as antigene enhancers: Unfused aromatic cations. *Biochem.* 32:10614–10621.

Wu, H. N., Lin, Y. J., Lin, F. P., Makino, S., Chang, M. F., and Lai, M. M. Human hepatitis delta virus RNA subfragments contain an autocleavage activity. *Proc. Natl. Acad. Sci. U.S.A.* 86(6):1831–1835.

Yoon, K., Hobbs, C. A., Koch, J., Sardago, M., Kutny, R., and Weis, A. L. 1992. Elucidation of the sequence-specific third-strand recognition of four Watson–Crick base pairs in a pyrimidine triple-helix motif: T:AT, C:GC, T:CG, and G.TA. *Proc. Natl. Acad. Sci. U.S.A.* 89:3840–3844.

Yu, M., Ojwang, J., Yamada, O., Hampel, A., Rapapport, J., Looney, D., and Wong-Staal, F. 1993. A hairpin ribozyme inhibits expression of diverse strains of human immunodeficiency virus type 1. *Proc. Natl. Acad. Sci. U.S.A.* 90(13):6340–6343.

Zaia, J. A., Chatterjee, S., Wong, K. K., Elkins, D., Taylor, N. R., and Rossi, J. J. 1992. Status of ribozyme and antisense-based developmental approaches for anti-HIV-1 therapy. *Ann. N.Y. Acad. Sci.* 660:95–106.

Zaug, A. J., Been, M. D., and Cech, T. R. 1986. The *Tetrahymena* ribozyme acts like an RNA restriction endonuclease. *Nature* (London) 324:429–433.

Zhao, J. J., and Pick, L. 1993. Generating loss-of-function phenotypes of the *Fushi tarazu* gene with a targeted ribozyme in *Drosophila. Nature* (London) 365:448–451.

Zuker, M. 1989. On finding all suboptimal foldings of an RNA molecule. *Science* 224:48–52.

Biotechnology and Safety Assessment, 2nd ed.
Edited by John A. Thomas
Copyright © 1998 Taylor & Francis

6

Food Allergens:
Implications for Biotechnology

Samuel B. Lehrer and Gerald Reese

Tulane University School of Medicine, New Orleans, Louisiana

As reviewed in previous chapters, new plant varieties that are potential food sources are being developed with recombinant DNA technologies (Harlander 1991; Kessler et al. 1992; Olempska-Beer et al. 1993). On the basis of these techniques, specific genetic modifications that could not have been accomplished through traditional breeding techniques have been achieved, including the expression of proteins into plants from other species. Several potential foods that have been derived from these new plant varieties by recombinant DNA technologies and are either commercially available or approaching commercial introduction (Table 1). Because genes governing these new traits code for proteins that are ordinarily not present in that plant, there is concern about the effects of the transferred proteins on individuals ingesting them (Harlander 1991; Kessler et al. 1992; Olempska-Beer et al. 1993). A major concern is the potential of these transferred proteins as food allergens (Kessler et al. 1992; Olempska-Beer et al. 1993).

FOOD ALLERGY

An adverse reaction to food is any clinically abnormal response attributed to exposure to a food or food additive; these reactions include those of both immunologic and nonimmunologic pathogenesis. Food allergy (food hypersensitivity) is a specific type of immunologically mediated adverse reaction resulting from the ingestion of a food or food additive. This reaction occurs only in some of the exposed subjects, requires prior exposure (sensitization), may occur only after a small amount of substance is ingested, and is unrelated to any physiological effect of the food or food additive (Sampson and Metcalf 1991). Most food allergies are mediated by IgE antibodies and have a rapid onset—typically within minutes of exposure (Sampson and Metcalf 1991). Other adverse reactions to foods are food intolerance, which is an abnormal physiologic response to an ingested food or food additive; food poisoning, which is basically a toxic reaction; and pharmacologic reaction to foods, which are due

TABLE 1. *Foods derived from new plant varieties created through recombinant DNA technology (final consultations under FDA's 1992 policy)*

Product	Company	Year of FDA evaluation
Cherry tomato with lowered ethylene content	DNA Plant Technology	Ongoing
Tomato with lowered ethylene content	DNA Plant Technology	
Pea with increased sweetness	DNA Plant Technology	
Improved texture sweet pepper	DNA Plant Technology	
Pepper with increased sweetness	DNA Plant Technology	
Glufosinate-tolerant corn	Agrevo	
High-oleic, low-linoleic-oil soybean	Du Pont	
Disease-resistant, increased dry ma potato	Frito Lay	
High-oleic oil corn	Du Pont	
High-protein lupin	Resource Seeds Inc.	
Male sterile corn	Pioneer Hi-Bred	
Glufosinate-tolerant corn	Pioneer Hi-Bred	
Glufosinate-tolerant sugar beet	Agrevo	
Lepidopterans resistance corn	Plant Genetics System	
Phosphinothricin-tolerant rice	Agrevo	
Bromoxynil-tolerant, resistance to c cotton	Calgene	
Colorado potato beetle resistant potato	Monsanto	
Colorado potato beetle resistant potato	Monsanto	
Sulfonylurea-tolerant flax	University of Saskatchewan	
Glyphosate-tolerant corn	Monsanto	
Canola with altered phytase activity	BASF	
Glufosinate-tolerant canola	Agrevo, Inc.	1997
Male sterile *Radicchio rosso*	Bejo Zaden BV	
Insect-protected corn	Dekalb Genetics Corp.	
High-oleic-acid soybean	Du Pont	
Virus-resistant squash	Seminis Vegetable Seeds	
Virus-resistant papaya	University Hawaii and Cornell University	
Modified fruit-ripening tomato	Agritope Inc.	1996
Glufosinate-tolerant corn	Dekalb Genetics Corp.	
Sulfonylurea-tolerant cotton	Du Pont	
Insect-protected potato	Monsanto Co.	
Insect-protected corn	Monsanto Co.	
Insect-protected corn	Monsanto Co.	
Glyphosate-tolerant/insect-protected corn	Monsanto Co.	
Insect-protected corn	Northrup King	
Male sterile/fertility restorer oilseed rape	Plant Genetic Systems	
Male sterile corn	Plant Genetic Systems	
Glufosinate-tolerant canola	AgrEvo Inc.	1995
Glufosinate-tolerant corn	AgrEvo Inc.	
Laurate canola	Calgene Inc.	
Insect-protected corn	Ciba-Geigy Corp.	
Glyphosate-tolerant cotton	Monsanto Co.	
Glyphosate-tolerant canola	Monsanto Co.	
Insect-protected cotton	Monsanto Co.	

TABLE 1. *Continued.*

Product	Company	Year of FDA evaluation
Virus-resistant squash	Asgro Seed Co.	1994
Flavr Savr tomato	Calgene Inc.	
Bromoxynil-tolerant cotton	Calgene Inc.	
Improved ripening tomato	DNA Plant Technology	
Glyphosate-tolerant soybean	Monsanto Co.	
Improved ripening tomato	Monsanto Co.	
Insect-protected potato	Monsanto Co.	
Delayed softening tomato	Zeneca Plant Science	

to chemicals present in foods that produce a drug-like effect. These last three re-actions are nonimmunologic (Sampson and Metcalf 1991). These reactions may be confused with true food allergy because they have similar symptoms (Sampson and Metcalf 1991).

In the induction of an allergic response (as depicted in Figure 1), proteins that are typically innocuous enter the body via mucosal surfaces and are processed by local

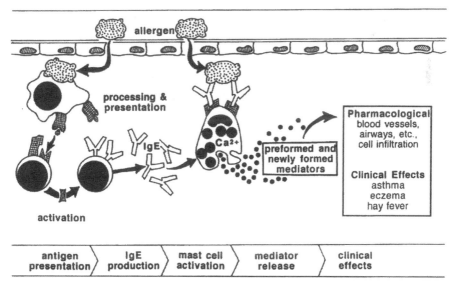

FIG. 1. Sensitization to food allergens. The process of IgE sensitization to food allergens is shown diagramatically. Innocuous protein molecules that cross the mucosal membrane barrier are processed and presented by antigen-presenting T cells. These cells, together with antigen, stimulate the production of IgE-specific antibodies. The IgE antibodies have the ability to fix to the surfaces of mast cells and basophils; subsequent interaction of allergen with cell-bound IgE antibody bridges these surface-bound IgE antibodies and stimulates the release of preformed and newly formed mediators that result in the clinical effects of allergy.

antigen-presenting cells that present peptide fragments of the allergen to specific cells called T lymphocytes or T cells. Recognition between the TH2 lymphocyte (a subset of T lymphocytes) and antigen-specific B lymphocytes (the cell that eventually produces antibody) results in the release of soluble factors, including interleukins 4, 5, and 13; these, in turn, stimulate B-cell proliferation and differentiation and the production of allergen-specific IgE (Romagnani 1991; de Vries et al. 1991; Wraith et al. 1989; Scherer et al. 1989). There is some evidence for genetic control of IgE antibody production to specific allergens (Marsh et al. 1974; Blumenthal et al. 1981). The portions of the protein molecule that interact with T cells are called T-cell epitopes, and those that interact with B cells or antibodies are called B-cell epitopes. The IgE antibodies have the ability to bind to specific receptors on mast cells or basophils, thus sensitizing these cells.

Subsequently, when an allergen reaches the sensitized mast cell, it cross-links surface-bound IgE, triggering the release of preformed and newly synthesized mediators. These mediators, in turn, elicit the clinical signs and symptoms of allergic diseases, including asthma, eczema, urticaria, hay fever, and anaphylaxis (Geha 1984; Roitt et al. 1985).

Precise figures concerning the prevalence of food allergies are difficult to obtain. In clinical surveys it has been estimated that about 500,000 to almost 1 million children younger than 6 years old have reproducible allergic reactions to foods. Studies in adults suggest that almost 2 to 4 million adult Americans are sensitive to foods or food additives. These figures contrast dramatically with the public's perception of the importance of allergic reactions to foods. It has been reported that at least one in four atopic adults believe that they have experienced adverse reactions following the ingestion or handling of foods. Similarly, parents believe that one of four of their children has experienced at least one adverse reaction to a food (Sampson and Metcalf 1991).

Factors that Affect the Development of Food Allergy

The quantity of a particular food ingested, which is influenced by the diet and culture of a region, can have a significant effect on the prevalence of a specific food allergy. The high prevalence of codfish allergy in Norway, rice and soy allergy in Japan, and peanut allergy in the United States are good examples. Route of exposure is another important factor in the development of food allergy. Not only are food allergens ingested, but exposure can also occur through skin contact and inhalation (O'Neil and Lehrer 1991). Gut permeability has long been thought to be a major factor in food allergy sensitization, but there has been very little investigation of this subject. Processing of food is another factor that can substantially alter food allergen content. Finally, intrinsic factors are clearly present; some foods apparently are very allergenic, whereas other foods are rarely if ever allergenic. This is discussed more fully in the next section on common food allergens that follows.

COMMON FOOD ALLERGENS

Food Allergens

Most foods contain literally thousands of different proteins to which an individual could potentially react. What makes some proteins within a particular food allergenic is not understood. Generally, allergens, including food allergens, can be classified as either major or minor. Major allergens are those to which 50% or more of a population sensitized to that particular food react (Lehrer and Salvaggio 1990). Reactions can be defined by skin test for IgE antibody response; however, more recently major or minor allergens are defined by reactivity to the individual allergen following testing of the sensitized subjects' sera by the immunoblot reaction. In addition to allergens being major or minor, they can also be isoallergens. Isoallergens are homologous proteins of similar molecular size, identical biological function, and at least 67% amino acid sequence identity. Isoallergens are homologous proteins that are identical immunologically yet may differ slightly in amino acid composition or carbohydrate content, which results in slightly different isoelectric points. Thus, several immunologically identical spots that are of similar yet distinct isoelectric points or molecular weights may occur during two-dimensional electrophoresis and appear to be distinct proteins.

Specific designations may be assigned to thoroughly characterized allergens according to the accepted taxonomic name of their source. The first three letters of the genus plus the first letter of the species name are used to indicate the source. An Arabic number is assigned according to the temporal order of allergen identification; however, generally the same number should be used to designate homologous allergens (King et al. 1994). For example, the first allergen described in brown shrimp, *Penaeus aztecus*, is designated Pen a 1, and the homologous molecule from Indian shrimp, *P. indicus*, is named Pen i 1. Examples of major food allergens and their names are shown in Table 2. The nomenclature system also provides rules for describing allergen genes, mRNAs, cDNAs, and, most importantly, recombinant and synthetic peptides of allergenic interest (King et al. 1994).

Foods that most commonly elicit allergic reactions are cereal grains, cow's milk, crustacea, fish, hen's eggs, legumes, mollusks, and tree nuts. The following section will briefly review some of the commonly allergenic foods and the allergenic proteins that have been identified from these foods. Generally, these foods have been identified as allergens based on the demonstration of positive skin test reaction or in vitro assay for specific IgE antibodies following a positive history; however, the preferred method of diagnosis is through double-blind, placebo-controlled food challenges (DBPCFC).

Animal-Derived Food Allergens

Although several foods from animal sources have been implicated in allergic reactions, there are only a few major food allergens that have been well studied. Many of these

TABLE 2. *Major food allergens identified*

Allergen source	Allergens (systematic and original names)	MW (kDa)	Sequence data	Reference[a]
Gadus callarias (cod)	Gad c 1; allergen M	12	C	Elsayed et al. 1976
Gallus domesticus (chicken)	Gal d 1; ovomucoid	28	C	Hoffman 1983
	Gal 2; ovalbumin	44	C	Langeland 1983
	Gal d 3; conalbumin (Ag 22)	78	C	Williams et al. 1982
	Gal d 4; lysozyme	14	C	Blake et al. 1965
Penaeus aztecus (brown shrimp)	Pen a 1; tropomyosin	36	P	Daul et al. 1993, 1994
Penaeus indicus (Indian shrimp)	Pen i 1; tropomyosin	34	P	Shanti et al. 1993
Metapenaeus enis (greasyback shrimp)	Met e 1; tropomyosin	34	C	Leung et al. 1994
Brassica juncea (oriental mustard)	Bra j 1; 25 albumin	14	C	Monslave et al. 1993
Hordeum vulgare (barley)	Hor v 1; BMAI-1	15	C	Mena et al. 1992
Sinapis alba (yellow mustard)	Sin a 1; 25 albumin	14	C	Menendez-Arias et al. 1988
Arachis hypogea (peanut)	Ara h 1	63.5	C	Stanley et al. 1995

SOURCE: Lehrer, Taylor et al. 1997, Table 58.6, p. 966.
[a]References refer to those where partial (P) or complete (C) sequence data are available.

foods (milk, eggs, fish, crustacea) and their allergens are described in more detail in the following paragraphs.

Several milk proteins have been identified as allergenic or immunogenic in humans. Many patients are allergic to more than one milk protein. Individuals allergic to cow's milk often have IgE antibodies directed against goat's or sheep's milk (Savilathi and Kuitunen 1992; Dean et al. 1993). Caseins and β-lactoglobulin appear to be the most frequently involved allergens in cow's milk sensitivity (Savilathi 1991; Savilathi and Kuitunen 1992; Amonette et al. 1993).

Egg allergy is one of the most frequently implicated causes of immediate food-allergic reactions in children in the United States and Europe (Crespo et al. 1994). Although there is extensive cross-reactivity among proteins from various birds, hen's egg proteins seem to be slightly more allergenic than those of duck eggs (Yunginger 1990). Egg white (albumin) appears to be more allergenic than yolk. Egg-white proteins have been extensively studied; most have been purified and their amino acid sequences determined (Li-Chan and Nakai 1989). Several major egg-white proteins have been identified as major allergens. Ovalbumin (Gal d 2) comprises more than 50% of egg-white protein. Ovotransferrin or conalbumin (Gal d 3) contributes 12% of the total protein of egg white. Ovomucoid (Gal d 1) comprises 10% of egg-white protein. Lysozyme (Gal d 4) is a small 14.3-kDa protein with isoelectric point (pI) of 10.7. It is a single polypeptide chain (with 129 amino acids [Canfield 1963]) cross-linked by four disulfide bonds. In addition to egg-white proteins, yolk proteins may also be allergenic. Apovitellenin I is a 9-kDa protein, and apovitellenin VI is a

170-kDa protein, the two of which comprise 2% of egg-yolk protein. Phosvitin (an iron-carrying molecule [Walsh et al. 1987, 1988]) comprises 10% of total egg-yolk protein. Livetins are derived from the blood of the hen and can also be found in the yolk (Yunginger 1991). Both phosvitin and livetins are potent allergens in some egg-sensitive individuals (Anet et al. 1985; Walsh et al. 1988).

The consumption of fish or inhalation of cooking vapors (Elsayed et al. 1972) from fish are frequent causes of IgE-mediated reactions. There have been no published reports on the prevalence of IgE-mediated reactions to a particular species of fish, for most studies refer only to cod or to "fish" in general. However, fish is one of the most commonly implicated allergenic foods and has been incriminated in fatal anaphylactic reactions (Yunginger et al. 1988). Although the overall prevalence of fish allergy is unknown, the incidence of fish hypersensitivity is observed to be higher in countries where fish consumption is above average.

One of the most comprehensive analyses of a food allergen has been the elegant work by Aas, Elsayed, and colleagues that resulted in purification and characterization of the codfish allergen, Gad c 1 (originally designated Allergen M). Gad c 1 belongs to a group of muscle tissue proteins known as parvalbumins (Elsayed and Bennich 1975) that control the flow of Ca^{2+} in and out of cells; parvalbumins are only found in the muscles of amphibians and fish. The existence of structurally related parvalbumins in different species of fish may explain cross-reactivity in fish-allergic individuals because Gad c 1 shares approximately 34% homology with similar proteins from hake, carp, pike, and whiting (Elsayad and Bennich 1975).

At least 30 edible species of crustacea are commonly consumed in the United States. Crustaceans (phylum Arthropoda, class Crustacea) include shrimp, prawns, crabs, lobsters, and crawfish (Yunginger 1991), which are frequent causes of food hypersensitivity. As is true for fish, a higher incidence of allergy to shellfish is seen in geographical areas where more is consumed on a regular basis. Shrimp is the most studied of the crustacean allergens. Daul et al. (1991, 1992, 1994) first isolated the major shrimp allergen (Pen a 1) from boiled brown shrimp (*P. aztecus*) and identified it as the shrimp muscle protein tropomyosin. Pen a 1 has a molecular weight of 36 kDa and is readily isolated from the boiling water (Lehrer et al. 1990) and meat of cooked shrimp. This allergen constitutes about 20% of water-soluble protein in crude cooked shrimp extract and inhibits up to 75% of IgE binding of pooled shrimp-sensitive subjects' serum to whole-body shrimp meat extract. The allergen-bound IgE in 28/34 (82%) sera from shrimp-sensitive individuals. Pen a 1 is composed of 284 amino acid residues and 2.9% carbohydrate; it has a pI of 5.2. Pen a 1 is the same as Pen i 1, the major allergen isolated from a different species of shrimp— *P. indicus* (Shanti et al. 1993). Leung et al. (1994) produced a recombinant shrimp allergen from a cDNA library of the greasyback shrimp, *M. enis*. That allergen has 284 amino-acid residues, is similar in amino acid composition to Pen a 1 and Pen i 1, and has a molecular weight of 34 kDa. In immunoblotting studies, recombinant Met e 1 allergen bound IgE in serum samples from all (8/8) individuals in the study with histories of anaphylactic reactions to shrimp. The authors confirmed the observations of other groups identifying the 34-kDa allergen as shrimp tropomyosin.

Food Allergens of Plant Origin

Several plant foods are major food allergens. These include legumes, particularly peanuts and soybeans; seeds and nuts; fruits; vegetables; grains; and spices. Several important food allergens of plant origin are discussed below.

Peanuts, although very popular, are the most allergenic food known. Peanut allergy is seldom outgrown, and allergic reactions to peanuts are often acute and severe. Peanut proteins have customarily been classified as albumins (water soluble) or globulins (saline soluble). Most peanut storage proteins are globulins; they make up 87% of the total protein (Johns and Jones 1916). Over the years, peanut proteins have been further fractionated and include albumins, arachin, and conarachin or nonarachin (Dechary et al. 1961; Altschul et al. 1964; Cherry et al. 1973; Dawson 1968, 1971). Globulins include two major proteins, α-arachin and α-conarachin (Cherry et al. 1973). A multiplicity of peanut allergens have been demonstrated (Sachs et al. 1981; Taylor et al. 1982; Barnett et al. 1983; Bush et al. 1983; Barnett and Howden, 1986; Meir-Davis et al. 1987; Burks et al. 1991; Burks et al. 1992) yet, with the exception of Ara h 1 and Ara h 2, have yet to be fully characterized and identified. Part of the problem in identification of peanut allergens lies in the large number of allergenic peanut proteins. Burks et al. (1991) identified a 63.5-kDa molecular weight glycoprotein peanut allergen using immunoblotting and enzyme-linked immunosorbent assay (ELISA) methods and sera from peanut-sensitive atopic dermatitis patients. This allergen (Ara h 1) has a pI of 4.55. In a later report (Burks et al. 1992), a second peanut allergen (Ara h 2) was identified and purified with a molecular weight of 17 kDa and an isoelectric point of 5.2. These studies of Burks and coworkers are the most advanced to date in the characterization of major plant-derived food allergens.

Globulins are also major proteins of the soybean. By adjusting the pH of the saline-soluble soybean protein fraction to 4.5, the globulins precipitate and leave a whey fraction that constitutes 6–8% of the protein (Rackins and Anderson 1964). The whey fraction contains haemagglutinin, trypsin inhibitors, and a urease (Shibasaki et al. 1980). Several soybean proteins have been studied as allergens. A study of Kunitz soybean trypsin inhibitor (KSTI) as an allergenic protein was prompted by a soy-allergic woman working with KSTI in an occupational setting (Moroz and Yang 1980). However, KSTI appears to be a relatively minor allergen. Herian et al. (1990) described a 20-kDa IgE binding protein from soybean and designated it S-II. Two serum samples from subjects allergic to soy showed IgE binding to a 20-kDa band, whereas no IgE binding was observed to pure KSTI. Roasting of the soybeans appeared to enhance IgE binding to the 20-kDa allergen. The allergen Gly m 30 K is described by Ogawa et al. (1991) as a 30-kDa-molecular-weight protein that is neither a soybean lectin nor the basic subunit of the 7S globulin but rather is a minor protein of the 7S globulin fraction. Sixty-five percent of the subjects in this study had specific IgE for Gly m Bd 30 K; however, these individuals did not experience severe or anaphylactic reactions to soy.

Rice *(Oryza sativa)* is a dietary staple for approximately one-half of the world's population. In Japan, rice has been found to aggravate atopic dermatitis frequently through IgE-dependent mechanisms. The major rice allergens consist of microhetero-geneous albumin proteins with molecular weights ranging from 14 to 60 kDa (Matsuda and Nakamura 1993). The rice allergens are encoded in a multigene family (Adachi et al. 1993). The nucleotide sequence of a cDNA coding for the major rice allergen has been identified (Izumi et al. 1992). The mature protein has a molecular weight of approximately 14 kDa.

Tree nuts are major plant food allergens. Brazil nuts are a cause of serious systemic anaphylaxis in some individuals. Arshad et al. (1991) found several allergenic fractions by immunoblotting using serum from allergic individuals to detect Brazil nut protein allergens. Another cause of significant nut-allergic reactions is almond. Bargman et al. (1992) used immunoblotting techniques to detect IgE binding proteins in almond extracts. An extensive number of proteins with molecular weights ranging from 38 to 70 kDa bound IgE. Two major allergens were identified. One is a 70-kDa heat-labile protein, whereas the other is a 45–50-kDa heat-stable protein.

Cross-Reactivity Among Food Allergens

Cross-reactivity can occur among food allergens. For example, within a given food group, clinically shrimp-allergic patients report reactivity to other crustacea such as crab, lobster, and crawfish in the absence of prior exposure (Daul et al. 1987). Cross-reacting allergens among crustacea have been confirmed by in vitro immunochemical studies (Lehrer 1986; Halmepuro et al. 1987). Fewer reports exist of cross-reactivity within vegetable food groups; however, one well-studied plant group is the legumes. Significant cross-reactivity, based on in vitro assays, is observed among peanuts, garden peas, soybeans, and chickpeas (Barnett et al. 1987; Taylor et al. 1987). In contrast to crustacea sensitivity, the clinical relevance of legume cross-reactivity has not been established because clinical hypersensitivity to one legume does not necessarily mean hypersensitivity to all legumes (Bernhisel-Broadbent and Sampson 1989).

In addition to cross-reactivity of foods within the same group, cross-reactivity has been described between foods and substances that are botanically unrelated or only distantly related. For example, it has been reported that ragweed pollen cross-reacts with melons and bananas (Anderson et al. 1970; Enberg et al. 1987), grass pollen cross-reacts with celery and a variety of other vegetables (Calkhoven et al. 1987; Wütrich and Dietschi 1985; Calkhoven et al. 1991), and birch pollen cross-reacts with several fruits (Calkhoven et al. 1991; Halmepuro et al. 1984). Recently it has been shown that latex allergens (newly identified proteins in natural rubber products) cross-react with fruits, including avocados, chestnuts, and bananas (Blanco et al. 1994; Lavaud et al. 1995). Studies from our laboratory have shown that marine animals belonging to different phyla such as oysters and crustacea (Lehrer and McCants 1988) or clams and shrimp (Desjardins et al. 1995) cross-react.

How can proteins from such distantly related substances cross-react? One explanation may be the presence of common structural or functional proteins. Profilins are ubiquitous cytoskeleton proteins that have a high degree of amino acid sequence homology and similar allergenicity (Hirschwehr et al. 1992; Valenta et al. 1992). Kraft and Valenta proposed profilins as a pan allergen (Valenta et al. 1993). This suggests that molecules of similar structure may have the same allergenicity. Our studies and those of others demonstrate that the principal shrimp allergen is tropomyosin, a major muscle protein (Daul et al. 1993; Daul et al. 1994; Shanti et al. 1993; Leung et al. 1994). However, in contrast to profilins, tropomyosins in beef, pork, and chicken are not allergenic, although they have at least 60% amino acid sequence homology with shrimp tropomyosin (Ayuso 1998). Thus, one must be careful when inferring cross-reactivity for all common structural proteins or proteins with similar amino acid sequences. The clinical significance of cross-reacting allergens is not entirely clear. However, cross-reacting allergens in different plants and animals poses serious concerns about the cross-reactivity of known allergens with transgenic products.

GENERAL CHARACTERISTICS OF FOODS AS ALLERGENS

Physical Properties

Generally, most food allergens are glycoproteins with acidic isoelectric points. This property is true for many antigens; thus, it is not a unique property for food allergens. Most known food allergens have molecular weights between 10 and 70 kDa (Taylor et al. 1987). Although smaller molecules could act as haptens, the molecular weight of 10 kDa probably represents the lower limit for the immunogenic response. The upper limit is probably a result of restricted mucosal absorption of larger molecules (Taylor et al. 1987). However, some allergens such as Ara h 1 (63.5 kDa) and Ara h 2 (17 kDa) exist in their native forms as parts of large protein polymers that are 200–300 kDa in size (Burks et al. 1991; Burks et al. 1992). It is not known whether these large molecules act as allergens or must be disassociated during the digestive process.

Valancy

Because all allergens must be able to bridge IgE molecules on the surface of mast cells to cause degranulation, they are constrained in their molecular dimensions. Allergens must contain at least two IgE antibody-reactive sites to trigger mediator release. However, because some monovalent allergens, such as the 21-residue venom peptide mellitin (King et al. 1993) or the ovalbumin-IgE monoclonal antiovalbumin experimental model (Böttcher and Hammerling 1978), can still elicit histamine release from basophils or mast cells or generate anaphylactic reactions in mice, allergens may have the ability to bind to surface IgE antibodies on basophils and mast cells and aggregate or aggregate and then bind, thus converting the monovalent allergens to polyligands that can trigger allergic reactions. It is not known whether this occurs in vivo in patients with allergic disease but is a substantial factor in allergic reactions.

Stability and Resistance

Allergens must reach the intestinal tract in an immunologically active form to exert their effects. Thus, conventional wisdom is that these proteins are comparatively resistant to heat or acid treatment as well as proteolysis and digestion; however, important exceptions do exist. Many allergenic food proteins are heat-resistant (i.e., peanuts, shrimp, cow's milk, fish). The application of heat promotes protein denaturation and the loss of conformational epitopes, but the resistance of many food allergens to heat denaturation suggests that conformational epitopes are not always critical for IgE binding. Although heat treatment leads to protein denaturation and loss of conformational epitopes, enzymatic or acidic cleavage of the polypeptide chains may destroy both conformational and linear epitopes. Resistance to digestion and other enzymatic processes are dependent upon the nature of the enzyme or hydrolytic system used in the experiments, the choice of systems for assessing the immunogenicity of the hydrolysis products and, to a lesser extent, the specific food allergen being evaluated.

Epitopes

Our knowledge of T-cell epitopes of food allergens is largely extrapolated from non-food allergens; however, food allergens may have unusual properties. Food allergens are recognized by gut-associated lymphoid tissue (GALT), and the immunoregulatory mechanisms of T-cell responses of this system have not been well defined; thus, there may be unique properties to T-cell epitopes in food allergens (King 1994b). However, two food allergens whose T-cell epitopes have been studied—Gal d 2 and Ara h 1—have not demonstrated any unique properties. On the basis of inducing nonresponsiveness and down-regulation of established immune responses in animals, there is considerable interest in peptides containing T-cell epitopes as a basis for immunotherapy in many allergies (de Vries and Lamb 1994). However, practical application of immunotherapy with T-cell-reactive peptides is still under study.

B-cell epitopes do not appear to have any unique or common pattern of amino acid sequence. The amino acid sequences of B-cell epitopes of four food allergens are summarized in Table 3. There are methods, although not totally successful, to predict B-cell epitopes. For example, algorithms based on the calculations of polar and nonpolar amino acid residues have been used to predict those residues located on the surface of the molecule. However, this information could be incomplete because host responsiveness may determine those epitopes to which an individual reacts.

When the biologic activities of various allergenic proteins have been compared, no consistent pattern representative for allergens in general or food allergens specifically has been evident. Comparisons of primary amino acid sequences of allergenic proteins or tertiary protein structure has not yielded any unique or typical pattern (King 1994a). When the primary structure of an allergen is compared with other proteins, however,

TABLE 3. *Sequence of some B-cell epitopes of four food allergens*

Food	Allergen	Sequence	Reference
Egg	Ovalbumin residue 323-339	LysGlyAlaGluAsnIleGluAla HisAlaAlaHisValAlaGlu SerIle	Johnson and Elsayed 1990
Cod	Parvalbumin residue 49-64	IleAlaAspGluAspLysGluGly PheIleGluGluAspGlu LeuLys	Elsayed et al. 1991
Milk	Lactoglobulin residue 124-134	ArgThrProGluValAspGlu AlaLeuGluAla	Adams et al. 1991
Shrimp	Tropomyosin residue 153-161	PheLeuAlaGluGluAlaAsp ArgLys	Shanti et al. 1993

SOURCE: Lehrer, Taylor et al. 1997, Table 58.4, p. 964.

amino acid sequence similarities occur with many proteins in our environment. When considered in the light of the evolution of all living organisms, this finding should not be surprising but rather serves to illustrate that there may be certain as yet undetected structural features of allergens that render them different from other proteins.

IMPLICATIONS FOR BIOTECHNOLOGY

Overview

As has been discussed in the previous chapters, recombinant DNA technology can have a profound effect in improving our food supply. Through these methods, crops can be improved by increasing stress, disease, and insect resistance and improving qualities desired by the consumers, including increased shelf life, better taste and flavor, and nutrition. Indeed, a variety of crops have been developed and approved by the Food and Drug Administration and are now available for use by consumers (Table 1). With regard to allergenicity, recombinant DNA technology can have both beneficial and potentially detrimental effects. The ability to alter the structure of proteins or suppress their expression has a potentially beneficial effect by reducing allergenicity for foods with known allergens. Conversely, expression of new proteins may result in new allergens or may quantitatively affect expression of homologous proteins that could be allergenic. Any of these possibilities may result in significant food allergy, and thus the allergenic activity of transgenic crops is an important concern.

Certainly it is known that recombinant proteins can be allergenic. Many recombinant allergens have been expressed and have been shown to retain their allergenic activity. This suggests that any recombinant protein allergen expressed in a transgenic crop has the potential to sensitize consumers who ingest this food. The implication for biotechnology when new food products are developed is that adequate testing for allergenic activity must be done before the release of transgenic foods to consumers.

When one considers the effects of biotechnology on altered foods, the following three issues are important:

1. Expression of proteins from known allergen sources and how to detect these allergens.
2. Expression of proteins from sources of unknown allergenicity and how the potential allergenicity of these new products can be assessed. This is the most difficult effect to monitor because the potential allergenic effects are unknown.
3. Potential beneficial effects of biotechnology on food products through the development of hypoallergenic products.

Measurement of Known Allergens by In Vitro Assays (IgE Binding)

The expression of recombinant proteins with known allergic activity in different foods is the least difficult issue to address. To date, most experience with recombinant allergens has been obtained from recombinant inhalant allergens, and for the most part these recombinant allergens bind IgE antibodies and thus retain their IgE-reactive epitopes. These findings suggest that recombinant proteins expressed in other species (i.e., altered food crops) will probably retain their allergenic activity and must be considered allergenic unless proven otherwise. Can such allergenic recombinant proteins be detected in new foods?

Recombinant allergens in genetically engineered or altered foods can be identified using traditional in vitro assays such as the radioallergosorbent test (RAST) inhibition or the enzyme-linked immunosorbent assay (ELISA) inhibition assays. Both assays are based on the competitive binding of IgE antibodies in patient serum between the test sample in a liquid phase and the solid-phase allergen. Testing increasing inhibitor concentrations can be used to measure unknown allergen quantities in assay samples.

These methods are well established, specific, sensitive, and reproducible. The only reagent of limited quantity is patients' sera. However, once sensitized individuals are identified, substantial quantities of sera can be obtained and serum pools produced. At least 1 L of plasma can be obtained from sensitized individuals by plasmapheresis, which can provide sufficient reagent for virtually unlimited allergen assays. An alternate possibility would be to use monoclonal antibodies raised against such allergens. This would provide a potentially unlimited quantity of the reagent. However, monoclonal antibodies may not bind to IgE reactive epitopes; thus, the only definitive way to identify immunologically active allergens is through demonstration of an IgE reactive epitope.

Examples of Allergen Detection in Transgenic Foods

The importance of testing recombinant proteins for allergenicity has been clearly demonstrated. For example, the methionine-rich 2S storage protein from the Brazil nut was expressed in soybean to improve the nutritional quality of soy meal as animal feed. Following genetic transformation, the 2S protein from the Brazil nut constituted a significant fraction of the soybean protein. Because Brazil nuts are allergenic, in vitro tests, including RAST and immunoblots, were used to test extracts of the transgenic soybeans for expression of Brazil nut allergens. Using sera from Brazil–nut-sensitive

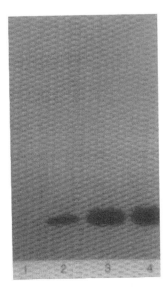

FIG. 2. Detection of Brazil nut allergen in transgenic soybean. A major Brazil nut allergen was detected with the serum of a Brazil–nut-allergic subject by immunoblot. Lane 1 contains normal soy, lane 2 transgenic soy expressing Brazil nut protein, lane 3 Brazil nut extract, lane 4 9-kDa Brazil nut allergen (Nordlee et al. 1996, Figure 3, p. 690).

subjects, 8/9 reacted to transgenic soybean extract (Fig. 2), which confirms that an immunologically functional Brazil nut allergen had been transferred to the soybeans (Nordlee et al. 1996). Although Brazil nut allergy is not common, if the allergens were widely dispersed in a commodity food, such as soybeans, exposure, and presumably reactivity, would increase.

Using RAST inhibition and immunoblotting, we assessed the allergenic potential of transgenic soybeans engineered to produce increased levels of oleic acid (McCants et al. 1997). Virtually identical inhibition of wild-type extract and an extract prepared from transgenic soybeans was demonstrated (Fig. 3). Results of immunoblotting

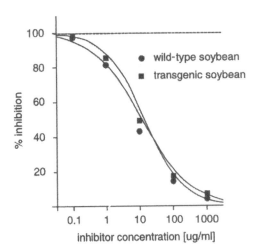

FIG. 3. Inhibition of wild-type soybean RAST with wild-type or transgenic soybean extracts. Increasing concentrations of either wild-type or transgenic soy extracts were tested for their ability to inhibit wild-type soy RAST. Statistical analysis does not demonstrate any significant difference in the inhibitory activity of either extract, which suggests quantitatively similar allergen content (Lehrer and Reese 1997, p. 123).

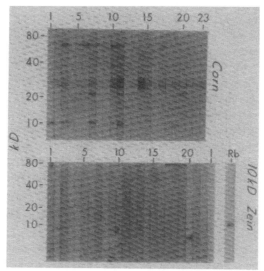

FIG. 4. The IgE antibody's reactivity to alcohol-soluble corn proteins and recombinant 10-kDa corn zein protein. Sera from 23 corn-reactive subjects demonstrated ongoing reactivity to alcohol-soluble corn proteins. None of the sera reacted to the 10-kDa recombinant serum protein.

demonstrated that both the wild-type and transgenic soybeans contained approximately 30 protein bands ranging in molecular weight from 14 to 100 kDa; intensity of the bands appeared to be identical (Fig. 4). The findings suggest that changes in soy proteins resulting in increased oleic acid content did not alter soy allergen levels and, thus, do not pose increased risk of allergy to consumers.

Lastly, we have investigated the allergenic activity of sulfur-rich corn proteins (Lehrer et al. 1997). Two zein proteins, 10 kDa and HSZ, have been identified; because efforts are under way to increase the expression of these proteins in corn or the seeds of other cereal grains to enhance the sulfur content of sulfur-poor crops, it is important to assess their potential allergenicity. Sera from 23 individuals demonstrated by skin test, RAST, or clinical history and immunoblot to be corn reactive were tested for IgE antibody reactivity to these two zein corn proteins using SDS-PAGE/immunoblotting. None of the sera from corn-reactive subjects demonstrated IgE reactivity against either the 10 kDa or HSZ zein proteins, which suggests that products encoding these genes do not pose an increased risk of allergy to consumers (Fig. 4).

Assessing Allergenicity in Genetically Engineered Foods Derived from Sources of Unknown Allergenicity

A much more difficult problem than the one outlined above is evaluation of the allergenic potential of foods engineered using proteins from sources of undetermined allergenicity. Predicting potential allergenicity is a major challenge for the food industry because there is no single predictive assay to assess the potential allergenicity of any protein.

A preliminary step, evaluation of amino acid sequence homology, may be useful in evaluating allergenicity of a transgenic protein. If sequence homologies are observed, particularly with regions containing IgE binding epitopes, it is essential that in vitro testing with RAST or ELISA should be performed to assess IgE reactivity. If no IgE antibody reactivity is detected, the second step should be to compare the physico-chemical and biologic characteristics—including molecular size, stability, solubility, and isoelectric point—of these proteins with major food allergens. Suspect proteins should again be tested to assess IgE reactivity. Clearly, these approaches may be ex-tremely difficult, if not impossible, to establish if IgE reactivity is not known. Even the most careful evaluation may not exclude a potential allergenic risk, and it may be prudent to follow new foods after their introduction into the marketplace.

Screening Amino Acid Sequence Homologies with Known Allergens

Because allergens from food and nonfood sources can cross-react, it is critical to compare amino acid sequences from all known allergens (not just with food aller-gens) with those from the genetically engineered food. Over 200 allergens have been identified, characterized, and sequenced; this information is available through public domain databases (i.e., Gen Bank, PIR, EMBL, and Swiss Prot). On the basis of the optimal peptide length for binding B-cell epitopes (8–12 amino acids), an immuno-logically significant sequence would require a match of at least eight contiguous identical or similar amino acids. Confirmation of such a sequence suggests that the transferred protein may be allergenic and in vitro testing, as described above, should be employed to assess further the allergenicity of the protein. Failure to find such a match strongly suggests that the introduced protein does not share linear epitopes with known allergens but does not exclude the possibility that the transferred protein is an allergen.

Linear epitopes are important; however, epitopes that are dependent on tertiary structure of the protein (conformational and discontinuous epitopes) are not as easily assessed and are not readily predicted based on the primary amino acid sequence of the allergen. Moreover, amino acid sequence similarity does not necessarily mean similar IgE reactivity (Reese et al. 1995), for the degree of homology must also be considered. Recently it has been shown that a single amino acid substitution in the major peanut allergen Ara h 2 could abate, diminish, or enhance IgE binding, depending on the substitution (Stanley et al. 1997). Thus, for a 10-residue region of a transgenic protein, an amino acid sequence identity of 90% with a known allergenic epitope may indicate possible allergenicity but does not prove clinically significant cross-reactivity.

Comparison of Physicochemical and Biological Characteristics

As stated above, food allergens generally share physicochemical and biologic char-acteristics, including molecular size, stability, solubility, and isoelectric point; thus, it is beneficial to compare these aspects in proteins of transgenic foods with known food allergens. Of these traits, stability to digestive processes may be an important factor when assessing potential allergenicity. Proteins that are stable under proteolytic and acidic conditions of the digestive tract are more likely to reach the intestinal mucosa and stimulate an immune response than more labile proteins. Thus, rapid degradation of proteins expressed in genetically engineered foods reduces the likelihood that the protein is an allergen.

Recently, an in vitro model to assess the stability of food allergens to digestion was developed (Astwood et al. 1996). In this model, known food allergens and other common plant proteins were exposed to simulated gastric fluid for varying periods of time. Food allergens were stable under these conditions for periods up to 2 min, and some major allergens were stable for more than 1 h; whereas, the few nonallergenic food proteins tested were degraded within 30 s. It should be emphasized that stability of a transgenic protein in simulated gastric acid is not synonymous with allergenicity because some allergens are rapidly degraded by proteolytic enzymes and many non-allergenic proteins can be very stable to proteolysis. Thus, there needs to be more detailed studies on the enzyme stability of more allergenic and nonallergenic food proteins in order for enzyme digestability to be considered as a major property of allergens.

Lastly, the prevalence of the protein in a food should also be considered. In plant foods, many food allergens are storage proteins that are present in large amounts; thus, many food allergens constitute a large proportion of total protein (1 to 80%) in offending foods (Yunginger 1990). Introduced proteins are usually expressed in plants in very low levels; from less than 0.01 to 0.4% (Astwood and Fuchs 1996). It is tempting to speculate that degree of exposure to allergens (i.e., the amount of an allergen in a food) is directly related to allergenic potential. However, the major allergen in codfish, Gad c 1, is not a predominant protein, and proteins that are major components of many foods, including myosin, tropomyosin, and actin from beef, pork, and chicken have not been identified as major allergens.

Biotechnologically Altered Foods

There can be a beneficial effect on allergenicity of transgenic foods in that foods eliciting allergenic responses can be rendered hypoallergenic through deletion of the allergen or modification of the allergenic epitopes. This technique is potentially very useful because it can substantially reduce a food's allergenic activity. Recently, genetic engineering of hypoallergenic rice has been accomplished (Adachi et al. 1993; Izumi et al. 1992; Watanabe 1993; Matsuda et al. 1993). A rice seed protein of approximately 16 kDa was isolated and identified as a major rice allergen. A cDNA clone encoding this protein was isolated from a cDNA library prepared from maturing rice seeds. On the basis of the nucleotide sequence of the cDNA, an antisense RNA strategy was applied to repress expression of this allergen in maturing rice seeds. Seeds from transgenic plants with the antisense gene have substantially reduced amounts of the allergen relative to the wild-type control (Fig. 5).

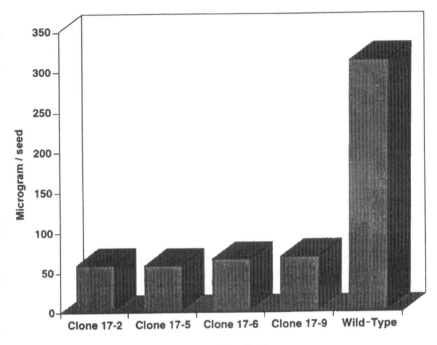

Rice Varieties

FIG. 5. Suppression of 16-kDa rice allergen using antisense technology. Rice allergen levels were quantified by ELISA for each genetically engineered rice variety (clones 17-2, 17-5, 17-6, 17-9) as compared with wild-type rice seeds (Astwood et al. 1997, Figure 4.4, p. 86).

Recently Burks and colleagues have investigated two major peanut allergens: Ara h 1 and Ara h 2. Four major IgE binding epitopes have been identified in Ara h 1, and three major IgE binding epitopes were identified in Ara h 2. Through a single amino acid substitution, as shown in Figure 6, IgE binding to peptides containing three of the four major epitopes of Ara h 1 and all three epitopes in Ara h 2 was abolished (Stanley et al. 1997). Current studies are directed at investigating the effects of such amino-acid substitutions on the entire allergen molecule. Together, these two studies are

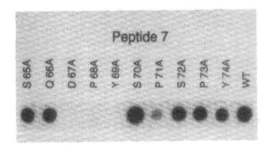

FIG. 6. Reduction of IgE binding capacity of peptide 7, which contains a major Ara h 2 IgE binding epitope through amino acid substitution. Peptides (S65A-S70A, P71A-P73A, Y74A) containing sequential amino acid substitution with alanine demonstrate altered IgE binding with a pool of peanut-allergic sera by immunoblot. Results indicate that some acid substitutions abolish IgE binding capacity (Stanley et al. 1997, Figure A, p. 249).

very exciting because they demonstrate that biotechnology can be used beneficially to suppress or alter expression of major allergens.

SUMMARY AND CONCLUSIONS

Recombinant techniques offer breeders, horticulturists, farmers, and consumers agricultural crops with improved qualities, including increased resistance to stress, insects, diseases, and herbicides. Because genes governing the coding of proteins conferring these new traits are not part of the plant's original genome, there are concerns for the safety of these newly engineered varieties. A major concern is the allergenic potential of transferred proteins. Systematic strategies to assess potential allergenicity of these proteins have been proposed; these strategies will be modified as more is known about the structural basis of allergens. The main problem is evaluating the allergenic potential of proteins from sources that have not been implicated as allergens. Current approaches include comparison of amino acid sequences of the transferred protein with those of known food and nonfood allergens and comparison of chemical and physical characteristics of the transferred protein with known food allergens. In vitro assays using sera from hypersensitive individuals or placebo-controlled double-blind food challenges should be used to confirm allergenicity of any transgenic foods before marketing.

Although this chapter has focused on problems arising from transferring known allergens or creating new allergens, bioengineering techniques also provide a unique opportunity to create hypoallergenic varieties of plants. Matsuda and coworkers (Tada et al. 1996) genetically engineered a hypoallergenic rice. By introducing genes in the antisense orientation, the levels of expression of the 16-kDa rice allergen were reduced when compared with the wild type. The exciting study by Stanley and associates (Stanley et al. 1997) suggests that altering IgE binding epitopes of the peanut allergen Ara h 2 could substantially reduce allergenicity of this ubiquitous food. Thus, biotechnology may be used to alter the allergenic potential of foods.

ACKNOWLEDGMENTS

The authors would like to acknowledge the word processing support of Pat Constant and Wendy Theard. Preparation of this manuscript was supported in part by funds provided by the Department of Medicine, Tulane University School of Medicine, and the National Fisheries Institute.

REFERENCES

Adachi, T., Izumi, H., Yamada, T., Tanaka, K., Takeuchi, S., Nakamura, R., and Matsuda, T. 1993. Gene structure and expression of rice seed allergenic proteins belonging to the α-amylase/trypsin inhibitor family. *Plant Mol. Biol.* 21:239–248.

Adams, S. L., Barnett, D., Walsh, B. J., and Pierce, R. J. 1991. Human IgE-binding synthetic peptides of bovine beta-lactoglobulin and alpha-lactalbumin. In vitro cross-reactivity of athe allergens. *Immunol. Cell Biol.* 68:191–197.

Altschul, A. M., Neucere, N. J., Woodham, A. A., and Dechary, J. M. 1964. A new classification of seed proteins: Application to the aleurins of *Arachis hypogaea*. *Nature* 203:501–504.

Amonette, M. S., Rosenfeld, S. I., and Schwartz, R. H. 1993. Serum IgE antibodies to cow's milk proteins in children with differing degrees of IgE-mediated cow's milk allergy; analysis by immunoblotting. *Pediat. Asthma Allergy Immunol.* 7:99–109.

Anderson, L. B., Dreyfuss, E. M., Logan, J., and Johnston, D. E. 1970. Melon and banana sensitivity coincident with ragweed pollinosis. *J. Allergy Clin. Immunol.* 45:310–319.

Anet, J., Back, J. R., Baker, R. S., Barnett, D., Burley, R. W., and Houden, M. E. H. 1985. Allergens in the white and yolk of hen's egg: A study of IgE-binding by egg proteins. *Int. Arch. Allergy Appl. Immunol.* 77:364–371.

Arshad, S. H., Malmberg, E., Krapt, K., and Hide, D. W. 1991. Clinical and immunological characteristics of Brazil nut allergy. *Clin. Exp. Allergy* 21:373–376.

Astwood, J. D., and Fuchs, R. L. 1996. Allergenicity of foods derived from transgenic plants. In *Highlights in Food Allergy*, eds. B. Wuthrich and C. Ortolani, *Allergy* 32:105–120, Monograph. Basel: Karger.

Astwood, J. D., Fuchs, R. L., and Lavrik, P. B. 1997. Food biotechnology and genetic engineering. In *Food allergy: Adverse reactions to foods and food additives.* eds., D. D. Metcalfe, H. A. Sampson, and R. A. Simon, 65–92. Cambridge, Massachusetts: Blackwell Science.

Astwood, J. D., Leach, J. N., and Fuchs, R. L. 1996. Stability of food allergens to digestion in vitro. *Bio Technol.* 14:1269–1273.

Ayuso, R., Reese G., Tanaka, L., Ibanez, M. D., Pascual, C., Burks, A. W., Sussman, G. L., Dalton, A. C., Lahoud, C., Lopez, M., and Lehrer, S. B. 1998. IgE antibody response to raw and cooked vertebrate meats and tropomyosins. (Abstract) *Ann. Allergy, Asthma Immunol.* 101:5239.

Bargman, T. J., Rupnow, J. H., and Taylor, S. L. 1992. IgE-binding proteins in almonds (Prunus amygdalus): Identification by immunoblotting with sera from almond-allergic adults. *J. Food Sci.* 57:717–720.

Barnett, D., Baldo, B. A., and Howden, M. E. H. 1983. Multiplicity of allergens in peanuts. *J. Allergy Clin. Immunol.* 72:61–68.

Barnett, D., Bonhan, B., and Howden, W. E. H. 1987. Allergenic cross-reactions among legume foods—in vitro study. *J. Allergy Clin. Immunol.* 79:433–438.

Barnett, D., and Howden, W. E. H. 1986. Partial characterization of an allergenic glycoprotein from peanut (*Arachis hypogaea* L.). *Biochim. Biophys. Acta* 882:97–105.

Bernhisel-Broadbent, J., and Sampson, H. A. 1989. Cross-allergenicity in the legume botanical family in children with food hypersensitivity. *J. Allergy Clin. Immunol.* 83:435–440.

Blake, C. C. F., Koenig, D. F., Mair, G. A., North, A. C. T., Phillips, D. C., and Sarma, V. R. 1965. Structure of hen egg-white lysozyme. *Nature* 206:757–761.

Blanco, C. Carrillo, T., Castillo, R., Quiralte, J., and Cuevas, U. 1994. Latex allergy: Clinical features and cross reactivity with fruits. *Ann. Allergy* 73(4):309–314.

Blumenthal, M. N., Namboodiri, K. Mendell, N., Glych, G., Elston, R. C., and Yunise, E. 1981. Genetic transmission of serum IgE levels. *Am. J. Med. Genet.* 10:219.

Böttcher, I., and Hammerling, G. 1978. Continuous production of monoclonal mouse antibodies with known allergenic specificity by a hybrid cell line. *Nature* 275:761–762.

Burks, A. W., Williams, L. W., Connaughton, C., Cockrell, G., O'Brien, T. J., and Helm R. M. 1992. Identification and characterization of a second major peanut allergen, *Ara h* II with use of the sera of patients with atopic dermatitis and positive peanut challenge. *J. Allergy Clin. Immunol.* 900:962–969.

Burks, A. W., Williams, L. W., Helm, R. M., Connaughton, C., Cockrell, G., and O'Brien, T. 1991. Identification of a major peanut allergen, *Ara h* I, in patients with topic dermatitis and positive peanut challenges. *J. Allergy Clin. Immunol.* 88:172–179.

Bush, R. K., Voss, M., Taylor, S. L., Nordlee, J., Busse, W., and Yunginger, J. W. 1983. Detection of peanut allergens by crossed radioimmuno-electrophoresis (CRIE) (abstract). *J. Allergy Clin. Immunol.* 71:95.

Calkhoven, P. G., Aalbers, M., Koshte, V. L., Pos, O., Oei, H. D., and Aalberse, R. C. 1987. Cross-reactivity among birch pollen, vegetables and fruits as detected by IgE antibodies is due to at least three distinct cross-reactive structures. *Allergy* 42:382–390.

Calkhoven, P. G., Aalbers, M. Koshte, V. L., Schilte, P. P., Yntema, J. L., Griffioen, R. W. Van Nierop, J. C., Oranje, A. P., and Aalberse, R. C. 1991. Relationship between IgG1 and IgE4 antibodies to foods and the development of IgE antibodies to inhalant allergens. II. Increased levels of IgG antibodies to foods in children who subsequently develop IgE antibodies to inhalant allergens. *Clin. Exp. Allergy* 21(1):99–107.

Canfield, R. E. 1963. The amino acid sequence of egg white lysozyme. *J. Biol. Chem.* 238:2698–2707.

Cherry, J. P., Dechary, J. M., and Ory, R. L. 1973. Gel electrophoretic analysis of peanut proteins and enzymes. I. Characterization of DEAE-cellulose separated fractions. *J. Agric. Food Chem.* 21:562–565.

Crespo, J. F., Pascual, C., Ferrer, A., Burks, A. W., Diaz Pena, J. M., and Esteban, M. M. 1994. Egg white-specific IgE level as a tolerance marker in the IgE level as a tolerance marker in the follow-up of egg allergy. *Allergy Proc.* 15:73–76.

Daul, C. B., Morgan, J. E., Waring, N. P., McCants, M. L., Hughes, J., and Lehrer, S. B. 1987. Immunological evaluation of shrimp-allergic individuals. *J. Allergy Clin. Immunol.* 80:716–722.

Daul, C. B., Slattery, M., Morgan, J. E., and Lehrer, S. B. 1991. Isolation and characterization of an important 36 kD shrimp allergen (abstract). *J. Allergy Clin. Immunol.* 87:192.

Daul, C. B., Slattery, M., Morgan, J. E., and Lehrer, S. B. 1992. Identification of a common major crustacea allergen (abstract). *J. Allergy Clin. Immunol.* 89:194.

Daul, C. B., Slattery, M., Morgan, J. E., and Lehrer, S. B. 1993. Common crustacea allergens: Identification of B cell epitopes with the shrimp specific monoclonal antibodies. In *Molecular biology and immunology of allergens*, eds. D. Kraft and A. Sehon, 291–293. Boca Raton, Florida: CRC Press, Inc.

Daul, C. B., Slattery, M., Reese, G., and Lehrer, S. B. 1994. Identification of the major brown shrimp (*Penaeus aztecus*) as the muscle protein tropomyosin. *Int. Arch. Allergy Clin. Immunol.* 105:49–55.

Dawson, R. 1968. The nutritive value of groundnut protein. 2. The correlation between electrophoretic and nutritive value. *Brit. J. Nutr.* 22:601–607.

Dawson, R. 1971. Comparison of fractionation of groundnut proteins by two different methods. *Anal. Biochem.* 41:305–313.

Dean, T. P., Adler, B. R., Ruge, F., and Warner, J. O. 1993. In vitro allergenicity of cow's milk substitutes. *Clin. Exp. Allergy* 23:205–210.

Dechary, J. M., Talluto, K. F., Evans, W. J., Carne, W. B., and Altschul, A. M. 1961. Alpha-conarachin. *Nature* 190:1125–1126.

Desjardins, A., Malo, J. L., L'Archevêque, J., Cartier, A., McCants, M., and Lehrer, S. B. 1995. Occupational IgE-mediated sensitization and asthma caused by clam and shrimp. *J. Allergy Clin. Immunol.* (abstract), 96:608–617.

de Vries, J. D., Gauchat, J. F., Aversa, G. G., Punnonen, J., Gascan, H., and Yssel, J. 1991. Regulation of IgE synthesis by cytokines. *Curr. Opin. Immunol.* 3:851–858.

deVries, J. E., and Lamb, J. R. 1994. Immunotherapy with allergen derived epitopes. *Allergy Clin. Immunol. News* 6:49–53.

Elsayed, S., Apold, S., Aas, K., and Bennich, H. 1976. The allergenic structure of allergen M from cod. Tryptic peptides of fragment TM1. *Int. Arch. Allergy Appl. Immunol.* 52:59–63.

Elsayed, S., Aas, K., Slette, K., and Johansson, S. G. O. 1972. Tryptic cleavage of a homogenous cod fish allergen and isolation of two active polypeptide fragments. *Immunochem.* 9:647–661.

Elsayed, S., Apold, J., Holen, E., and Vik, H. 1991. The structural requirements of epitopes with IgE binding capacity demonstrated by three major allergens from fish, egg and tree pollen. *Scand J. Clin. Lab. Invest. Suppl.* 204:17–31.

Elsayed, S., and Bennich, H. 1975. The primary structure of allergen M from cod. *Scand. J. Immunol.* 4:203–208.

Enberg, R. N., Leickley, F. E., McCollough, J., Bailey, J., and Ownby, D. R. 1987. Watermelon and ragweed share allergens. *J. Allergy Clin. Immunol.* 79:867–875.

Geha, R. S. 1984. Human IgE. *J. Allergy Clin. Immunol.* 74:109–120.

Halmepuro, L., Salvaggio, J., and Lehrer, S. B. 1987. Crawfish and lobster allergens: Identification and structural similarities with other crustacea. *Int. Arch. Allergy Appl. Immunol.* 82:213–220.

Halmepuro, L., Vuontela, K., Kalimo, K., and Bjorksten, F. 1984. Cross-reactivity of IgE antibodies with allergens in birch pollen, fruits and vegetables. *Int. Arch. Allergy Appl. Immunol.* 74:235–240.

Harlander, S. K. 1991. Biotechnology—A means for improving our food supply. *Food Technol.* 45(4):84–95.

Herian, A. M., Taylor, S. L., and Bush, R. K. 1990. Identification of soybean allergens by immunoblotting with sera from soy-allergic adults. *Int Arch. Allergy Appl. Immunol.* 92:193–198.

Hirschwehr, R., Valenta, R., Ebner, C., Ferreira, F., Sperr, W. R., Valent, P., Rohac, M., Rumpold, H., Scheiner, O., and Kraft, D. 1992. Identification of common allergenic structures in hazel pollen and hazelnuts; A possible explanation for sensitivity to hazelnuts in patients allergic to tree pollens. *J. Allergy Clin. Immunol.* 90:927–936.

Hoffman, D. R. 1983. Immunochemical identification of the allergens in egg white. *J. Allergy Clin. Immunol.* 71:17–22.

Izumi, H., Adachi, T., Fujii, N., Matsuda, T., Nakamura, R., Tanaka, K., Urisu, A., and Kurosawa, Y. 1992. Nucleotide sequence of a cDNA clone encoding a major allergenic protein in rice seeds. Homology of the deduced amino acid sequence with member of α-amylase/trypsin inhibitor family. *FEBS Lett.* 302:213–216.

Johns, C. O., and Jones, D. B. 1916. The proteins of the peanut, *Arachis hypogaea*: I. The globulins arachin and conarachin. *J. Biol. Chem.* 28:770–787.

Johnson, G., and Elsayed, S. 1990. Antigenic and allergenic determinants of ovalbumin. III MHC Ia-binding peptide (OA 323–229) interacts with human and rabbit specific antibodies. *Mol. Immunol.* 27:821–827.

Kessler, D. A., Taylor, M. R., Maryanski, J. H., Flamm, E. L., and Kahl, L. S. 1992. The safety of foods developed by biotechnology. *Science* 256:1747–1749 and 1832.

King, T. P. 1994a. Antigenic determinants: B cells. Paper presented at Conference on Scientific Issues Related to Potential Allergenicity in Transgenic Food Crops at Annapolis, MD.

King, T. P. 1994b. T cell epitopes. In *Proceedings, conference on scientific issues related to potential allergenicity in transgenic food crops*. 275–290, FDA Docket No. 94N-0053. Washington D.C.: Food and Drug Administration.

King, T. P., Coscia, M. R., and Kochoumian, L. 1993. Structure-immunogenicity relationship of a peptide allergen, mellitin. In *Molecular biology and immunology of allergens*, eds. D. Kraft and A. Sehon, 11–20. Boca Raton, Florida: CRC Press, Inc.

King, T. P., Hoffman, D., Lowenstein, H., Marsh, D. G., Platts-Mills, T. A. E., Thomas, W. 1994. Allergen nomenclature. *Allergy Clin. Immunol. News* 6:38–44.

Langeland, T., and Harbitz, O. 1983. A clinical and immunological study of allergy to hen's egg. V. Purification and identification of a major allergen (antigen 22) in hen's egg white. *Allergy* 38:131–139.

Lavaud, F., Prevost, A., Cossart, C., Guerin, A., Bernard, J., and Kochman, S. 1995. Allergy to latex, avocado pear, and banana: Evidence for 30 kD antigen immunoblotting. *J. Allergy Clin. Immunol.* 95:557–564.

Lehrer, S. B. 1986. The complex nature of food antigens: Studies of cross-reacting crustacea allergens. *Ann. Allergy* 57:267–272.

Lehrer, S. B., Ibanez, M. D., McCants, M. L., Daul, C. B., and Morgan, J. E. 1990. Characterization of water soluble shrimp allergens released during boiling. *JACI* 85:1005–1013.

Lehrer, S. B., and McCants, M. L. 1988. Reactivity of IgE antibodies with crustacea and oyster allergens: Evidence for common antigenic structures. *J. Allergy Clin. Immunol.* 76:803–809.

Lehrer, S. B., and Reese, G. 1997. Recombinant proteins in newly developed foods: Identification of allergenic activity. *Int. Arch. Allergy Immunol.* 113:122–124.

Lehrer, S. B., and Salvaggio 1990. Allergens: Standardization and impact of biotechnology—A review. *Allergy Proc.* 11:197–208.

Lehrer, S. B., Reese, G., Ortega, H., El-Dahr, J. M., Goldberg, B., and Malo, J.-L. 1997. IgE antibody reactivity to aqueous-soluble, alcohol-soluble and transgenic corn proteins. *J. Allergy Clin. Immunol.* 99:S147.

Lehrer, S. B., Taylor, S. L., Hefle, S. L., and Bush, R. K. 1997. *Food allergens*. Vol. 2, *Allergy and Allergic Diseases*, ed. A. B. Kay, 961–980. Oxford: Blackwell Science Ltd.

Leung, P. S. C., Chu, K. H., Chaw, W. K., Aftab A., Bandea, C. I., Kwan, H. S. Nagy, S. M., and Gershwin, M. E. 1994. Cloning, expression, and primary structure of *Metapenaeus enis* tropomyosin, the major heat stable shrimp allergen. *J. Allergy Clin. Immunol.* 92:837–845.

Li-Chan, E., and Nakai, S. 1989. Biochemical basis for the properties of egg white. *Crit. Rev. Poultry Sci.* 2:21–58.

Marsh, D. G., Bias, W. B., and Ishizaka, K. 1974. Genetic control of basal serum IgE level and its effect on specific reaginic sensitivity. *Proc. Natl. Acad. Sci. U.S. A* 71:3588.

Matsuda, T., Alvarez, A. M., Tada, Y., Adachi, T., and Nakamura, R. 1993. Gene engineering for hypeallergenic rice: Repression of allergenic protein synthesis in seeds of transgenic rice plants by antisense RNA. In *Proceedings of the international workshop on life science in production and food-consumption of agricultural products session-4*. Tsukuba, Japan: Tsukuba Center.

Matsuda, T., and Nakamura, R. 1993. Molecular structure and immunologic properties of food allergens. *Trends Food Sci. Tech.* 4:289–293.

McCants, M., Lehrer, S. B., Reese, G., and Tracey, D. 1997. Allergy assessment of high oleic acid transgenic soy beans (G94-1). *J. Allergy Clin. Immunol.* 99:S479.

Meier-Davis, S., Taylor, S. L., Nordlee, J., and Bush, R. K. 1987. Identification of peanut allergens by immunoblotting (abstract). *J. Allergy Clin. Immunol.* 79:218.

Mena, M., Sanchez-Monge, R., Gomez, L., Salcedo, G., and Carbonero, P. 1992. A major barley allergen associated with Baker's asthma disease is a glycosylated monomeric inhibitor of insect α-amylase: cDNA cloning and chromosomal location of the gene. *Plant Mol. Biol.* 20:451–458.

Menendez-Arias, L., Moneo, I., Dominguez, J., and Rodriguez, R. 1988. Primary structure of the major allergen of yellow mustard (*Sinapis alba*) seed. Sin a I. Eur. *J. Biochem.* 177:159–166.

Monsalve, R. I. Gonzalez de la Pena, M. A., Menendez-Arias, L., Lopez-Otin, C., Billalba, M., and Rodriguez, R. 1993. Characterization of a new mustard allergen, Bra j IE. Detection of an allergenic epitope. *Biochem. J.* 293:625–632.

Moroz, L. A., and Yang, W. H. 1980. Kunitz soybean trypsin inhibitor, a specific allergen in food anaphylaxis. *N. Engl. J. Med.* 302:1126–1128.

Nordlee, J. A., Taylor, S. L., Townsend, J. A., Thomas, L. A., and Bush, R. K. 1996. Identification of a Brazil nut allergen in transgenic soybeans. *N. Engl. J. Med.* 334:688–692.

Ogawa, T., Bando, N., Tsuji, H., Okajima, H., Nishikawa, K., and Sasoka, K. 1991. Investigation of the IgE-binding proteins in soybeans by immunoblotting with the sera of soybean-sensitive patients with atopic dermatitis. *J. Nutr. Sci. Vitamin* 37:555–565.

Olempska-Beer, Z. S., Kuznesof, P. M. DiNovi, M., and Smith, M. J. 1993. Plant biotechnology and food safety. *Food Technol.* 47(12):64–72.

O'Neil, C. E., and Lehrer, S. B. 1991. Occupational reactions to food allergens. In *Food allergy: Adverse reactions to foods and additives*, D. D. Metcalfe, eds. H. A. Sampson and R. A. Simon, 207–235. Boston: Blackwell Scientific Publications.

Rackins, J. J., and Anderson, R. L. 1964. Isolation of soybean trypsin inhibitors by DEAE cellulose chromatography. *Biochem. Biophys. Res. Commun.* 15:230–235.

Reese, G., Tracey, D., Daul, C. B., and Lehrer, S. B. 1995. IgE and monoclonal antibody reactivities to the major shrimp allergen Pen a 1 and (tropomyosin) and vertebrate tropomyosins. Advances in Experimental Medicine and Biology. In *Proceedings of the international symposium on molecular biology of allergens and the atopic immune response.* In *New Horizons in Allergy Immunology*, eds. A. Sehon, K. T. Kayglass, and D. Kraft, 225–230. New York: Plenum Press.

Roitt, I. M., Brostoff, J., and Male, D. K. 1985. *Immunology.* London, New York: Gower Medical Publishing.

Romagnani, S. 1991. Human T_H1 and T_H2 subsets: Doubt no more. *Immunol. Today* 12:256.

Sachs, M. I., Jones, R. T., and Yunginger, J. W. 1981. Isolation and partial characterization of a major peanut allergen. *J. Allergy Clin. Immunol.* 67:27–34.

Sampson, H. A., and Metcalfe, D. D. 1991. Immediate reactions to foods. In *Food allergy: Adverse reactions to foods and food additives*, eds. D. D. Metcalfe, H. A. Sampson, and R. A. Simon, 99–112. Oxford: Blackwell Scientific Publications.

Savilathi, E. 1991. Cow's milk allergy. *Allergy* 36:73–88.

Savilathi, E., and Kuitunen, M. 1992. Allergenicity of cow milk proteins. *J. Pediatr.* 121:S12–S20.

Scherer, M. T., Chan, B. M., Smith, J. A., Perkins, D. L., and Gefter, M. L. 1989. Control of cellular and humoral immune responses by peptide containing T cell epitopes. *Cold Spring Harbor Symp. Quant. Biol.* 54:479–504.

Shanti, K. N., Martin, B. M., Nagpal, S., Metcalfe, D. D., and Sabba-Rao, P. V. 1993. Identification of tropomyosin as the major shrimp allergen and characterization of its IgE binding epitopes. *J. Immunol.* 151:5354–5363.

Shibasaki, M., Suzukim, S., Tajima, S., Nemoto, H., and Kuroume T. 1980. Allergenicity of major component proteins of soybeans. *Int. Arch. Allergy Appl. Immunol.* 61:441–448.

Stanley, J. S., King, N., Burks, A. W., Huang, S. K., Sampson, H., Cockrell, G., Helm, R. M., West, C. M., and Bannon, G. A. 1997. Identification and mutational analysis of the immunodominant IgE binding epitopes of the major peanut allergen Ara h 2. *Arch. Biochem. Biophys.* 342(2):244–253.

Tada, Y., Nakase, M., Adachi, T., Nakamura, R., Shimada, H., Takahashi, M., Fujimura, T., and Matsuda, T. 1996. Reduction of 14-16 kDa allergenic proteins in transgenic rice plants by antisense gene. *FEBS Lett.* 391:341–345.

Taylor, S. L., Lemanski, R. F., Bush, R. K., and Busse, W. W. 1987. Food allergens: Structure and immunologic properties. *Ann. Allergy* 59:93–99.

Taylor, S. L., Nordlee, J. A., Yunginger, J. W., Jones, J. T., Sachs, M. I., and Bush, R. K. 1982. Evidence for the existence of multiple allergens in peanuts (abstract). *J. Allergy Clin. Immunol.* 69:128.

Valenta, R., Duchene, M., Ebner, C., Valent, P., Aillaber, C., Deviller, P., Ferreira, F., Tejkl, M., Edelmann, H., Kraft, D., and Scheiner, O. 1992. Profilins constitute a novel family of functional plant pan-allergens. *J. Exp. Med.* 175:377–385.

Valenta, R., Duchene, M., Sperr, W. R., Valent, P., Vrtala, S., Hirschwehr, R., Ferreira, F., Kraft, D., and Scheiner, O. 1993. Profilin represents a novel plant pan-allergen. In *Molecular biology and immunology of allergens.* eds. D. Kraft and A. Sehon, 47–51. Boca Raton, Florida: CRC Press.

Walsh, B. J., Barnett, D., Barley, R. W., Elliott, C., Hill, D. J., and Howden, M. E. H. 1988. New allergens from hen's egg white and egg yolk. *Int. Arch. Allergy Appl. Immunol.* 37:81–86.

Walsh, B. J., Elliott, C., Baker, R. S., Barnett, D., Burley, R. W., Hill, D. J., and Howden, M. E. H. 1987. Allergenic cross-reactivity of egg-white and egg-yolk proteins. *Int. Arch. Allergy Appl. Immunol.* 84:228–232.

Watanabe, M. 1993. Hypoallergenic rice as a physiologically functional food. *Trends Food. Sci. Technol.* 4:125–128.

Wraith, D. C., Smilek, D. E., Mitchell, D. J., Steinman, L., and McDevitt, H. O. 1989. Antigen recognition in autoimmune encephalomyelitis and the potential for peptide-mediated immunotherapy. *Cell* 59:247–255.

Yunginger, J. W. 1990. Classical food allergens. *Allergy Proc.* 11:7–9.

Yunginger, J. W. 1991. Food antigens. In *Food allergy: Adverse reactions to foods and food additives*, eds. D. D. Metcalfe, H. A. Sampson, and R. A. Simon, 36–51. Boston: Blackwell Scientific Publications.

Yunginger, J. W., Sweeney, K. G., Sturner, W. Q., Giannandrea, L. A., Teigland, J. D., Bray, M., Benson, P. A., York, J. A., Biedrzycki, L., Squillace, D. L., and Helm, R. M. 1988. Fatal food-induced anaphylaxis. *J. Am. Med. Assoc.* 260:1450–1452.

Biotechnology and Safety Assessment, 2nd ed.
Edited by John A. Thomas
Copyright © 1998 Taylor & Francis

7

The Pharmacokinetics and Toxicity of Phosphorothioate Oligonucleotides

Arthur A. Levin, Richard S. Geary, Janet M. Leeds,
David K. Monteith, Rosie Yu, Mike V. Templin, and Scott P. Henry

Isis Pharmaceuticals, Inc., Carlsbad, California

THE SCIENCE OF ANTISENSE

Antisense therapeutic agents are now moving from the research laboratory to the clinic. The first generation of antisense drugs will be marketed before the new millennium. This promising new class of therapeutic agents will revolutionize the way that disease processes will be treated and perhaps even how we come to classify diseases. Antisense drugs are being used in a broad range of disease indications from viral infections, to cancer, to inflammatory diseases. The molecular targets range from viral transcription factors, and oncogenes, to cellular adhesion molecules. Many drugs are well beyond Phase I clinical trials, with the data on the safety and efficacy of these agents now being analyzed.

The first generation of antisense agents are modified from endogenous oligodeoxynucleotides by substituting a phosphorothioate linkage between deoxyribonucleotides for the phosphodiester linkage of DNA. A phosphorothioate linkage differs from the phosphodiester linkage by the substitution of a sulfur for one of the nonbridging oxygens in the phosphodiester bond (Fig. 1). This substitution markedly increases the nuclease resistance of the oligonucleotide (Hoke et al. 1991).

Antisense oligonucleotides function by selectively hybridizing with a targeted messenger RNA (mRNA). Hybridization with the target results in a reduction in the expression of the target gene either by destabilizing the mRNA, physically blocking translation, or altering mRNA processing (Crooke 1995). No matter what the exact mechanism of action, the end result is a selective knockdown of the expression of the targeted gene (Fig. 2). What makes this class of therapeutic agent unique is the high specificity for a single gene target even when that gene is part of a complex multigene family. This is a class of agents that makes it possible to inhibit specific isozymes or even specific gene variants such as mutated oncogenes (Monia et al. 1992; Dean and McKay 1994).

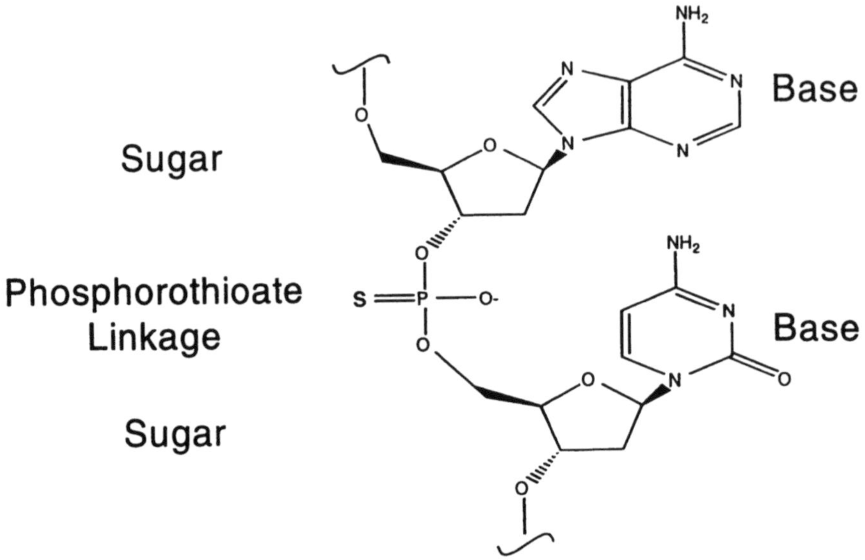

FIG. 1. The structure of a phosphorothioate oligonucleotide. These compounds differ from endogenous DNA in the sulfur in the linkage between nucleotides. Note that the substitution of the sulfur for the oxygen results in a chiral center and enhances the nuclease resistance.

Selectivity is defined by the well-known rules of Watson and Crick base-pairing. To design an antisense oligonucleotide, one simply needs the sequence of the target mRNA, the ability to make a series of different oligonucleotides complementary to different regions on the mRNA target, and the ability to test them for activity either in vivo or in vitro. Activity should include an assessment of the target gene expression (protein levels) or mRNA levels, or both (Bennett 1998). Making a series of oligonucleotides, each hybridizing to a different portion of the mRNA, ensures that one or more of the oligonucleotides will have activity. Only a fraction of the oligonucleotides that are complementary to the mRNA will be active antisense compounds (Chiang et al. 1991; Bennett 1994; Dean and McKay 1994; Bennett et al. 1996; Miraglia et al. 1996; Monia et al. 1996; Bennett et al. 1997) because not all parts of the mRNA are accessible to the antisense oligonucleotides. Thus, only some of the complementary sequences can hybridize to the target. This apparent fastidiousness is related to secondary structure of mRNA and perhaps even secondary structure of the antisense oligonucleotides. Like other macromolecules, mRNA has secondary structure, and it is this secondary structure that can limit access of an antisense molecule to its target (Lima et al. 1992; Crooke 1995). If the antisense oligonucleotide is blocked from hybridizing by secondary structure of the mRNA, the antisense activity is muted or absent. In practice, it is generally necessary to synthesize a few dozen oligonucleotides that span the length of the mRNA (i.e., 5'-UTR, coding region, and 3'-UTR) to find active compounds. All of the existing data suggest that simply making a single 20-mer oligonucleotide

How Antisense Works

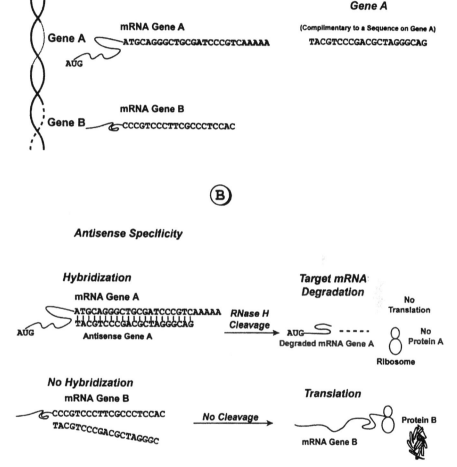

FIG. 2. How antisense oligonucleotides work. (**A**) Two different gene products have two different mRNAs. An antisense oligonucleotide complementary to a portion of Gene A is introduced. (**B**) Hybridization of the antisense for Gene A with the mRNA for Gene A. The antisense for Gene A does not hybridize with the mRNA for Gene B. The antisense-Gene A mRNA complex is a substrate for RNase H, and the message is degraded. No Protein A is translated. The mRNA for Gene B is unaffected by the presence of the antisense oligonucleotide and is therefore translated to Protein B.

to an arbitrary portion of a target has a low probability of producing an active compound. This phenomenon probably explains why some researchers fail to find antisense activity with one or two randomly selected oligonucleotides (Bennett 1998).

Like small molecules, it is possible and generally necessary to assess pharmacologic activity on the basis of dose-response relationships. It is also useful to think in terms of traditional pharmacological agents when describing antisense activity. Antisense compounds should behave like other compounds with potencies that reflect binding affinities. In this case, the binding affinity is the hybridization affinity with the mRNA target. This process is completely analogous to the familiar ligand–receptor kinetics so well characterized for small molecules.

By synthesizing a series of oligonucleotides with increasing numbers of mismatched bases, it is possible to relate binding affinity with pharmacologic activity. These types of studies are an effective tool for demonstrating sequence specificity. Because the highest hybridization affinities require perfect matches to the target sequences, intentionally including one or more mismatched bases results in an antisense molecule with reduced affinity for its mRNA target. Carefully planned in vivo or in vitro experiments can generate a series of dose–response curves that reflect the binding affinities (Anderson et al. 1996; Monia et al. 1996). Thus, in some ways it is possible to think of antisense oligonucleotides in terms very similar to those used for small molecules. However, in other aspects antisense oligonucleotides need to be considered as very different from traditional molecules. Two of those areas are pharmacokinetics and toxicity.

One of the most remarkable differences between small molecules and antisense oligonucleotides is that many of the toxicologic and pharmacokinetic properties are independent of sequence. As a result, an antisense oligonucleotide targeted to an oncogene will be very similar in toxicologic or pharmacokinetic profile to an oligonucleotide targeting a viral mRNA or an mRNA in the inflammatory cascade, if one assumes that the knockdown of the expression of the target gene is not deleterious. This property is highly advantageous to the development of these novel therapeutic agents because information gained with one molecule is applicable to all other oligonucleotides in that class (Levin et al. 1998). The reason for the similarity between oligonucleotides is that virtually all phosphorothioate oligonucleotides contain the same four nucleotides. The only distinctions between an anticancer drug and an antiviral drug are differences in sequence; that is, differences in the order of the nucleotides and perhaps small differences in abundance of some nucleotides. All oligonucleotides share common physical and chemical properties because they are so closely related. All phosphorothioate oligonucleotides are polyanions that have generally similar protein-binding affinities, and they are all metabolized by exonucleases in plasma and tissues at similar rates. Thus, in the discussion that follows, we will employ several different oligonucleotides as examples, but what is true for one is generally true for most. To date, only one significant biologic difference (other than pharmacologic activity) has been identified for different sequences, and that will be discussed in a section on the immunostimulatory effects of phosphorothioate oligonucleotides.

PHARMACOKINETICS OF PHOSPHOROTHIOATE OLIGONUCLEOTIDES

Pharmacokinetic considerations have directed the design of antisense drugs. Synthetic phosphodiester oligodeoxynucleotides would be attractive as antisense therapeutic agents because they would be metabolized by all of the normal nucleotide metabolic pathways, and metabolites would be innocuous. Phosphodiester DNA fragments cannot be used as antisense agents because endogenous nucleases rapidly degrade them. After intravenous administration, phosphodiester oligonucleotides are degraded and eliminated within minutes (Emelen and Mannik 1978; Wickstrom 1986; Goodchild and Kohli 1991; Sands et al. 1994; Srinivasan and Iversen 1995). Clearly, with rapid degradation like this it is not possible to deliver significant concentrations of oligonucleotides to sites of pharmacologic activity.

There is one area where there has been a successful use of phosphodiester oligonucleotides as antisense agents. Apparently, after direct injections into the brain, it is possible to demonstrate antisense activity of phosphodiester compounds directed toward genes that express neurotransmitter receptors (Chiasson et al. 1994; Pasternak and Standifer 1995). Why the central nervous system is different is not completely clear, but two factors come to mind; the first is that the drug is being delivered at the site of action and therefore yields high local concentrations, and the second factor is possibly that the brain is lacking nucleases. With this exception, phosphodiester oligonucleotides have been not useful as antisense agents.

By comparison, phosphorothioate oligonucleotides have an attractive pharmacokinetic profile in both animals and humans (Geary et al. 1997a; Glover et al. 1997; Leeds et al. 1997a). Currently, these compounds are being administered by intravenous infusions, subcutaneous injections, subcutaneous infusions, intradermal injections, and intravitreal injections. With the exception of intravitreal injections (Leeds et al. 1997b), extensive systemic exposure and broad tissue distribution occur with all of the these routes of administration. (In the future, with advances in medicinal chemistry, orally available antisense drugs will enter clinical trials.) In contrast to the rapid degradation of phosphodiester oligonucleotides, the first generation of phosphorothioate oligonucleotides are present in tissues for days after injection (Cossum et al. 1993; Cossum et al. 1994; Geary et al. 1997a; Leeds et al. 1997a). The kinetics of oligonucleotides in plasma are quite similar between animals and man (half-lives in the range of 30 to 60 min), and they scale from one species to the next on the basis of body weight, not surface area (Geary et al. 1997a). It is possible to show that, when dosed on the basis of body weight, the concentrations of oligonucleotides administered by a 2-h constant infusion are similar between humans and monkeys (Geary et al. 1997a; Leeds et al. 1997a). Thus, it has been possible to predict plasma concentrations in humans from nonclinical data. Because acute toxicities of this class of compound are related to peak plasma concentrations, the ability to predict plasma levels is critical for designing dose regimens.

Plasma clearance is a combination of nuclease degradation (metabolism) and distribution out of the plasma into tissues and, to a much lesser extent, elimination through urinary excretion. Although tissue distribution accounts for the majority of plasma clearance, phosphorothioate oligonucleotides are metabolized by exonucleases in plasma (Leeds et al. 1996; Cummins et al. 1997; Nicklin et al., n.d.). These enzymes cleave single nucleotides from the 3' or 5' termini to yield a shortened oligonucleotide and a mononucleotide. These cleavage events occur repeatedly and result in a progressively shortened oligonucleotide. Capillary gel electrophoresis of plasma samples demonstrates a series of peaks that comigrate with oligonucleotides shortened by one, two, three, or more nucleotides (Fig. 3).

Close examination of plasma concentrations and metabolite profiles from animal studies or clinical trials reveals that the kinetics of the appearance of metabolites in plasma are different than for small molecules. The metabolism of these phosphorothioate oligonucleotides, although appearing sequential on capillary gel electrophoresis, does not share the same kinetics as sequential metabolism of small molecules. With some compounds, sequential metabolism leads to the sequential appearance and disappearance of metabolites as metabolism progresses. For example, nitrobenze is reductively metabolized by gut flora to the nitroso, hydroxylamine, and then amine forms, and each metabolite rises and falls in sequence (Levin and Dent 1982). In contrast, with the sequential metabolites of phosphorothioate oligonucleotides, it seems that metabolism is rapid enough that metabolites shortened by one, two, or three nucleotides are present in plasma almost immediately after dosing (Glover et al. 1997; Geary et al. 1997b). Yet the sequential rise and fall of metabolites so typical in sequential metabolites are not apparent. Remarkably, despite the rapid appearance of a full complement of metabolites, after 4 h the parent compound is still the most abundant oligonucleotide species. Metabolism of oligonucleotides proceeds rapidly but does not completely deplete the parent compound. The rates of disappearance of parent compound and the major oligonucleotide metabolites are essentially parallel in mice, monkeys, and humans following intravenous administration (see human data in Fig. 4).

The disappearance of parent oligonucleotide from plasma does not proceed at a constant rate. In fact, the disappearance of parent compound seems to slow after an initial rapid burst of metabolic activity (Geary et al. 1997b). Plotting the rate of disappearance of parent oligonucleotide with time yields a biphasic curve with an initial steep slope indicating rapid disappearance and a second shallower slope indicative of slower metabolism (Fig. 5). One possible explanation for this slowdown relates to competition for binding and cleavage by exonucleases. When an oligonucleotide first enters the circulation, the full-length compound is the only substrate for exonucleases. However, after the initial round of cleavage events there is an ever-increasing pool of metabolites competing at the site of metabolism. Because each of the metabolites may compete equally well for binding to the active site, with each round of metabolism the probability that the parent compound will be cleaved diminishes. It is also possible that, once an oligonucleotide is bound to the active site of the exonuclease, there could be a sequential cleavage of that one molecule. Thus, the formation of oligonucleotides shortened by multiple cleavage events may actually be favored over the single-cleavage metabolite.

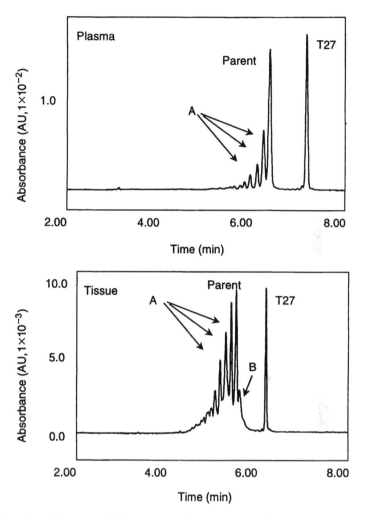

FIG. 3. An electropherogram of plasma and renal cortex extracts from a monkey treated with a phosphorothioate oligonucleotide at 3 mg/kg qod for 28 days. Plasma and tissue were extracted (Leeds et al. 1996) and run on CGE. The peak for the internal standard is labeled T27. The intact oligonucleotide is labeled "parent," and the arrows point out the metabolites that have been shortened by one, two, or three nucleotides (n-1, n-2, n-3, etc.). A metabolite with a slower migration time designated by the letter B is found in tissues. This metabolite is only found in tissues and may represent an oligonucleotide increased by one nucleotide (Cummins et al. 1997).

A second explanation for the apparent change in the rate of disappearance of parent compound is related to the stereochemistry of the oligonucleotides. Each of the phosphorothioate linkages has a chiral center. Thus, a 20-mer phosphorothioate oligonucleotide antisense drug is actually a mixture of 2^{19} stereoisomers. Nuclease resistance of the phosphorothioate linkage differs between the R and S forms in that

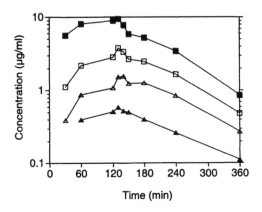

FIG. 4. Plasma profile in humans treated with a 2-h infusion of a 20-mer oligonucleotide at 2 mg/kg. The parent 20-mer (filled squares), the 19-mer (open squares), the 18-mer (open triangles), and the 17-mer (filled triangles) metabolites are depicted.

the latter is more nuclease resistant (Koziolkiewicz et al. 1997). The synthesis of phosphorothioate oligonucleotide is not stereoselective, which results in a nearly equal distribution of *R* and *S* stereoisomers at each linkage. This random distribution of nuclease-resistant and labile bonds may also serve to explain the apparent changes in the rate of disappearance of parent compound. The initial rapid rate of disappearance of parent phosphorothioate oligonucleotide may represent the nuclease cleavage

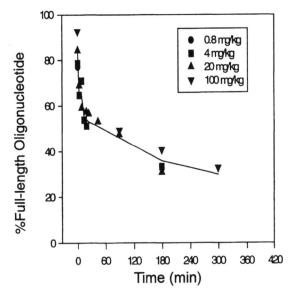

FIG. 5. The percent full-length oligonucleotide versus time in mice treated with various doses of a 20-mer phosphorothioate oligonucleotide. The line represents a best fit to the data. (From Geary et al. 1997b, Figure 7a. Used with permission of the publishers.)

of the more labile linkages. Cleavage at each of the linkages would proceed rapidly until the nuclease reaches a more resistant S linkage, at which time the rate of cleavage would decrease. Thus, over time there would tend to be an enrichment of oligonucleotides with S linkages at the 3' and 5' termini. More importantly, this stereospecific metabolism would result in metabolic kinetics similar to those observed in vivo. Pharmacokinetic models made by assuming a 50% faster rate of metabolism of one of the stereoisomers predict a metabolic pattern that closely mimics the patterns seen in vivo. The factors that control the pattern of metabolism remain to be determined, but it is clear that the metabolic profile of this class of compounds is distinctive.

Following injections or infusions of phosphorothioate oligonucleotides there is a rapid clearance of parent oligonucleotide as well as the chain-shortened metabolites (Fig. 4). Although metabolism is clearly playing a role in the reduction in plasma concentration, because the parent compound after 4 or more hours is often the most abundant oligonucleotide species indicates metabolism is not the major factor in plasma clearance. These data suggest that there may be some other factor driving the disappearance of oligonucleotide from plasma. Renal and fecal elimination of intact oligonucleotide is relatively limited. One of the factors that modulates urinary excretion is binding to plasma proteins. Oligonucleotides bind to plasma proteins with affinities in the micromolar to millimolar range (Leeds et al. 1994b). Binding to plasma proteins prevents the filtration of oligonucleotide in the kidney. When the binding affinity of the oligonucleotide or oligonucleotide metabolite decreases, as it would in metabolism, excretion in urine increases. This hypothesis is supported by data that show that when oligonucleotides are modified in such a way as to reduce protein binding, there is a resultant increase in urinary excretion (Crooke et al. 1996). Urinary excretion is also altered under conditions where protein binding is saturated. At high plasma concentrations, when plasma protein binding is saturated, there is an increase in the amount of intact oligonucleotide detected in urine (Fig. 6).

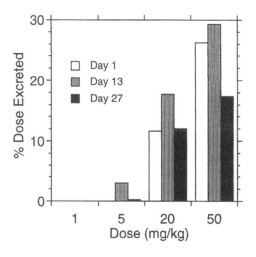

FIG. 6. The urinary excretion of intact oligonucleotide after intravenous administration of various doses to mice. Mice were administered a phosphorothioate oligonucleotide intravenously at the indicated dose every other day for the indicated number of days. Urinary excretion of pooled urine samples from groups of 3 mice were collected for 24 h after dosing. Urinary excretion was determined by capillary gel electrophoresis of urine extracts.

The absence of significant excretion suggests that the driving force in the plasma clearance of oligonucleotides is distribution to tissues (Agrawal et al. 1991; Cossum et al. 1993; Cossum et al. 1994; Agrawal et al. 1995; Crooke et al. 1996; Geary et al. 1997a; Geary et al. 1997b; Leeds et al. 1997a; Phillips et al. 1997; Nicklin et al. 1998). Oligonucleotide can be detected in tissues within minutes of administration (Cossum et al. 1993; Geary et al. 1997b; Phillips et al. 1997). As stated above, all phospho-rothioate oligonucleotides share a similar pharmacokinetic profile, and tissue and organ distribution is no exception. Phosphorothioate oligonucleotides are distributed to most tissues rapidly, but the organs that contain the highest concentrations are the liver and kidneys. As much as 40% of administered dose can be found in these organs alone, which supports the conclusion that tissue distribution is the primary determi-nant of plasma clearance. The spleen, bone marrow, and lymph nodes are also sites of significant distribution, but the concentrations attained by these tissues are much lower than the liver and kidney. After parenteral administration, phosphorothioate oligonucleotides can be found in nearly all tissues, but the brain, adipose tissue, and testes have very low levels of oligonucleotides (Agrawal et al. 1991; Cossum et al. 1993; Cossum et al. 1994; Geary et al. 1997b; Leeds et al. 1997a; Phillips et al. 1997). The half-lives of oligonucleotides in tissues range from 20 to more than 120 h, and the liver and kidneys have the longest half-lives (Cossum et al. 1993; Cossum et al. 1994; Levin et al. 1998).

The cellular distribution of phosphorothioate oligonucleotides has been studied us-ing autoradiography, immunohistochemistry, and fluorescent-tagged oligonucleotide (Plenat et al. 1994; Sands et al. 1994; Oberbauer et al. 1995; Rappaport et al. 1995; Rifai et al. 1996; Butler et al. 1997). Within organs there are distinct cell populations that accumulate oligonucleotide. In the liver for example, parenchymal cells and non-parenchymal cells can be shown to take up oligonucleotide, but it is the Kupffer cells that appear to have highest concentrations (Butler et al. 1997). In the kidney, oligonucleotide can be detected in both the medulla and the cortex, but cells within the proximal tubules accumulate the greatest concentrations. At the high doses used in repeat dose toxicity studies (10–100 mg/kg), Kupffer cells and proximal tubu-lar epithelium cells accumulate sufficient concentrations of oligonucleotide to result often in granules containing oligonucleotide visible within the cytoplasm (Sarmiento et al. 1994; Henry et al. 1997a; Henry et al. 1997c).

Like other more commonly used therapeutic agents, there is still more to be learned about the uptake of oligonucleotides into cells. Some cellular uptake results from the pinocytotic and phagocytic activity of cells. This mechanism may be playing a significant role in the uptake by Kupffer cells, and possibly even proximal tubule cells. Binding to the scavenger receptor may also be a mechanism of transmem-brane movement of oligonucleotide. Coadministration of scavenger receptor ligands and oligonucleotides causes an alteration in the disposition of the phosphorothioate oligonucleotides (Sawai et al. 1996; Kuipers et al. 1997). Another mechanism for cellular uptake may be related to transit across membranes by a shuttling mechanism whereby an oligonucleotide transits a membrane by shuttling from one membrane protein to another. Of all these mechanisms, the least is known about shuttling, but it may be the most important. Regardless of the mechanism, there is now clear and compelling evidence for the movement of oligonucleotides into cells.

Once inside the cell, oligonucleotide can bind the target mRNA and produce the desired biologic activity. The fate of the oligonucleotide once inside the cell is similar to that in plasma. There are exonucleases present within cells that begin cleaving nucleotides off the 3' and 5' termini. Note that in plasma the predominant exonuclease activity is at the 3' end. In tissues there is evidence for both 5' and 3' exonuclease activity (Gaus et al. 1997; Nicklin et al. 1998). Each cleavage event leads to the formation of two products. The first is an oligonucleotide shortened by a single nucleotide, and the second is a phosphorothioate mononucleotide. The shorter oligonucleotide can then serve as a substrate for further cleavage events, each liberating yet a shorter oligonucleotide and phosphorothioate mononucleotide (Fig. 7). This process continues until the oligonucleotide is completely degraded or until the shortened oligonucleotide is itself excreted. Extensively shortened oligonucleotides have reduced protein binding and are more likely to be excreted in the urine.

The phosphorothioate mononucleotides liberated by exonuclease cleavage are exactly like endogenous nucleotides except that a sulfur is substituted for one of the oxygens in the phosphate group. Although a complete analysis of the fate of these phosphorothioate mononucleotides has not been performed, there is sufficient data at this time to infer their ultimate fate. A scheme for the metabolism of phosphorothioate oligonucleotides is presented in Figure 7. A key piece of data was obtained by Cossum and coworkers (Cossum et al. 1993) when they performed a mass balance experiment with a phosphorothioate oligonucleotide radiolabeled with ^{14}C at the C-2 position of thymidine. Radiolabeling the oligonucleotide in this manner allowed for the fate of both the oligonucleotide and the thymidine residue to be followed. After intravenous injection, approximately 15% of the radiolabel was found in the urine in the form of low-molecular-weight components. The low-molecular-weight nature of the metabolites in urine supports the concept that metabolism to low-molecular-weight components is essential for urinary excretion. The low-molecular-weight metabolites in urine may represent the radiolabeled mononucleotide, nucleoside, or other small-molecular-weight metabolites. The same study demonstrated that a significant portion of the radiolabel was excreted in expired air in the form of carbon dioxide. The production of CO_2 from the thymidine in a phosphorothioate oligonucleotide is strong evidence that the mononucleotides are metabolized by the same pathways as the endogenous nucleotides. More than 50% of the nucleotides liberated by the nuclease cleavage of the oligonucleotide are catabolized to CO_2. Assuming the nucleotide follows the normal catabolic route, we can infer that the first step in nucleotide degradation is the removal of the thiophosphate, which leaves the nucleoside and liberates a thiophosphate ion. The thiophosphate would then be diluted in the vast compartment of phosphate pools or spontaneously oxidized to phosphate, thus liberating a sulfate. The nucleoside could then be a substrate for a nucleoside phosphorylase yielding the ribose sugar and the base that can ultimately be metabolized to CO_2. This pathway

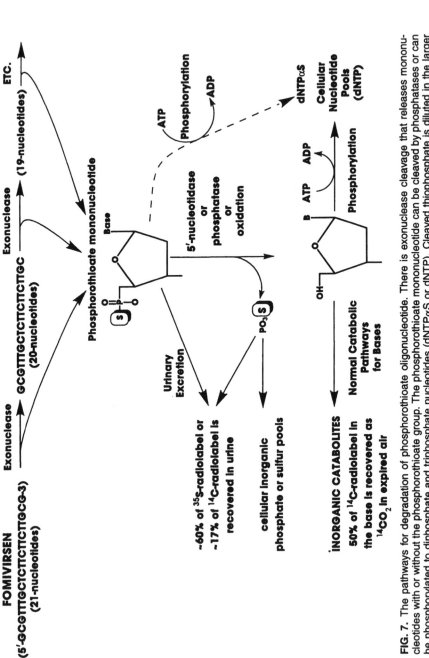

FIG. 7. The pathways for degradation of phosphorothioate oligonucleotide. There is exonuclease cleavage that releases mononucleotides with or without the phosphorothioate group. The phosphorothioate mononucleotide can be cleaved by phosphatases or can be phosphorylated to diphosphate and triphosphate nucleotides (dNTPαS or dNTP). Cleaved thiophosphate is diluted in the larger phosphate pools or excreted in the urine. At any time the phosphorothioate linkage can be oxidized to phosphate with the subsequent release of sulfate. The sulfate would be excreted or incorporated like endogenous sulfate.

suggests that for each $^{14}CO_2$ liberated from the thymine there should be a release of a thiophosphate moiety. Because more than half of the radiolabel is detected as CO_2, at least half of the molecule, including the sulfur, may be metabolized by normal pathways to yield innocuous inorganic metabolites. When a phosphorothioate oligonucleotide is radiolabeled with ^{35}S in the thioate linkage, more than half of the radiolabel is detected in urine as small-molecular-weight metabolites (Agrawal et al. 1991; Agrawal et al. 1995). These data are consistent with the ^{14}C data, and suggest that the ultimate fate of the sulfur would be in small-molecular-weight fractions like thiophosphates, sulfate, or mono- or di-deoxynucleotides.

There are other possible pathways for the thiophosphate mononucleotides released by exonuclease cleavage. One pathway is the oxidation of the thiophosphate moiety to phosphate-liberating sulfate. Oxidation of the phosphorothioate commonly occurs even under storage conditions. Alternatively, the thiophosphate mononucleotide could be dephosphorylated by the ubiquitous phosphatases, thus releasing thiophosphate whose fate was described above.

The last possible pathway for the thiophosphate mononucleotide could be into the nucleotide mono-, di-, and tri-phosphate pools. Thiophosphoate mononucleotides are substrates for various kinases and thus can be incorporated in these pools. Note that the thiophosphate would still be susceptible to oxidation to a phosphodiester in any of these states. As part of the triphosphate pool, the nucleotide thiotriphosphate (dNTPαS) could be incorporated in nucleic acids. However, the incorporation is stereospecific, and only one of the stereoisomers (the S form) of the dNTPαS is a substrate for the polymerases (Eckstein et al. 1972; Eckstein and Thomson 1995). There are at least two potential consequences of this incorporation. First, the thioate is oxidized to a phosphodiester, and sulfate is released. Second, the molecule is stable and hybridizes in a sequence-dependent manner with the appropriate complementary nucleotide on the opposite strand. Although we have not characterized this, if there is incorporation of oligonucleotide-derived material in DNA, it is apparent from extensive experience in vitro that these compounds bind to complementary nucleotides with high fidelity. It is also clear from assays of the genotoxicity with this class of compounds that, even under conditions whereby cells are incubated with 5000 μg/ml concentrations of the phosphorothioate oligonucleotide, there is no evidence for genotoxicity. Because the in vitro cytotoxicity of these compounds is low (Crooke et al. 1992; Azad et al. 1993; Crooke 1993), it is highly unlikely that the nucleotide thiotriphosphates are interfering with energy storage or production by altering adenosine triphosphate (ATP) levels or regulation.

This proposed scheme describing the fate of all of the possible metabolites suggests that the products of metabolism are innocuous small-molecular-weight metabolites or are endogenous molecules. This information is critical because metabolism is the major mechanism of clearance of phosphorothioate oligonucleotides. If these pathways are in fact the actual routes of metabolism (and there is no reason to think that they are not), it is not surprising that these compounds have a highly acceptable toxicity profile.

TOXICITY OF PHOSPHOROTHIOATE OLIGONUCLEOTIDES

Phosphorothioate oligonucleotides have been administered to patients in Phase I and Phase II clinical trials that extend out to 1 year in cancer trials, and months of treatment have been administered to patients in clinical trials for other indications. The side-effect and adverse-event profiles are consistent with existing nonclinical data in mice, rats, rabbits, dogs, and monkeys. The scope of the nonclinical evaluations that have been performed with phosphorothioate oligonucleotides includes acute, chronic, and subchronic administration, segment 1 and 2 reproductive toxicity studies, and a full battery of genotoxicity assays. The remainder of this chapter will address the results of these studies and the relationship between exposure and toxicity.

To explore the toxicity of antisense oligonucleotides fully, it is absolutely essential to use sequences that have activity in the animal models. Because the sequences of target genes are different in the various species, to characterize the toxicity (if any) associated with the inhibition of the target gene expression, an oligonucleotide active in the test species should be used whenever possible. At Isis Pharmaceuticals we routinely test both the human sequence and the rodent sequence in toxicity studies (Bennett et al. 1997; Henry et al. 1997e). For primates we have generally observed that there is complete or sufficient homology to have hybridization. In rodents, comparing the toxicities associated with the human sequence (inactive) with the rodent sequence (active) allows us to understand which effects are related to the class of compound and which toxicities are related to the knockdown of the specific genes.

Phosphorothioate oligonucleotides can be envisioned to produce toxicity by several different mechanisms. The most obvious mechanism is a toxicity that is related to the knockdown expression of the target protein: exaggerated pharmacologic activity. Judicious selection of antisense targets is the best way to avoid toxicities related to the intended activity of the drug. Antisense therapeutic agents under development at this time hybridize with viral genes, oncogenes, or genes that are part of multigene families. Reductions in the expression of these genes should have little or no consequence on normal cellular function. Most researchers have avoided selecting protein targets from among the proteins that are necessary for cell survival. However, Sadler and coworkers (Sadler et al. 1995; Augustine et al. 1995) have used antisense oligonucleotides in vitro to knockdown the expression of genes crucial for embryonic development and have in the process produced developmental abnormalities.

A second potential mechanism of toxicity is the unintended hybridization of the antisense oligonucleotide with an inappropriate target. It is statistically unlikely there would be a perfect match in the genome for a 15- to 20-mer oligonucleotide, which suggests the chance of inappropriate hybridization is remote. Note also that with phosphorothioate oligonucleotides, hybridization affinity falls off markedly with sequence mismatches (Anderson et al. 1996; Monia et al. 1992; Monia et al. 1996), and thus for significant pharmacologic activity there needs to be a perfect match. It is also important to realize that not every oligonucleotide, even those with perfect matching sequences, will inhibit gene expression. As with any other therapeutic agents there are strong agonists and weak agonists. Because of secondary structure of the receptor, mRNA, not all oligonucleotides can hybridize with all regions (Lima et al. 1992; Lima et al. 1997a; Lima et al. 1997b). Like other receptor-binding phenomena, there are steric restrictions preventing the agonist (the antisense construct) from having access to the binding site. Thus, hybridization to an inappropriate mRNA target is disfavored because of statistical improbability of finding a perfect match and the rare chance that any random match would be accessible for hybridization.

Toxicity could also result from nonspecific interactions with cellular or extracellular components like proteins (Henry et al. 1997b; Henry et al. 1997d). Phosphorothioate oligonucleotides are polyanionic molecules that are known to bind to plasma proteins and may also bind to intracellular protein components. Toxicities that result from protein binding are less likely to be related to the specific sequence and more likely to be related to the chemical class. In fact there are several toxicities that have been observed in nonclinical studies that are virtually identical from sequence to sequence, irrespective of the intended target (reviewed in Henry et al. 1997c and Levin et al. 1998). In the section on the systemic toxicities of phosphorothioate oligonucleotides we will show that, although toxicities are nearly identical from sequence to sequence, there are markedly different toxicities that vary by species.

Systemic Toxicities of Phosphorothioate Oligonucleotides

Genotoxicity

At least five phosphorothioate oligonucleotides have been examined in a range of genotoxicity assays (Table 1). These assays were performed according to the appropriate guidelines with the appropriate positive controls run side-by-side with the

TABLE 1. *Summary of genetic toxicity assays for the various phosphorothioate oligodeoxynucleotides*

Assay	ISIS 2105	ISIS 2302	ISIS 2922 (fomivirsen)	ISIS 3521	ISIS 5132
Bacterial mutagenesis					
Salmonella	Neg[a]	Neg	Neg	Neg	Neg
Escherichia coli		Neg	Neg	Neg	Neg
In vitro assays					
In vitro cytogenetics	Neg	Neg	Neg		
Mouse lymphoma mammalian mutagenesis		Neg	Neg		
CHO/HGPRT mammalian mutagenesis	Neg				
In vitro rat hepatocyte UDS	Neg				
In vivo assay					
In vivo mouse micronucleus	Neg		Neg		

[a]Negative results. Positive controls had the appropriate response in all cases.

oligonucleotide. In vitro assays were performed at limit concentrations of 5000 μg/ml and in vivo micronucleus assays were performed at 80% of the LD_{50}. All of the assays, both in vivo and in vitro, have been negative, suggesting that phosphorothioate oligonucleotides are nongenotoxic. This class of compounds is metabolized to endogenous nucleotides and other small water-soluble metabolites like sulfate and thiophosphate. It is unlikely that these compounds or endogenous nucleosides or nucleotides would accumulate at concentrations sufficient to produce genotoxicity. Other mechanisms of genotoxicity could be envisioned like homologous recombination. However, the oligonucleotide itself is single-stranded and would not be expected to participate in homologous recombination events. From both theoretical considerations and the collection of data on 5 compounds in 16 assays, it appears that phosphorothioate oligonucleotides are not genotoxic.

Cytotoxicity

In vitro studies demonstrate that this class of compounds does not have a high cytotoxic potential. Data from in vitro genotoxicity studies suggest that concentrations of up to 5000 μg/ml produce some cytotoxicity, as indicated by decreased proliferation rates. Biochemical indices of cytotoxicity have been examined in detail, and at concentrations of 100 μM there are reductions in amino acid incorporation in HeLa cells (Crooke et al. 1992). Significant alterations in the cell viability as measured by biochemical assays only occur at concentrations above 200 μM (Azad et al. 1993). If one assumes a molecular weight of 7000 for a phosphorothioate, these concentrations translate into the range of 1000 μg/g of cells—levels that are well beyond those predicted to occur in clinical settings and are similar to concentrations observed in the liver and kidneys at toxic dose levels.

Acute Toxicity

Acute toxicity studies in mice have been performed as dose range finding studies for the mouse micronucleus test. Single-dose LD_{50}s have been determined for three different sequences and range from 720 to >1000 mg/kg (Levin et al. 1998). The exact cause of lethality was not determined.

In monkeys, acute toxicities have been extensively characterized. Rapid infusions of phosphorothioate oligonucleotides (10 mg/kg infused over 10 minutes) produce a series of physiologic responses that under some circumstances can lead to cardiovascular collapse and death (Cornish et al. 1993; Galbraith et al. 1994; Henry et al. 1997b; Henry et al. 1997c). These hemodynamic changes are considered to be secondary to activation of the complement cascade. Complement activation in these animals is indicated by the release of C3a, C5a, and Bb, but not C4a. This pattern of complement split products is characteristic of activation of complement through the alternative pathway, not the classical pathway (Henry et al. 1997b). The biochemical event that causes complement activation has not been determined unambiguously, but there are some working hypotheses. One hypothesis is that the oligonucleotide, because of its polyanionic nature, binds to one or more of the factors that normally deactivate the low level of constitutive activation of the alternative pathway, which allows the constitutive activity to trigger the cascade. Another hypothesis is that binding of the polyanionic oligonucleotide with one of the deactivating factors results in sequestration or loss of the deactivating factors, which leads to a triggering of the cascade (Henry et al. 1997b). The activation of complement is transient and associated with peak plasma concentrations of the oligonucleotide. Thus, when plasma levels fall after injections or infusions there is a diminished potential for complement activation. Repeated administration of phosphorothioate oligonucleotides does not alter the sensitivity of the monkeys to this effect.

The split products released by complement activation are bioactive compounds, like the anaphylatoxins C3a and C5a, which can produce profound cardiovascular events, such as hypotension and circulatory collapse. Because of the potentially life-threatening consequence of complement activation, the relationship between plasma concentrations and complement activation has been extensively characterized. It has been possible to establish a clear relationship between plasma concentrations and complement activation (Henry et al. 1997b). Correlating plasma concentrations produced by infusions of different doses at different dose rates with the appearance of complement split products demonstrates that complement activation is a threshold phenomenon. When one plots plasma concentrations of oligonucleotide versus the appearance of the complement split product Bb, a threshold is apparent. Complement is only activated when plasma concentrations exceed a threshold of 40–50 μg/ml. This threshold is similar for all 20-mer phosphorothioates we have examined (Henry et al. 1997c; Levin et al. 1998). Below this threshold there is little or no complement activation. In general, plasma concentrations in clinical studies are in the range of 10 to 25 μg/ml. Note that plasma concentrations in existing clinical trials are maintained well below the threshold values for complement activation, and thus, complement activation has not been observed in clinical studies performed to date.

In addition to complement activation, there is also a prolongation of activated partial thromboplastin time (APTT) (Henry et al. 1997a; Henry et al. 1997c; Henry et al. 1997b). The prolongation of APTT is generally not associated with prolongation of prothrombin time (PT) and has not been associated with significant hemorrhage. The prolongation of APTT is transient, and the inhibitory effects are directly proportional to plasma concentrations of oligonucleotide. Within a short period after the administration of oligonucleotide, APTT returns to normal limits as plasma is cleared of oligonucleotide. In clinical trials, the magnitude of the increase in APTT at the end of a 2-h infusion is in the range of a 1-s increase per μg/mL in plasma. Thus, in patients with plasma concentrations in the 8–15 μg/ml range, there would be an 8- to 15-s increase in APTT that would rapidly reverse as plasma was cleared of oligonucleotide (Glover et al. 1997). In monkey studies and in clinical trials, the rise and fall of APTT are nearly identical on the first day of dosing as well as on the last, which indicates no cumulative effects of treatment. These small increases in APTT that rapidly reverse are not generally considered clinically significant. In animal studies, there have been observations of increased superficial bruises at the dose of 50 mg/kg every other day for a month, but no evidence of clinically significant hemorrhage (Henry et al. 1997a). Bruising has not been observed in clinical trials with these drugs.

Subchronic Toxicities of Phosphorothioates

As was observed with the acute toxicities, there are also marked species differences in the subchronic toxicities. Treatment of rodents with phosphorothioate oligonucleotides results in a series of changes, including diffuse mononuclear cell infiltrations in various organs, splenomegaly, and lymphoid hyperplasia (Sarmiento et al. 1994; Henry et al. 1997c; Henry et al. 1997e; Monteith et al. 1997). Together this pattern of changes is generally regarded as immunostimulation. Primates are considered less sensitive to immune stimulation, for they demonstrate only mild lymphoid hyperplasia (Henry et al. 1997a); immune stimulation has not been noted in clinical trials.

After subchronic administration of oligonucleotide to rodents, mononuclear or histiocytic infiltrates can be observed in numerous organs and tissues, including the liver, spleen, heart, pancreas, and salivary glands (Sarmiento et al. 1994; Henry et al. 1997c; Henry et al. 1997e; Monteith et al. 1997). The effect is dose-related, and these changes are much more apparent at doses that exceed 20 mg/kg; however, even at lower doses it is possible to demonstrate immunostimulation with some sequences. The histiocytic infiltrates occur in the absence of fibrotic changes, and if treatment is discontinued the infiltrates gradually reverse (Henry et al. 1997e). The splenomegaly that occurs is partially the result of increases in the number of lymphocytes in the B-cell areas (Sarmiento et al. 1994; Henry et al. 1997c; Henry et al. 1997e; Monteith et al. 1997) and partially the effect of increased extramedullary hematopoiesis. There does not appear to be a proliferative effect in T-cell areas.

In vitro studies confirm the sensitivity of B-cells to the proliferative effects of phosphorothioate oligonucleotides. Treatment of B-cells with various phosphorothioate oligonucleotides in vitro results in a marked increase in the incorporation of tritiated thymidine. The increase in thymidine incorporation with some oligonucleotide sequences exceeds the stimulatory effects of lipopolysaccharides. Although all phosphorothioate oligonucleotides produce this effect to some extent, the severity of immunostimulatory effects is sequence-dependent. However, the sequence-related effects are not the result of hybridization. A series of investigations by several different laboratories have demonstrated that there are specific sequence motifs that are particularly immunostimulatory. The sequence AACGTT has been reported to be one of the more immunostimulatory motifs (Yamamoto et al. 1994; Krieg et al. 1995; Krieg et al. 1996). Alterations in this motif, particularly a change in the CG, results in reduced potency. It is clear from in vivo studies that the CG was not the only sequence motif with immunostimulatory potential because there are examples of oligonucleotides without CG sequences that show significant immunostimulatory activity (Monteith et al. 1997).

In vivo investigations have demonstrated that the immunostimulatory effects of phosphorothioate oligonucleotides are polyclonal. The increase in B-cell numbers results in a general expansion of all clones, and there is a subsequent increase in total antibody production (Branda et al. 1993; Branda et al. 1996; Liang et al. 1996). Note that analyses of plasma samples from laboratory animals and humans treated with repeated administration of phosphorothioate oligonucleotides demonstrates that there are no antibodies directed against the phosphorothioate oligonucleotide (Levin et al. 1998). In fact these compounds are very weak antigens and need to be conjugated to a hapten to elicit an immune response. Recent in vivo studies have demonstrated that there are increases in cytokine production associated with the administration of these compounds (Yi et al. 1996). Circulating concentrations of IL-6, IL-12, MCP-1, and MIP-1β were increased up to thirtyfold over control values for 24 to 48 h after a single administration of a highly potent immunostimulatory oligonucleotide. The cytokines IL-2, IL-4, IL-10, and IFN-γ were not increased over control values (Zhao et al. 1997). The time course of the changes in the circulating levels of cytokines is consistent with the time course of immunostimulation observed in vivo in which immunostimulatory effects are observed for some time even after a single high dose in rodents.

The clinical relevance of immunostimulation is questionable because it is not a prominent response in primates treated with phosphorothioate oligonucleotides (Henry et al. 1997a; Henry et al. 1997c), nor has it been observed in clinical trials to date.

Subchronic Toxicities and Hematologic Changes

Treatment of animals with phosphorothioate oligonucleotide results in some hematologic changes. Rodents treated with phosphorothioate oligonucleotides have dose-related increases in monocytes and other white-blood-cell types (Sarmiento et al. 1994; Henry et al. 1997e). These findings are predictable based on the increased cytokine concentrations and the increased numbers of histiocytes described above. Also consistent with the immune stimulation described in rodents and the increase in spleen size and activity of the reticular endothelium system is a reduction in platelet numbers in rodents. This reduction has a similar time-course and dose-response relationship to the splenomegaly. Reductions in, or alterations of, the megakaryocyte populations in mice have been reported in mice treated with very high doses of phosphorothioate oligonucleotide (Sarmiento et al. 1994), but in other studies it is possible to detect alterations in circulating platelets in the absence of marrow effects (Henry et al. 1997e and unpublished observations). It appears that the major factor in the thrombocytopenia in rodents is an increase in destruction, not necessarily an effect on production.

In primates infused with phosphorothioate oligonucleotides there are occasional observations of transiently reduced platelet numbers during the infusion. This reduction rapidly reverses as plasma is cleared of oligonucleotide. The incidence of this is sporadic, but the reductions in circulating platelets during infusion can be impressive (up to 60%) in some individuals. Within hours after the end of infusion there is recovery, which suggests that the effect is independent of production, and probably destruction, but may be related to sequestration (Templin and Levin, unpublished observations).

Subchronic Toxicities in Organs and Tissues

Toxicities associated with repeated administration of phosphorothioate oligonucleotides were correlated with organs containing the highest concentrations. The only significant histologic lesions observed in the connective tissue, the central nervous system, the cardiovascular system, the gastrointestinal system, the integument, the reproductive tracts, and the respiratory system were mononuclear or histiocytic cell infiltrates. These changes were only observed in rodents and attributed to an immuno-stimulatory effect of phosphorothioate oligonucleotides (reviewed in Henry et al. 1997c and Levin et al. 1998).

The sites of highest accumulation of oligonucleotide, the liver and the kidneys, are potential targets for toxicity. In rodents, the liver appears to be a target organ for phosphorothioate oligonucleotide toxicity. Repeated administration of oligonucleotides at doses greater than 20 mg/kg qod produce changes in hepatic morphology and functional changes. As dose is increased above 20 mg/kg, there is a greater incidence of hepatocellular changes. These changes are generally characterized as Kupffer-cell hypertrophy and occasional single-cell necrosis that is often accompanied by increases in serum transaminases and reductions in albumin and cholesterol. Slight increases in serum transaminases have occasionally been observed in primates but in the absence of hepatocellular changes. Note that for both rodents and primates treatment with phosphorothioate oligonucleotides results in increases in Kupffer-cell granulation (Sarmiento et al. 1994; Henry et al. 1997e). This specific hepatocellular effect is related to the uptake of oligonucleotide in this cell type (Butler et al. 1997).

The kidney is the site of the highest concentration of oligonucleotide in rodents and monkeys (Geary et al. 1997b; Philips et al. 1997; Leeds 1997a). Histologic examination of the kidney cortex demonstrates that in some toxicity studies the uptake of oligonucleotide is sufficient to make it possible to visualize granules containing oligonucleotide in the proximal tubular epithelium (Butler et al. 1997; Henry et al. 1997a). In primates, doses of 10 mg/kg administered every other day for a month produce a mild and focal change in the height of the proximal tubular epithelium and occasional cells with active-looking nuclei. Together the effects have been described as a minimal atrophic–regenerative change, but this change is not associated with any alterations in renal functional assays (Horner et al. 1997). At doses greater than 50 mg/kg, tubular degeneration and tubular hemorrhage occur, but these doses are manyfold greater than those used clinically. The concentrations of oligonucleotide associated with the minimal and focal morphologic changes in the kidney are generally greater than 1000 μg/g of tissue. Concentrations of total oligonucleotide in the renal cortex of primates dosed with clinically relevant doses are less than 500 μg/g

(Geary et al. 1997a; Levin et al., 1998). Because pharmacokinetic parameters determined in monkeys have been predictive of kinetics seen in patients, the rate of renal clearance derived from monkey studies was used to predict the concentrations of oligonucleotide in the renal cortex of humans treated with various dose regimens. On the assumption of a dose regimen of 2 mg/kg 3 times per week, these predictions yield renal concentrations that are lower than the concentrations associated with even the minimal changes noted in toxicity studies.

Fertility and Reproductive Toxicity

In subchronic and chronic toxicity studies, there have been no observations of testicular effects following treatment with phosphorothioate oligonucleotides. This finding is supported by pharmacokinetic data indicating that little or no oligonucleotide reaches the testes. In female rodents treated with phosphorothioate oligonucleotides at doses that produce immunostimulation (>20 mg/kg) there are increases in mononuclear cell infiltrates in the uterus and ovaries, but in Segment 1 reproductive toxicity assays there has been no indication of reduced fertility or reproduction performance. In mice, administration of an antisense oligonucleotide directed against a developmentally significant gene, *Sry*, yielded no abnormal pups or alteration in male–female birth ratio (Gaudette et al. 1993). Full Segment-2 teratogenicity studies in mice have been negative with an antisense oligonucleotide against human ICAM 1 and an antisense oligonucleotide directed against the mouse ICAM gene (Templin, Henry, and Levin, unpublished observations). Studies are in progress to define the transplacental kinetics of this class of drugs. Exposure of the pregnant dam to antisense oligonucleotides, however, has not been associated with direct teratogenic effects. In contrast, direct injection of antisense oligonucleotides into developmentally significant genes in embryo culture has demonstrated sequence-specific effects (Augustine et al. 1995; Sadler et al. 1995; Bavik et al. 1996), but in vivo maternal exposure to oligonucleotides has not been associated with a specific reduction in an embryonic gene expression and terata. Conclusive studies to this effect have yet to be reported.

The role of possible reductions in target gene expression in the embryo and the role that knockdown of expression of maternal genes in the outcome of pregnancies makes it necessary to perform teratogy testing for antisense oligonucleotides directed at mammalian genes. When an antisense oligonucleotide is directed toward a viral gene that does not exist in the human or rodent genome, it is unlikely that there will be anything but class-related toxicities. Clearly the developmental effects of inhibition of target protein expression will have to be examined for each antisense oligonucleotide for each target mRNA.

Carcinogenicity Studies

As of this writing, no 2-year carcinogenicity studies have been performed with phosphorothioate oligonucleotides. It is already apparent that if these studies are performed their interpretation will be complex because of the underlying proliferative response of the immune system to phosphorothioate oligonucleotides in rodents. Given underlying differences in responses of rodents and primates, the relevance of any findings in rodents will have to be considered carefully.

CONCLUSIONS

Although the studies performed to characterize the toxicity and pharmacokinetics of phosphorothioate oligonucleotides are typical of other pharmaceutical agents, there are several technical aspects related to the antisense mechanism or chemical class that are distinct from either small molecules or other biotechnology products. Because of the relative ease for identifying active oligonucleotides, these are among the few biotechnology products for which it is possible to isolate and synthesize species-specific analogs easily. In this way, it is much easier to characterize the biologic effects of an active form of antisense oligonucleotides than it is to clone, grow, and purify most protein-based biotechnology products, many of which are also highly species-specific.

Antisense oligonucleotides have the distinct advantage of having well-defined class effects that differ little from sequence to sequence. The pharmacokinetic profile is likewise independent of sequence and is remarkably similar from species to species. This cross-species predictability taken together with the ability to study active oligonucleotides in the species of interest are significant advantages to the development of these compounds as therapeutic agents.

ACKNOWLEDGMENTS

The authors would like to thank John Barnes for his extraordinary efforts in preparing and reviewing this manuscript. Without his efforts this chapter would not have been possible.

REFERENCES

Agrawal, S., Temsamani, J., Galbraith, W., and Tang, J.Y. 1995. Pharmacokinetics of antisense oligonucleotides. *Drug Disposition* 28:7–16.

Agrawal, S., Temsamani, J., and Tang, J. Y. 1991. Pharmacokinetics, biodistribution, and stability of oligodeoxynucleotide phosphorothioates in mice. *Proc. Natl. Acad. Sci. U.S.A.* 88:7595–7599.

Anderson, K. P., Fox, M. C., Brown-Driver, V., Martin, M. J., and Azad, R. F. 1996. Inhibition of human cytomegalovirus immediate-early gene expression by an antisense oligonucleotide complementary to immediate-early RNA. *Antimicrob. Agents Chemother.* 40(9):2004–2011.

Augustine, K. A., Liu, E. T., and Sadler, T. W. 1995. Antisense inhibition of engrailed genes in mouse embryos reveals roles for these genes in craniofacial and neural tube development. *Teratol.* 51:300–310.

Azad, R., Driver, V. B., Tanaka, K., Crooke, R. M., and Anderson, K. P. 1993. Antiviral activity of a phosphorothioate oligonucleotide complementary to RNA of the human cytomegalovirus major immediate-early region. *Antimicrob. Agents Chemother.* 37:1945–1954.

Bavik, C., Ward, S. J., and Chambon, P. 1996. Developmental abnormalities in cultured mouse embryos deprived of retinoic acid by inhibition of yolk-sac retinol binding protein synthesis. *Proc. Natl. Acad. Sci. U.S.A.* 93:3110–3114.

Bennett, C. F. 1994. Inhibition of cell adhesion molecule expression with antisense oligonucleotides: Activity in-vitro and in-vivo. *Clin. Chem.* (Winston-Salem, N.C.) 40:644–645.

Bennett, C. F. 1998. Antisense oligonucleotides: Is the glass half full or half empty? *Biochem. Pharmacol.* 55:9–19.

Bennett, C. F., Dean, N., Ecker, D. J., and Monia, B. P. 1996. Pharmacology of antisense therapeutic agents—Cancer and inflammation. In *Methods in Molecular Medicine: Antisense Therapeutics*, ed. S. Agrawal. Totowa, NJ: Humana Press, 13–46.

Bennett, C. F., Kornbrust, D., Henry, S., Stecker, K., Howard, R., Cooper, S., Dutson, S., Hall, W., and Jacoby, H. I. 1997. An ICAM-1 antisense oligonucleotide prevents and reverses dextran sulfate sodium-induced colitis in mice. *J. Pharmacol. Exp. Ther.* 280:988–1000.

Branda, R. F., Moore, A. L., Lafayette, A. R., Mathews, L., Hong, R., Zon, G., Brown, T., and McCormack, J. J. 1996. Amplification of antibody production by phosphorothioate oligodeoxynucleotides. *J. Lab. Clin. Med.* 128:329–338.

Branda, R. F., Moore, A. L., Mathews, L., McCormack, J. J., and Zon, G. 1993. Immune stimulation by an antisense oligomer complementary to the rev gene of HIV-1. *Biochem. Pharmacol.* 45:2037–2043.

Butler, M., Stecker, K., and Bennett, C. F. 1997. Cellular distribution of phosphorothioate oligodeoxynucleotides in normal rodent tissues. *Lab. Invest.* 77:379–388.

Chiang, M.-Y., Chan, H., Zounes, M. A., Freier, S. M., Lima, W. F., and Bennett, C. F. 1991. Antisense oligonucleotides inhibit intercellular adhesion molecule 1 expression by two distinct mechanisms. *J. Biol. Chem.* 266:18162–18171.

Chiasson, B. J., Armstrong, J. N., Hooper, M. L., Murphy, P. R., and Robertson, H. A. 1994. The application of antisense oligonucleotide technology to the brain. *Cell. Mol. Neurobiol.* 14:507–521.

Cornish, K. G., Iversen, P., Smith, L., Arneson, M., and Bayever, E. 1993. Cardiovascular effects of a phosphorothioate oligonucleotide with sequence antisense to p53 in the conscious rhesus monkey. *Pharmacol. Commun.* 3:239–247.

Cossum, P. A., Sasmor, H., Dellinger, D., Troung, L., Cummins, L., Owens, S. R., Markham, P. M., Shea, J. P., and Crooke, S. 1993. Disposition of the ^{14}C-labeled phosphorothioate oligonucleotide ISIS 2105 after intravenous administration to rats. *J. Pharmacol. Exp. Ther.* 267:1181–1190.

Cossum, P. A., Troung, L., Owens, S. R., Markham, P. M., Shea, J. P., and Crooke, S. T. 1994. Pharmacokinetics of a ^{14}C-labeled phosphorothioate oligonucleotide, ISIS 2105, after intradermal administration to rats. *J. Pharmacol. Exp. Ther.* 269:89–94.

Crooke, R. M. 1993. In vitro and in vivo toxicology of first generation analogs. In *Antisense research and applications*, eds. S. T. Crooke and B. Lebleu, 471–492. Boca Raton, Florida: CRC Press.

Crooke, R. M., Hoke, G. D., and Shoemaker, J. E. E. 1992. In vitro toxicological evaluation of ISIS 1082, a phosphorothioate oligonucleotide inhibitor of herpes simplex virus. *Antimicrob. Agents Chemother.* 36:527–532.

Crooke, S. T. 1995. *Therapeutic applications of oligonucleotides.* Austin, Texas: R. G. Landes Company.

Crooke, S. T., Graham, M. J., Zuckerman, J. E., Brooks, D., Conklin, B. S., Cummins, L. L., Greig, M. J., Guinosso, C. J., Kornbrust, D., Manoharan, M., Sasmor, H. M., Schleich, T., Tivel, K. L., and Griffey, R. H. 1996. Pharmacokinetic properties of several novel oligonucleotide analogs in mice. *J. Pharmacol. Exp. Ther.* 277:923–937.

Cummins, L. L., Winniman, M., and Guas, H. G. 1997. Phosphorothioate oligonucleotide metabolism: Characterization of the "N+"-mer by CE and HPLC-ES/MS. *Bioorg. Med. Chem. Lett.* 7:1225–1230.

Dean, N. M. and McKay, R. 1994. Inhibition of protein kinase C-α expression in mice after systemic administration of phosphorothioate antisense oligodeoxynucleotides. *Proc. Natl. Acad. Sci. U.S.A.* 91:11762–11766.

Eckstein, F., Schulz, H. H., Ruterjans, H., Haar, W., and Maurer, W. 1972. Stereochemistry of the transesterification step of ribonuclease T1. *Biochem.* 11:3507–3512.

Eckstein, F., and Thomson, J. B. 1995. Phosphate analogs for study of DNA polymerases. *DNA Replica.* 262:189–202.

Emelen, W., and Mannik, M. 1978. Kinetics and mechanisms for removal of circulating single-stranded DNA in mice. *J. Exp. Med.* 147:684–699.

Galbraith, W. M., Hobson, W. C., Giclas, P. C., Schechter, P. J., and Agrawal, S. 1994. Complement activation and hemodynamic changes following intravenous administration of phosphorothioate oligonucleotides in the monkey. *Antisense Res. Dev.* 4:201–206.

Gaudette, M. F., Hampikian, G., Metelev, V., Agrawal, S., and Crain, W. 1993. Effect on embryos of injection of phosphorothioate-modified oligonucleotides into pregnant mice. *Antisense Res. Dev.* 3:391–397.

Gaus, H. J., Owens, S. R., Winniman, M., Cooper, S., and Cummins, L. L. 1997. On-line HPLC electrospray mass spectrometry of phosphorothioate oligonucleotide metabolites. *Anal. Chem.* 69:313–319.

Geary, R. S., Leeds, J. M., Henry, S. P., Monteith, D. K., and Levin, A. A. 1997a. Antisense oligonucleotide inhibitors for the treatment of cancer: 1) Pharmacokinetic properties of phosphorothioate oligodeoxynucleotides. *Anticancer Drug Design* 12:383–393.

Geary, R. S., Leeds, J. M., Fitchett, J., Burckin, T., Troung, L., Spainhour, C., Creek, M., and Levin, A. A. 1997b. Pharmacokinetics and metabolism in mice of a phosphorothioate oligonucleotide antisense inhibitor of C-raf-1 kinase expression. *Drug Metab. Dispos.* 25(11):1272–1281.

Glover, J. M., Leeds, J. M., Mant, T. G. K., Amin, D., Kisner, D. L., Zuckerman, J., Levin, A. A., and Shanahan, W. R. 1997. Phase 1 safety and pharmacokinetic profile of an ICAM-1 antisense oligodeoxynucleotide (ISIS 2302). *J. Pharmacol. Exp. Ther.* 282:1173–1180

Goodchild, J. and Kohli, V. 1991. Ribozymes that cleave an RNA sequence from human immunodeficiency virus: The effect of flanking sequence on rate. *Arch. Biochem. Biophys.* 284:386–391.

Henry, S. P., Bolte, H., Auletta, C., and Kornbrust, D. J. 1997a. Evaluation of the toxicity of ISIS 2302, a phosphorothioate oligonucleotide, in a 4-week study in cynomolgus monkeys. *Toxicol.* 120:145–155.

Henry, S. P., Giclas, P. C., Leeds, J., Pangborn, M., Auletta, C., Levin, A. A., and Kornbrust, D. J. 1997b. Activation of the alternative pathway of complement by a phosphorothioate oligonucleotide: Potential mechanism of action. *J. Pharmacol. Exp. Ther.* 281:810–816.

Henry, S. P., Monteith, D. K., and Levin, A. A. 1997c. Antisense oligonucleotide inhibitors for treatment of cancer: 2) Toxicologic properties of phosphorothioate oligodeoxynucleotides. *Anticancer Drug Design* 12:395–408.

Henry, S. P., Novotny, W., Leeds, J., Auletta, C., Crooke, R., and Kornbrust, D. J. 1997d. Inhibition of coagulation by a phosphorothioate oligonucleotide. *Antisense and Nucleic Acid Drug Dev.* 7:503–510.

Henry, S. P., Taylor, J., Midgley, L., Levin, A. A., and Kornbrust, D. L. 1997e. Evaluation of the toxicity profile of ISIS 2302, a phosphorothioate oligonucleotide, following repeated intravenous administration, in a 4-week study in CD-1 mice. *Antisense Nucleic Acid Drug Dev.* 7:473–481.

Hoke, G. D., Draper, K., Freier, S. M., Gonzalez, C., Driver, V. B., Zounes, M. C., and Ecker, D. J. 1991. Effects of phosphorothioate capping on antisense oligonucleotide stability, hybridization and antiviral efficacy versus herpes simplex virus infection. *Nucleic Acids Res.* 19:5743–5748.

Horner, M. J., Monteith, D. K., Gillett, N. A., Butler, M., Henry, S. P., Bennett, C. F., and Levin, A. A. 1997. Evaluation of the renal effects of phosphorothioate oligonucleotides in monkeys. *Fund. Appl. Toxicol.* 36:147.

Koziolkiewicz, M., Wozjcik, M., Kobylanska, A., Rebowski, B., Guga, P., and Stec, W. J. 1997. Stability of stereoregular oligo(nucleoside phosphorothioate)s in human plasma diastereoselectivity of plasma 3'-exonuclease. *Antisense and Nucleic Acid Drug Dev.* 7:43–48.

Krieg, A. M., Matson, S., and Fisher, E. 1996. Oligonucleotide modifications determines the magnitude of B-cell stimulation by CpG motifs. *Antisense and Nucleic Acid Drug Dev.* 6:133–139.

Krieg, A. M., Yi, A.-K., Matson, S., Waldschmidt, T. J., Bishop, G. A., Teasdale, R., Koretzky, G. A., and Klinman, D. M. 1995. CpG motifs in bacterial DNA trigger direct B-cell activation. *Nature* (London) 374:546–549.

Leeds, J. M., Geary, R. S., Henry, S. P., Glover, J., Shanahan, W., Fitchett, J., Burckin, T., Truong, L., and Levin, A. A. 1997a. Pharmacokinetic properties of phosphorothioate oligonucleotides. *Nucleosides and Nucleotides* 16(7–9):1689–1693.

Leeds, J. M., Graham, M. J., Truong, L., and Cummins, L. L. 1996. Quantitation of phosphorothioate oligonucleotides in human plasma. *Anal. Biochem.* 235:36–43.

Leeds, J. M., Henry, S. P., Truong, L., Zutsi, A., Levin, A. A., and Douglas, K. J. 1997b. Pharmacokinetics of a potential human cytomegalovirus therapeutic, a phosphorothioate oligonucleotide, after intravitreal injection in the rabbit. *Drug Metab. Dispos.* 25(8):921–926.

Leeds, J. M., Truong, L., Cossum, P., Prowse, C., Crooke, S. C., and Kornbrust, D. 1994b. Interaction of phosphorothioate oligonucleotides with plasma proteins. *Pharm. Res.* 11:S-352.

Levin, A. A., and Dent, J. G. 1982. Comparison of the metabolism of nitrobenzene by hepatic microsomes and cecal microflora from Fischer-344 rats in vitro and the relative importance of each in vivo. *Drug Metab. Dispos.* 10:450–454.

Levin, A. A., Monteith, D. K., Leeds, J. M., Nicklin, P. L., Geary, R. S., Butler, M., Templin, M. V., and Henry, S. P. 1998. Toxicity of oligodeoxynucleotide therapeutic agents. In *Handbook of experimental pharmacology: Antisense research and applications,* ed. Crooke, S. T. Heidelberg: Springer–Verlag.

Liang, H., Nishioka, Y., Reich, C. F., Pisetsky, D. S., and Lipsky, P. E. 1996. Activation of human B cells by phosphorothioate oligodeoxynucleotides. *J. Clin. Invest.* 98:1119–1129.

Lima, W. F., Brown-Driver, V., Fox, M., Hanecak, R., and Bruice, T. W. 1997a. Combinatorial screening and rational optimization for hybridization to folded hepatitis C virus RNA of oligonucleotides with biological antisense activity. *J. Biol. Chem.* 272(1):626–638.

Lima, W. F., Mohan, V., and Crooke, S. T. 1997b. The influence of antisense oligonucleotide-induced RNA structure on E. *coli* RNase H1 activity. *J. Biol. Chem.* 272(29):18191–18199.

Lima, W. F., Monia, B. P., Ecker, D. J., and Freier, S. M. 1992. Implication of RNA structure on antisense oligonucleotide hybridization kinetics. *Biochem.* 31(48):12055–12061.

Miraglia, L., Geiger, T., Bennett, C. F., and Dean, N. 1996. Inhibition of interleukin-1 type receptor expression in human cell-lines by an antisense phosphorothioate oligonucleotide. *Int. J. Immunopharmacol.* 18:227–240.

Monia, B. P., Johnston, J. F., Ecker, D. J., Zounes, M., Lima, W. F., and Freier, S. M. 1992. Selective inhibition of mutant Ha-ras mRNA expression by antisense oligonucleotides. *J. Biol. Chem.* 267:19954–19962.

Monia, B. P., Sasmor, H., Johnston, J. F., Freier, S. M., Lesnik, E. A., Muller, M., Geiger, T., Altmann, K. H., Moser, H., and Fabbro, D. 1996. Sequence-specific antitumor activity of a phosphorothioate oligodeoxyribonucleotide targeted to human C-raf kinase supports an antisense mechanism of action in vivo. *Proc. Natl. Acad. Sci. U.S.A.* 93:15481–15484.

Monteith, D. K., Henry, S. P., Howard, R. B., Flournoy, S., Levin, A. A., Bennett, C. F., and Crooke, S. T. 1997. Immune stimulation—A class effect of phosphorothioate oligonucleotides in rodents. *Anticancer Drug Design.* 12:421–432.

Nicklin, P. L., Craig, S. J., and Phillips, J. A. 1998. Pharmacokinetic properties of phosphorothioates in animals—Absorption, distribution, metabolism and elimination. In *Handbook of experimental pharmacology: Antisense research and applications*, ed. Crooke, S. T. Heidelberg: Springer–Verlag.

Oberbauer, R., Schreiner, G. F., and Meyer, T. W. 1995. Renal uptake of an 18-mer phosphorothioate oligonucleotide. *Kidney Int.* 48:1226–1232.

Pasternak, G. W., and Standifer, K. M. 1995. Mapping of opioid receptors using antisense oligodeoxynucleotides: Correlating their molecular biology and pharmacology. *Trends Pharmacol. Sci.* 16:344–350.

Phillips, J. A., Craig, S. J., Bayley, D., Christian, R. A., and Nicklin, P. L. 1997. Pharmacokinetics, metabolism and elimination of a 20-mer phosphorothioate oligodeoxynucleotide (CGP 69846A) after intravenous and subcutaneous administration. *Biochem. Pharmacol.* 54:657–668.

Plenat, F., Klein-Monhoven, N., Marie, B., Vignaud, J.-M., and Duprez, A. 1994. Cell and tissue distribution of synthetic oligonucleotides in healthy and tumor-bearing nude mice. *Am. J. Pathol.* 147:124–135.

Rappaport, J., Hanss, B., Kopp, J. B., Copelend, T. D., Bruggeman, L. A., Coffman, T. M., and Klotman, P. E. 1995. Transport of phosphorothioate oligonucleotide in kidney: Implications for molecular therapy. *Kidney Int.* 47:1462–1469.

Rifai, A., Brysch, W., Fadden, K., Clark, J., and Schlingensiepen, K. H. 1996. Clearance kinetics, biodistribution, and organ saturability of phosphorothioate oligodeoxynucleotides in mice. *Am. J. Pathol.* 149:717–725.

Sadler, T. W., Liu, E. T., and Augustine, K. A. 1995. Antisense targeting of *Engrailed*-1 causes abnormal axis formation in mouse embryos. *Teratol.* 51:292–299.

Sands, H., Gorey-Feret, L. J., Cocuzza, A. J., Hobbs, F. W., Chidester, D., and Trainor, G. L. 1994. Biodistribution and metabolism of internally 3H-labeled oligonucleotides. I. Comparison of a phosphodiester and phosphorothioate. *Mol. Pharmacol.* 45:932–943.

Sarmiento, U. M., Perez, J. R., Becker, J. M., and Narayanan, R. 1994. In vivo toxicological effects of rel A antisense phosphorothioates in CD1 mice. *Antisense Res. Dev.* 4:99–107.

Srinivasan, S. K., and Iversen, P. 1995. Review of in vivo pharmacokinetics and toxicology of phosphorothioate oligonucleotides. *J. Clin. Lab. Anal.* 9:129–137.

Wickstrom, E. 1986. Oligodeoxynucleotide stability in subcellular extracts and culture media. *J. Biochem. Biophys. Meth.* 13:97–102.

Yamamoto, T., Yamamoto, S., Kataoka, T., and Tokunaga, T. 1994. Ability of oligonucleotides with certain palindromes to induce interferon production and augment natural killer cell activity is associated with their base length. *Antisense Res. Dev.* 4:119–122.

Yi, A.-K., Chace, J. H., Cowdery, J. S., and Krieg, A. M., 1996. IFN-γ promotes IL-6 and IgM secretion in response to CpG motifs in bacterial DNA and oligodeoxynucleotides. *J. Immunol.* 156:558–564.

Zhao, Q., Temsamani, J., Zhou, R.-Z., and Agrawal, S. 1997. Pattern and kinetics of cytokine production following administration of phosphorothioate oligonucleotides in mice. *Antisense and Nucleic Acid Drug Dev.* 7:495–502.

Biotechnology and Safety Assessment, 2nd ed.
Edited by John A. Thomas
Copyright © 1998 Taylor & Francis

8

Biotechnology in Agriculture and the Environment: Benefits and Risks

Christine McCullum and David Pimentel

Cornell University, Ithaca, New York

Maurizio G. Paoletti

Padova University, Padova, Italy

Technological advances in biotechnology, including genetic engineering, have enabled transfer of genetic traits both within species and between entirely different plant and animal species. Currently, biotechnology techniques are being used in various fields, including agriculture, veterinary medicine, pharmaceutical development, forestry, energy conservation, and waste treatment (BIO 1990). These techniques, if applied responsibly, have the potential to increase productivity in crops and livestock, control pests, produce new food and fiber crops, and develop effective medicines (Paoletti and Pimentel 1996). Potential environmental and economic benefits from biotechnology include the reduction of fossil fuel in agriculture and forestry through improved nutrient availability in crops and livestock, use of fewer artificial inputs (e.g., synthetic nitrogen fertilizers, insecticides, and fungicides), and more cost-effective and environmentally friendly waste management practices, such as bioremediation.

If realized, these improvements will help protect ecological systems by reducing habitat degradation (Paoletti and Pimentel 1996). In addition, some of the biotechnology techniques should improve the economics of agricultural and forestry production systems. Although genetic engineering can be expected to provide major benefits to agriculture and the environment, risks with the use of this technology should also be recognized. In this chapter, we assess the environmental, health, and socioeconomic benefits and risks of biotechnology, including genetic engineering, in agricultural systems.

DISEASE-RESISTANT CROPS

Engineering Virus-Resistance in Crops

Resistance against crop disease in plants, caused by viruses, bacteria, and fungi is now being explored through biotechnology and genetic engineering techniques as a way to reduce the loss of crops (Moffat 1992; Gasser and Fraley 1992). Because viruses in the field cannot easily be treated, the production of genetically engineered, virus-resistant crops is agronomically significant (Mannion 1995). In addition, few antibacterial chemicals are available to control bacterial diseases (Schroth and McCain 1991). It has been estimated that viruses, bacteria, and fungi are collectively responsible for significant crop losses estimated at 12%, or nine hundred million tons, of preharvest yield worldwide (Cramer 1967; Krimsky and Wrubel 1996).

More than 350 field tests of genetically engineered disease-resistant plants have been approved in the United States since 1987, and the majority of these have been created to produce disease-resistant, genetically-engineered crops impervious to viral infections (Krimsky and Wrubel 1996). Success in engineering virus resistance in tobacco, alfalfa, potato, cucumber *(Cucumis sativus)*, melon *(Cucumis melo)*, alfalfa, and tomato plants have been reported by Cuozzo et al. (1988), Hill et al. (1991), Truve et al. (1993), Gonsalves et al. (1992), Dong et al. (1991), and Xue et al. (1994), respectively. (See Table 1.)

Field trials with tobacco containing the gene from the mosaic virus for the production of the coat protein have shown that resistance can be transgenically induced. For example, in China, field trials of tobacco that contains the tobacco mosaic virus and tomatoes with cucumber mosaic virus are under way (Chen and Gu 1993). Efforts are also being aimed at rice because of its importance as a staple crop in this region. In Japan, a method for producing fertile transgenic rice plants using an electroporation system has been developed (Shimamoto et al. 1989). Transgenic rice plants expressing the rice stripe virus–coat protein (RSV–CP) have been developed to fight the rice stripe virus, one of the major viruses of rice plants in Japan, Korea, China, and Taiwan (Hayakawa et al. 1992).

The findings of a 3-year biosafety study of ecological risks have demonstrated that expressing the introduced gene (RSV–CP) in a japonica rice variety (Kinuhikari) resulted in transgenic rice plants that did not: (1) affect morphological and ecological traits with the exception of some somaclonal variations, (2) hybridize with closely grown rice plants, (3) exhibit the tendency to become weeds, (4) produce any detectable toxic substances, and (5) have any observable effects on subsequent cultivation, microorganisms in soil, insects in florae, or on surrounding plants (Yahiro et al. 1994). However, these results will need to be followed up with longer-term biosafety assessment in the future.

In the United States, squash and the papaya are two of the more recent models of crop engineering for virus resistance. In 1994, the genetically engineered, virus-resistant squash developed by Asgrow seed company for resistance to zucchini yellow mosaic virus (ZYMV) and watermelon mosaic virus II (WMV II) was one of the first

TABLE 1. *Plants genetically engineered for virus resistance that have been approved for field tests in the United States from 1987 to July 1995*

Crop	Disease(s)	Research organization
Alfalfa	Alfalfa mosaic virus, Tobacco mosaic virus (TMV), Cucumber mosaic virus (CMV)	Pioneer Hi-Bred
Barley	Barley yellow dwarf virus (BYDV)	USDA
Beets	Beet necrotic yellow vein virus	Betaseed
Cantelope and squash	CMV, papaya ringspot virus (PRV)	Upjohn
	Zucchini yellow mosaic virus (ZYMV), Watermelon mosaic virus II (WMVII)	
	CMV	Harris Moran Seed
	ZYMV	Michigan State University
	ZYMV	Rogers NK Seed
	Soybean mosaic virus (SMV)	Cornell University
	SMV, CMV	New York State Experiment Station
Corn	Maize dwarf mosaic virus (MDMV) Maize chlorotic mottle virus (MCMV), Maize chlorotic dwarf virus (MCDV)	Pioneer Hi-Bred
	MDMV	Northup King
	MDMV	DeKalb
	MDMV	Rogers NK Seed
Cucumbers	CMV	New York State Experiment Station
Lettuce	Tomato spotted wilt virus (TSWV)	Upjohn
Papayas	PRV	University of Hawaii
Peanuts	TSWV	Agracetus
Plum Trees	PRV, plum pox virus	USDA
Potatoes	Potato leaf roll virus (PLRV), Potato virus X (PVX), Potato virus Y (PVY)	Monsanto
	PLRV, PVY, late blight of potatoes	Frito-Lay
Potatoes	PLRV	Calgene
	PLRV, PRY	University of Idaho
	PLRV, PVY	USDA
	PVY	Oregon State University
Soybeans	SMV	Pioneer Hi-Bred
Tobacco	ALMV, tobacco etch virus (TEV), Tobacco vein mottling virus	
	TEV, PVY	University of Florida
	TEV, PVY	North Carolina State University
	TMV	Oklahoma State University
	TEV	USDA
Tomatoes	TMV	Monsanto
	CMV, tomato yellow leafcurl virus	
	TMV, ToMV	Upjohn
	ToMV	Rogers NK Seed
	CMV	PetoSeed
	CMV	Asgrow
	CMV	Harris Moran Seeds
	CMV	New York State Experiment Station
	CMV	USDA

SOURCE: Krimsky and Wrubel 1996.

genetically engineered crops commercialized in the United States. (Biotechnology Information Center 1996). Researchers have also developed two genetically engineered papaya lines by utilizing rDNA techniques to isolate and clone a papaya ringspot virus (PRV) that encodes for the production of the viral coat protein (Fitch et al. 1992). Papaya, one of the three largest crops in Hawaii, has been decimated in recent years by PVR. Hawaiian papaya growers believe these two lines of genetically engineered, disease resistant papaya could save the $45 million Hawaiian papaya industry from extinction (McCandless 1996). However, it should be noted that, although the mild strain of PRV displayed excellent resistance to PRV isolates from Hawaii, it showed only moderate to no protection to isolates from different geographic regions (e.g., Bahamas, Mexico, Florida, Australia, China, Guam, Brazil, Thailand, and Ecuador) (Tennant et al. 1994; Gonsalves et al. 1994).

Engineering Bacterial and Fungi Resistance in Crops

Harms (1992), who has reviewed recent developments in the production of resistance to fungal and bacterial diseases via genetic engineering points out that, although there has been much research on how to incorporate such resistance into crop plants, few improvements have been made in this area. Chen and Gu (1993) have described efforts that are being made to combat bacterial blight, which can reduce rice yields by as much as 10%, through genetic engineering.

Developing disease-resistant crops should also receive high priority secondary to the large amounts of fungicides that are currently applied to fruit and vegetable crops (Pimentel et al. 1993). Aspelin et al. (1994) reported that in 1993 131 million pounds of pesticidal active ingredient was applied at a cost of $584 million. Fungicides are sometimes harmful to beneficial insects and toxic to earthworms and many other beneficial soil biota (Edwards and Bohlen 1992; Paoletti et al. 1988, 1991). The number and activity of these soil biota are important in maintaining soil fertility over time because they recycle nutrients in organic matter and aid in water percolation and soil aeration (Crossley et al. 1992). Furthermore, fungicides rank highest for carcinogenicity of all pesticides applied to agriculture and account for approximately 70% of human health problems associated with pesticide use (NAS 1987).

One way to reduce crop losses to fungi and the external application of fungicides is to introduce genes that encode proteins with antifungal properties into crop plants. Several genes have been identified so far in fungi, bacteria, and plants that are effective for the engineering of resistance to fungi based on their ability to produce enzymes, such as chitinase, that attack the cell wall of fungi. Transgenic tobacco plants with a chitinase gene from beans produced elevated levels of chitinase in roots and leaves compared with control plants in greenhouse experiments. Both experimental and control plants were grown in soil inoculated with the fungal pathgen *Rhizoctonia solani*. A positive association was also found between the level of chitinase expressed in the experimental plants and survival (Broglie et al. 1991). Broglie et al. (1993) and Lin et al. (1995) also reported some success has been achieved with engineering resistance to the stem rot pathogen (*Rhizoctonia solani*) in oilseed rape or canola (*Brassica napus*) and rice, respectively. The engineering of resistance to the fungus *Fusarium oxysporum* in the tomato has also been successful (Van den Elzen 1993).

Risks

Creation of New Weeds

In terms of risks, it has been proposed that large-scale cultivation of plants expressing viral and bacterial genes could lead to novel ecological risks. The most significant ecological risk would be gene transfer via pollination from cultivated crops to wild relatives. For example, it has been postulated that the virus-resistant squash commercialized in 1994 could transfer its newly acquired virus-resistance genes to wild squash (*Cucurbita pepo*), which is native to the southern United States, where it is an agricultural weed. If the virus-resistance genes were to spread, newly disease-resistant wild squash could become a hardier, more abundant weed. Moreover, because the United States is the origin for squash, changes in the genetic make-up of wild squash could lessen its value to squash breeders (Goldburg 1995).

Another area of concern is the production of virus-resistant sugar beets, which is likely to result in exchange of genes between cultivated and wild populations of beets (*Beta vulgaris L.*) because production areas contain wild or weed beet populations, or both, separated by only a few kilometers. A genetic exchange could take place owing to wind pollination, biotic pollination, or the common gynoniecy of wild beets (Boutin et al. 1987; Cuguen 1994). A genetic introgression from seed beet to wild beet popluations has already been observed in Europe (Santoni and Berville 1992).

Viruses that Infect New Hosts

Some plant pathologists also hypothesize that development of virus-resistant crops may allow viruses to infect new hosts through transencapsidation. This may be especially important for certain viruses (e.g., luteoviruses) for which possible heterologus encapsidation of other viral RNAs with the expressed coat protein is known to occur naturally. With other viruses, such as the PRV, risk of heteroencapsidation is thought to be mininal because the papaya itself is infected by very few viruses (Gonsalves et al. 1994).

Creation of New Viruses

Virus-resistant crops may also lead to the creation of new viruses through an exchange of genetic material or recombination between RNA virus genomes. Recombination between RNA virus genomes requires infection of the same host cell with two or more viruses. Several authors have pointed out that recombination may also occur in genetically engineered plants expressing viral sequences on infection with a single virus and that large-scale cultivation of such plants may lead to increased possibilities of combinations (Hull 1990; Palukaitis 1991; de Zoeten 1991; Tepfer 1993). It has recently been shown that an RNA transcribed from a transgene can recombine with an infecting virus to produce highly virulent new viruses (Greene and Allison 1994).

An overall strategy of risk assessment utilizing an incremental approach entails: (1) identifying potential hazards, (2) determining frequency of recombination between homologous but nonidentical sequences, and (3) determining whether such recombinants can have selective advantage (Tepfer et al. 1994). Fernandez-Cuartero et al. (1994) have already demonstrated that, even though a particular pseudorecombinant strain was at a competitive disadvantage relative to a parent cucumber mosaic virus (CMV) strain, a spontaneous recombinant that arose from the pseudorecombinant (resulting from pseudorecombination or the situation in which gene components of one virus are exchanged with the proteins of another coat) had enhanced fitness relative to either of the other original strains.

HERBICIDE-RESISTANT CROPS (HRCs)

At the moment at least 4 engineered crops for target herbicide resistance are on the market, and 13 among the key crops in world production have been extensively tested in field trials (Table 2). In addition, some crops (e.g., corn) are being engineered to contain both herbicide (Glyphosate) and insecticide-resistance (BT δ-endotoxin) (*Gene Exchange* 1997). The potential benefits and risks of herbicide-resistant crops (HRCs) are discussed in this section.

Potential Benefits Associated with the Use of HRCs

Possible Reduced Use of Herbicides

Proponents have argued that reduction of herbicides adopted for HRC crops occurs primarily because these "new" herbicides are needed in lower doses (if compared for instance with atrazine, 2,4-D, and alachlor) and are applied later in crops, postemergence. However, higher resistance of the crop to the target herbicide would, in practice, suggest to the farmer to adopt a higher rate than advised to ensure that all weeds are burned in one tractor trip with the targeted broad spectrum herbicide (Pimentel and Paoletti 1996).

Improved Integrated Pest Management (IPM)

Integrated pest management IPM could benefit from some HRCs if alternative nonchemical methods were applied first to control weeds and the target herbicide were used later, only when and where the threshold of weeds is surpassed, in postemergence

TABLE 2. *Herbicide-resistant crops (HRCs) approved for field tests in the United States from 1987 to July 1995*

Crop	Herbicide	Research organization
Alfalfa	Glyphosate	Northrup King
Barley	Glufosinate/Bialaphos	USDA
Canola (oilseed rape)	Glufosinate/Bialaphos	University of Idaho
		Hoechst–Roussel/AgrEvo
	Glyphosate	InterMountain Canola
		Monsanto
Corn	Glufosinate/Bialaphos	Hoechst–Roussel/AgrEvo
		ICI
		Upjohn
		Cargill
		DeKalb
		Holdens
		Pioneer Hi-Bred
		Asgrow
		Great Lakes Hybrids
		Ciba–Geigy
		Genetic Enterprises
	Glyphosate	Monsanto
		DeKalb
	Sulfonylurea	Pioneer Hi-Bred
		Du Pont
	Imidazolinone	American Cyanamid
Cotton	Glyphosate	Monsanto
		Dairyland Seeds
		Northrup King
	Bromoxynil	Calgene
		Monsanto
		Rhone Poulenc
	Sulfonylurea	Du Pont
		Delta and Pine Land
	Imidazolinone	Phytogen
Peanuts	Glufosinate/Bialaphos	University of Florida
Potatoes	Bromoxynil	University of Idaho
		USDA
	2,4-D	USDA
	Glyphosate	Monsanto
	Imidazolinone	American Cyanamid
Rice	Glufosinate/Bialaphos	Louisiana State University
Soybeans	Glyphosate	Monsanto
		Upjohn
		Pioneer Hi-Bred
		Northrup King
		Agri-Pro
	Glufosinate/Bialaphos	Upjohn
		Hoechst/AgrEvo
	Sulfonylurea	Du Pont
Sugar beets	Glufosinate/Bialaphos	Hoechst–Roussel
	Glyphosate	American Crystal Sugar
Tobacco	Sulfonylurea	American Cyanmid
Tomatoes	Glyphosate	Monsanto
	Glufosinate/Bialaphos	Canners Seed
Wheat	Glufosinate/Bialaphos	AgrEvo

SOURCE: Adapted from Krimsky and Wrubel 1996 and *Gene Exchange* 1997.

(Krimsky and Wrubel 1996). If HRCs could be implemented in IPM programs without underestimating the induction of weed resistance and adopting all the available nonchemical alternatives to manage weed control, this technology would be a step toward more sustainable agriculture (Kirmsky and Wrubel 1996). However, in practice, insufficient work of extension outreach and appropriate protocols promoted by the producers could only lead to a further link of the farmland to the producers and their marketing policies aimed at increasing their sales of the targeted herbicides aside to their HRCs seeds.

Benefits to Developing Countries

Although the majority of HRCs currently on the market and under development belong to key crops in Western agriculture (Krimsky and Wrubel 1996), a few innovations have been proposed that would help developing countries. For example, HRCs have been proposed for improved control of parasitic flowering seeds such as *Orobanche* and *Stringa*, both which can severely reduce grain yields. The HRCs would permit more effective herbicide action against the soil parasitic weed without damaging the target crop. Trials on boomrape have demonstrated that the engineered plants can overproduce at a rate at least double the reproduction of the control plants. However, the authors observed that this technology can only be used with weeds that do not have the potential to interbreed with wild relatives that could themselves become weeds (Joel et al. 1995). For example, in northern African countries, most crops such as sorghum, wheat, and canola (oilseed rape) have their wild relatives nearby, which therefore increases the risk that genes from the herbicide-resistant crop varieties can be transferred to wild relatives (Mikkelsen 1996; BSTID 1996). The same gene escape risks are possible for tomato, corn, and potato in South America.

Potential Risks Associated with Use of HRCs

Building Weed Resistance

The risk that herbicide-resistant genes from a transgenic crop variety can be transferred via pollination into weedy relatives has been demonstated for canola (oilseed rape) and sugar beet. Mikkelsen et al. (1996) and Brown and Brown (1996) have shown that herbicide-resistant genes from a transgenic canola move quickly into wild relative weedy populations. Boudry et al. (1994) also noted consistent gene flow between the cultivated sugar beets and weed beet populations.

Repeated use of the same herbicide in the same area creates problems of plant resistance to the target herbicide. This concern has consistent bases in the recent history of herbicides (Wrubel and Gressell 1994). For instance, if glyphosate from an actual few million hectares of crops were allowed to associate with HRC·crops, the resulting acreage could be around 70 million ha, and pressure on weeds to evolve resistant biotypes could be pronounced (Gressel 1992; Krimsky and Wrubel 1996).

Sulfonylureas and imidazolinones, to be targeted in HRC crops, are particularly prone to rapid evolution of resistant weeds and have already resulted in several cases of resistance (LeBaron and McFarland 1990; Wrubel and Gressel 1994). Extensive adoption of HRCs will increase the acreage and surface treated, thereby exacerbating the resistance problems (Krimsky and Wrubel 1996).

Environmental Risks

Even if less environmentally persistent than previous herbicides (e.g., Alachor, 2,4-D, atrazine), the "environmentally friendly" HRCs have, as do most chemical pesticides, consistent or severe environmental effects (Dekker and Comstock 1992).

Bromoxynil (Commercial Name: Buctril)

Bromoxynil has been targeted in HRC cotton by Calgene and Monsanto (Table 2). This herbicide has traditionally been used in winter cereals, cotton, corn, sugarbeets, and onions to control large-leaf weeds. Drift has been observed that has resulted in damage to nearby grapes, cherries, alfalfa, and roses (Al Khatib et al. 1992). In addition, leguminous plants can be very sensitive to this herbicide (Abd Alla and Omar 1993), and potatoes can be damaged as well. Consistent residues over the accepted standards have been detected in soil and groundwater (Miller et al. 1995) and as fallout (Waite et al. 1995). Rodents tested have demonstrated some mutagenic responses (Rogers et al. 1995). Stafilinid beetles have been shown to have reduced survival and egg production at suggested dosages (Samsoe-Petersen 1995). Crustaceans *(Daphnia magna)* have also been severely affected (Buhl et al. 1993).

Glufosinate/Bialaphos (Commercial Name: Basta)

Many crops have been modified for this herbicide-resistance PAT (phosphinothricin acetyl transferase) gene, which has been introduced into alfalfa, corn, barley, wheat, rice, canola, peanuts, soybeans, sorghum, tomatoes, and sugarbeets (Table 2). Turf-grass *(Agrostisis stolonifera)* and other components have been engineered for resistance but need appropriate environmental risk assessment before being marketed (Lee 1996). This herbicide has been on the market since 1984 as a synthetic development of a natural pathogen toxin of *Streptomyces viridochromogenes* (Charudattan et al. 1996; Duke and Abbot 1995). The amount of active ingredient used is 200–900 g/ha postemergence on the engineered plants (AgrEvo 1996).

Although detrimental effects on users and consumers seem unlikely under recommended doses (Hack et al. 1994), toxic effects on humans and animals have been reported (Cox 1996). For example, the Basta surfactant (sodium polyoxyethylene alkylether sulfate) has been shown to have strong vasodilative effects in humans and cardiostimulative effects in rats (Koyama et al. 1997). Treated mice embryos exhibited specific morphological defects (Watanabe and Iwase 1996).

Imidiazolinone (Commercial Name: Imazethapyr)

Imidiazolinone is applied at low doses (38–50 g/ha) in beans and soybeans postmergence (Burnside et al. 1994; Perucci and Scarponi 1994). It has been observed that, at the field rate of 50 g/ha, there is no effect in laboratory and field microbial biomass. However, higher doses induce catalase and hydrolase activity and increase the risk for monocultural practices (Pertucci and Scarponi 1994) and reduced mycelial growth in *Sclerotinia trifoliorum* (Reichard et al. 1997).

Sulfonylurea (Commercial Name: Safari)

Sulfonylurea is used as a herbicide on wheat, barley, sugarbeets, cotton, maize, potatoes, and soybeans postmergence. Drift from very low amounts (5–30 g/ha) can damage cultures, and potential losses may result in several crops, wild plants, and nontarget invertebrates (Cox 1992).

Glyphosate (Commercial Name: Roundup)

Most HRCs have been engineered for glyphosate resistance. Although adverse effects of herbicide-resistant soybeans have not been observed on feeding animals such as cows, chickens, and catfish; genotoxic effects have been demonstrated in other nontarget organisms (Cox 1995 a,b). Earthworms have been shown to be severely damaged by the glyphosate herbicide at 2.5–10 l/ha. (Rebanova et al. 1996). For example, *Allolobophora caliginosa*, the most common earthworm in European, North American, and New Zealand fields is damaged by this herbicide (Mohamed et al. 1995; Springett and Gray 1992). In addition, aquatic organisms, including fish, can sometimes be severely damaged (Henry et al. 1994; WHO 1994). The prevalence of the nematode *Steinerema feltiae*, a useful biological control organism, is reduced by 19–30% by use of this herbicide (Forschler et al. 1990).

Health Risks

The unknown health risks associated with the use of herbicides (as well as most xenobiotics) involve the effects of low-level chronic exposures (Wilkinson 1990). Although most research has addressed cancer risk, much less research has been focused on neurological, immunological, developmental, and reproductive effects (Krimsky and Wrubel 1996). Much of this problem reflects the dearth of methodologies and diagnostic tests at the disposal of scientists necessary to evaluate the risks caused by exposure to many chemials, including herbicides, properly.

Although industry often stresses the desirable characteristics of their HRCs, environmental and alternative agriculture groups, as well as some scientists, disagree, with the contention that these products are safe. For example, research has shown that the application of glyphosate can increase the level of plant estrogens in the bean *Vicia faba* (Sandermann and Wellman 1988). Feeding experiments have shown that cows fed transgenic glyphosate-resistant soybeans had a statistically significant difference in daily milk fat production compared with other test groups (Hammond et al. 1996).

Increased Costs Associated with Use of HRCs

Because the herbicides for which HRCs are being designed are almost all under patent, they will be more expensive than many of the herbicides they are intended to replace (Fig. 1). In addition, although analysts project that, for example, switching to bromoxynil for broadleaf weed control in cotton could result in savings of $37 million each year from reductions in herbicide purchases of 40% to 50%, few economic product evaluations that demonstrate cost savings with the use of HRCs have been published (Krimsky and Wrubel 1996). Furthermore, recent problems with use of glyphosate-resistant cotton in the Mississippi Delta region (crop losses resulting in up to $500 thousand of this year's cotton crop) suggest this technology needs to be perfected further before some farmers will reap economic benefits (Fox 1997).

Some scientists suggest that use of HRCs will cause a shift to the use of fewer broad-spectrum herbicides, which will reduce the amount of sprays and the amount of herbicide used per application (Hayenga et al. 1992), others contend that the overall use of HRCs will actually increase herbicide use and thereby increase costs associated with their application (Goldburg et al. 1990; Rissler and Mellon 1993; Paoletti and Pimentel 1995; Paoletti and Pimentel 1996).

Bacillus thuringiensis (BT) for Insect Control

More than 40 BT crystal protein genes have been sequenced and 14 distinct genes identified and classified into 6 major groups based on amino acids and insecticidal activity (Krimsky and Wrubel 1996). Many crop plants have been engineered with the BT δ-endotoxin, including alfalfa, corn, cotton, potatoes, rice, tomatoes, and tobacco (Table 3). The amount of toxic protein expressed inside the modified plant is 0.01–0.02% of the total soluble proteins (Strizhov et al. 1996).

Potential Economic and Environmental Benefits

Corn

Supporters of BT usage cite current trials demonstrating a high level of efficacy in controlling corn borer damages on plants. Corn engineered with BT δ-endotoxin has the potential to recover 5–15% of corn borer damage in 70 million acres in the United States with a projected benefit of $50 million annually (Steffey 1995). However, without careful management tactics based on university agriculture extension and technical expertise, growers and seed suppliers could see unforseen environmental risks.

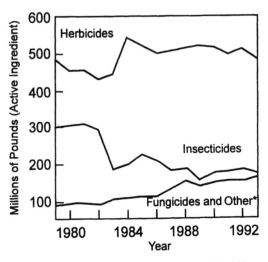

* Other = rodenticides, fumigants and molluscicides

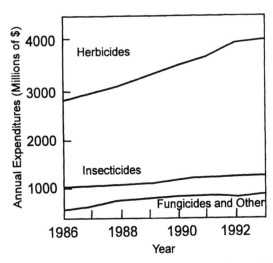

FIG. 1. The choice of herbicide resistance as the main target for engineered crops seems related to the promising trend of herbicide sales and returns rather than to other environmental strategies such as potential reduction of pesticides in the environment (from Aspelin et al. 1992).

Cotton

Cotton was the first crop plant engineered with the BT δ-endotoxin released into the market. It has been estimated that caterpillar pests, including the cotton bollworm and budworm, cost U.S. farmers about $171 million/year as measured in yield losses and insecticide costs (Head 1991). Benedict et al. (1992) predicted that the widespread

TABLE 3. *Transgenic insect-resistant crops containing BT endotoxins. Approved field tests in United States from 1987 to July 1995*

Crop	Research organization
Alfalfa	Mycogen
Apples	Dry Creek
	University of California
Corn	Asgrow
	Cargill
	Ciba–Geigy
	Dow
	Genetic Enterprises
	Holdens
	Hunt–Wesson
	Monsanto
	Mycogen
	NC+Hybrids
	Northrup King
	Pioneer Hi-Bred
	Rogers NK Seed
Cotton	Calgene
	Delta and Pineland
	Jacob Hartz
	Monsanto
	Mycogen
	Northrup King
Cranberry	University of Wisconsin
Eggplant	Rutgers University
Poplar	University of Wisconsin
Potatoes	USDA
	Calgene
	Frito-Lay
	Michigan State University
	Monsanto
	Montana State University
	New Mexico State University
	University of Idaho
Rice	Louisiana State University
Spruce	University of Wisconsin
Tobacco	Auburn University
	Calgene
	Ciba–Geigy
	EPA
	Mycogen
	North Carolina State University
	Roham & Haas
Tomatoes	Campbell
	EPA
	Monsanto
	Ohio State University
	PetoSeeds
	Rogers NK Seeds
Walnuts	University of California, Davis
	USDA

SOURCE: Adapted from Krimsky and Wrubel 1996 and *Gene Exchange*, 1996.

use of BT-cotton could reduce insectide use and thereby decrease costs by as much as 50 to 90%, which would save farmers $86 to $186 million/year, respectively.

Potatoes

Genetically engineered Russet Burbank potatoes with *Bacillus thuringiensis tene-brionis* (the δ-endotoxin is represented inside the engineered potatoes at 0.05–0.1% CryIIIA as a percentage of total protein) have been shown to be successful (Perlak et al. 1993). This formulation is also suggested as a base to develop an effective and sustainable IPM program for potatoes.

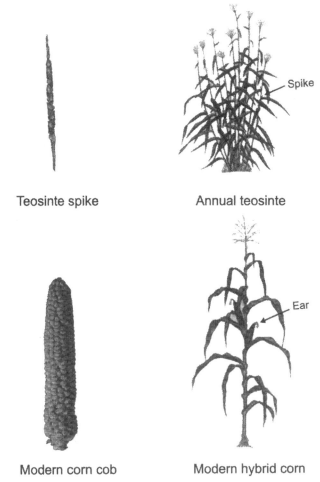

Teosinte spike   Annual teosinte

Modern corn cob   Modern hybrid corn

FIG. 2. Traditional vertical breeding has sometimes resulted in very different crops from the original wild relatives. This may be illustrated by comparing modern corn with its wild relative teosinte. However horizontal transfer of genes has been not possible in traditional breeding and is a new and revolutionary trend in breeding.

TABLE 4. *The numbers of native and introduced plants and insects in Florida*

Type	Plant	Insects
Native species	2,525	11,512
Immigrant species and established in nature	0	946
Introduced species and established in nature	925	42
Cultivated species but not present in nature	25,000	5

SOURCE: After Frank and McCoy 1995.

Eggplants

The δ-endotoxin Cry3b-engineered eggplants have demonstrated consistent effects against potato beetles (Iannacone et al. 1997). However, resistance is one harrowing prospect, and proper resistance management has to be considered.

Potential Environmental Risks

Insects that develop resistance to transgenic crop varieties are one of the possible risks associated with the use of BT δ-endotoxin in genetically engineered crop varieties. Resistance to BT has already been demonstrated in the cotton budworm and bollworm (Tabashnik 1992; Bartlett 1995). If BT-engineered plants become resistant, a key insecticide that has been utilized successfully in IPM programs, could be lost (Paoletti and Pimentel 1995). Therefore, proper resistance-management strategies with the use of this new technology are imperative. Another potential risk is that the BT δ-endotoxin could be harmful to nontarget organisms (Goldburg and Tjaden 1990; Jepson et al. 1994). For example, it is not clear what potential effect BT δ-endotoxin residues that are incorporated into soils will have against an array of nontarget useful invertebrates living in the rural landscape (Jepson et al. 1994; Paoletti and Pimentel 1995).

RECYCLING OF TOXIC WASTES AND IMPROVED WASTE MANAGEMENT

Advances in technology have resulted in the availability of a variety of chemicals, many of which have increased ecosystem pollution. Currently, in the United States, some 70,000 different chemicals are released into the environment through soil, water, and air (Newton and Dillingham 1994); an estimated 100,000 chemicals are used worldwide (Nash 1993). Cleanup of hazardous wastes by conventional technologies is projected to cost between $400 and $750 billion in the United States alone on the basis of estimates obtained from a variety of federal and private sources (Salt et al. 1995; USGS 1995). Various strategies that are utilizing genetic engineering and biotechnological methods to deal more efficiently with waste removal and management include bioremediation, phytoremediation, and the production of biodegradable plastics.

Bioremediation

Biotechnology and genetic engineering may help reduce environmental pollution through bioremediation, which is the application of biological treatments using microbes to degrade chemical materials at polluted sites effectively. Because bioremediation provides continuous cleanup of contaminated sites, such as pesticide residues in agricultural ecosystems, it has significant advantages over other techniques. Furthermore, a marked degree of self-regulation is present in such systems because the added microbes survive by consuming and degrading chemicals but die off when the nutrient source is reduced or eliminated (Pimentel et al. 1997).

Investigations into genetically modified bacteria for bioremediation have produced a strain of *Pseudomoas cepacia* that degrades a wide spectrum of chemical pollutants, including vinyl chloride, dichloroethylene, phenol, toluene, xylene, and creosol (EPA 1991). Most importantly, this bacterium produces an enzyme that is capable of degrading trichloroethylene, a persistent industrial degreaser. Bioremediation techniques are growing in importance and diversity as new approaches are developed to use both wild and genetically engineered biota for chemical pollution abatement.

Phytoremediation

Phytoremediation, or the use of specially selected and engineered metal-accumulating plants for environmental cleanup, is an emerging technology that may serve as a cost-effective approach to treating soils and groundwater contaminated with toxic metals (Salt et al. 1995). Currently, cleanup of sites in the United States contaminated with heavy metals and organic mixtures can alone cost $42.5 billion. Phytoextraction, or the use of metal accumulating in plants to remove toxic metals from soils, could be utilized to clean up sandy loam soil to a depth of 50 cm, which would cost $60,000–$100,000 compared with at least $400,000 for excavation and storage alone using traditional soil removal methods (Salt et al. 1995). Rhizofiltration, or the use of plant roots to remove toxic metals from polluted waters, may offer an advantage in water treatment because of the ability of plants to remove 60% of their dry weight as toxic metals, thus markedly reducing the generation and disposal costs of hazardous radioactive residue. Rhizofiltration could also be a cost-competitive technology in the treatment of groundwater containing low concentrations of toxic metals.

Genes encoding the cadmium-binding protein, methallothionein, have been recently shown to be expressed in plants. A research group at Peking University has engineered a human gene encoding methallothionein (a protein that binds heavy metals) into tobacco. The genetically engineered plants have survived exposure to cadmium concentrations that were 25 times greater than the dose that killed control plants. More importantly, these genetically engineered plants have also been shown to absorb cadmium from the soil. Researchers are also now engineering the gene into weeds in the hope of using the transgenic weeds to reduce heavy metal contamination on farmland (Chen and Gu 1993).

Biodegradable Plastics

Another area of waste management biotechnology and genetic engineering could play a role is in the production of biodegradable plastics. Currently, plastics account for 20% by volume of all municipal solid wastes (Stein 1992). Estimates of the current global market for biodegradable plastics range up to 1.3 billion kg per year (Lindsay 1992). However, until recently, the cost of these biodegradable plastics, such as polyhydroxyalkanoates (PHAs), polymers made entirely by bacterial fermentation, has been one of the major limiting factors in their use (Poirier et al. 1992, 1995). Producing PHAs in plants would significantly reduce the expense of manufacture compared with fermentation and make these biopolymers competitive with petroleum-based plastics for low-cost uses (Poirier et al. 1995; Nawrath and Somerville 1995).

Recently, efforts to produce PHAs in plants have been successful. Bacterial genes that are necessary to synthesize polyhydroxybutyrate (PHB), a type of PHA, were transferred to *Arabidopsis thaliana* plants. The genetically engineered plants accumulated up to 14% dry weight of PHB without deleterious effects on plant growth or fertility and with a level of polymer yield that is considered commercially practical (Nawrath, Poirier, and Somerville 1994). Several companies are now pursuing development of PHA-producing genetically engineered oilseed rape (canola) crops (Poirier et al. 1995; Wrage 1995).

Environmental Risks

In initial prerelease testing, the interactions of such new genetically engineered organisms with nontarget organisms in the soil communities and contaminated landscapes must be carefully monitored to avoid potentially deleterious side effects (Paoletti and Pimentel 1996). Also, it will be important to monitor the safe environmental disposal of plants that accumulate toxic materials.

NOVEL FOOD PRODUCTS

The objective of biotechnology and genetic engineering is to improve the food supply by increasing crop and livestock productivity, enhancing nutrient composition and availability, and improving food characteristics such as size, taste, and texture. The following are examples of the types of applications that have been achieved or are under way.

Increased Crop and Livestock Productivity

In the past, scientific breeders of plants and animals have utilized the rich genetic resources of cultivars and land races in crop improvement programs. It has been estimated that at least half of the increase in agricultural productivity realized this century may be directly attributable to "artificial selection, recombination, and intraspecific gene transfer procedures" (Woodruff and Gall 1992). Use of biotechnology and genetic engineering may result in even greater productivity gains through high-yield crop varieties, improved pest and disease control, the production of herbicide-resistance in crops, enhanced nutrient availability, and tolerance to a variety of environmental stressors (Mannion 1995).

Biotechnology is also being used as a way to increase livestock productivity by enhancing milk and meat production. This is achieved by isolating the gene for the protein hormone (e.g., somatotropin) made in the anterior pituitary gland of animals and inserting it into bacterial cells, which results in an increased production of the hormone. Applications of this technology include the injection of recombinant bovine somatotropin (rBST) into dairy cows to increase milk production and injection of recombinant porcine somatotropin (rPST) into growing pigs to increase carcass leanness and decrease fattiness (Etherton 1994). Some authors have speculated that deceasing the quantities of feed consumed per unit of output may benefit the environment by decreasing environmental pollution through reducing the quantity of feed provided; reducing the quantity of fertilizer and other inputs associated with growing, processing, and storing animal feed; and reducing animal waste products (Etherton 1994; van Weerden and Verstegen 1989).

Biotechnology may also play a role in improving crop plants that, in turn, enhance secondary productivity in livestock (Mannion 1995). Researchers at the Commonwealth Scientific and Industrial Research Organization's division of plant industry in Canberra have found a way to insert a sunflower gene into the subterranean clover, a major constituent of Australian sheep pastures. The new clover, with genes from sunflower that code for a protein in sulfur amino acids, provides sheep with a sulfur-rich diet that is necessary for wool production. Creating such a transgenic clover resulted in about a hundredfold increase in the sulfur-rich protein. However, it has been estimated that (i.e., an increase in wool production by 5–10%), the level of sulfur-rich protein needs to be boosted another tenfold to have a substantial impact on wool production. The group is testing gene promoters that could achieve this. Producing the same amount of wool from fewer sheep could reduce soil erosion and other environmental pollution produced by these animals (Thwaites 1993).

Developments in biotechnology and genetic engineering may also be able to improve ruminant nutrition by modifying the microbes that are involved in ruminal fermentation. The objective will be to find suitable foreign bacterial genes that can be inserted into ruminal bacterial organisms (Wallace 1994). Techniques of genetic engineering are also playing an important role in increasing animal productivity by improving and developing vaccines and pharmaceuticals (e.g., fertility hormones). Hybridoma technology, which results in the generation of monoclonal antibodies by cell fusion procedures, will be increasingly useful in diagnosing specific diseases as well as in disease prevention and treatment (Woodruff and Gall 1992). The broad range of potential vaccines for control of various diseases is especially promising because of their low environmental risks and excellent socioeconomic benefits (Paoletti and Pimentel 1996).

Enhanced Nutrient Composition and Availability

Enhanced Nutrient Composition

Another application of biotechnology in agriculture is intended to improve the food supply by enhancing the nutritional composition of foods. In 1992, Monsanto was able to produce a genetically engineered potato successfully with an increased starch content. A higher starch content reduces oil absorbtion during frying and thereby lowers the cost of frying certain food products (e.g., French fries, potato chips) and reduces oil content in the finished product (Zinnen and Voichick 1994). Genetically engineered strains of oilseed rape (canola) are undergoing development at Calgene, where applications include high-stearic-oil-content margarine and edible canola oil with reduced saturated fat content (Krimsky and Wrubel 1996).

Through biotechnology, scientists also hope to create foods that protect against cancer, heart disease, osteoporosis, and other life-threatening illnesses. For example, the National Cancer Institute (NCI) is currently working on increasing the amount of phytochemicals, which are linked to cancer prevention, in foods such as garlic, parsley, and citrus fruits. Government, university, and industry scientists are also developing cereal grains with increased amounts of both soluble and insoluble fiber to lower cholesterol and fight digestive cancers, respectively; milk with improved calcium bioavailability to help protect against osteoporosis; and vegetables with boosted levels of antioxidants (e.g., super carrots with five times the amount of β-carotene) to help reduce the risk of cancer (Rohlfing 1991).

Enhanced Nutrient Availability

Biotechnology and genetic engineering can also be employed to enhance nutrient availability in agricultural systems. Genetic engineering has been applied to the problem of nitrogen fixation with a specific focus on the genetic makeup of organisms that fix nitrogen from the atmosphere and the genetic basis for the relationship between leguminous species (e.g., peas, beans, alfalfa, clover, peanuts) and the nitrogen-fixing bacteria that occupy their nodules (Mannion 1995). Scientists in China have recently developed recombinant strains of nitrogen-fixing bacteria that have a higher nitrogen fixation efficiency than traditional bacteria. These recombinant strains of bacteria have been spread over more than a million hectares of rice and soybean fields, and preliminary results show crop yield increases of 5–10% (Chen and Gu 1993).

In addition, a mechanism of biological nitrogen-fixation similar to that of natural legumes eventually may be genetically engineered into plants such as wheat and corn. If this is achieved, the need for commercial nitrate fertilizers could be significantly reduced (Mannion 1995). However, because the molecular mechanisms required for symbiotic nitrogen fixation are complex (involving at least 17 genes), achieving this will require an investment in research over several decades (Woodruff and Gall 1992).

Other Improved Food Characteristics

Other applications of biotechnology to improve the food supply include the improvement of certain food characteristics such as size, ripeness, acidity or sweetness, taste, and texture. Genetic modifications have enabled the production of fruit that may have better taste and an enhanced shelf life through delayed pectin degradation (Bennett et al. 1989) or altered responses to the plant hormone ethylene (Bleecker 1989). Among the first such novel products on the market was Calgene's slow-ripening Flavr Savr® tomato, which was approved for sale by the Food and Drug Administration in May 1994. This tomato has been transfected with an "antisense" gene responsible for the enzyme polygalacturonase, which solubilizes pectin (Holden 1989). Because pectin degradation increases fruit ripening and decreases shelf-life, preventing translation of the message (by having the antisense message present) for polygalacturonase should retard fruit ripening and increase prepurchase and postpurchase shelf life (AMA 1991). DNA Plant Technology (DNAP) corporation's Endless Summer® tomato, which like Calgene's Flavr Savr has an antibiotic-resistance marker gene encoded into it, also incorporates a gene-splicing technique to retard ripening (Cummins 1995).

Risks

Environmental Risks

The production of plants and animals to suit specific environments could lead to the transformation of more land presently occupied by natural ecosystems into agricultural land. The ecosystems of the tropics on the land not suited for agriculture of any kind are particularly vulnerable. Thus, the removal of yet more natural ecosystems would further deplete biodiversity. Genetic engineering may also favor monocultures that will threaten the global centers of ordered biodiversity (Hopkins et al. 1991; Rissler and Mellon 1993; Third World Network 1995).

Health Risks

One potential health risk is related to the use of genetically engineered animal hormones. One can illustrate using the case of rBST. The Food and Drug Administration has ruled that the presence of rBST in milk is safe for children and adults (FDA 1995). Even so, there have been questions regarding the impact of this technology on animal and human health (Broom 1995). Use of rBST in dairy cows increases the chances of bacterial infections and mastitis and also reduces the reproductive cycle in treated dairy cows (GA0 1992; Burton et al. 1994; Broom 1995). Milestone et al. (1994) report that increased infections in cattle will require treatment with antibiotics. Although not all antibioitics appear in milk, some do. Thus, if more antibiotics are used, an indirect risk to humans may arise because some residues may remain in the milk (GAO 1992).

A second potential human health concern is the risk of introducing allergens into the food supply. A recent study by Nordlee and colleagues demonstrated the transfer of a major food allergen from Brazil nuts to transgenic soybeans during the development of a genetically engineered crop variety (Nordlee et al. 1996). Furthermore, although only a dozen foods may produce allergic reactions—mainly protein foods—biotechnology allows for nontraditional food proteins (e.g., moths, insects) to be present for which no knowledge currently exists regarding their allergenic or nonallergenic qualities (Nordlee et al. 1996).

A third potential human risk relates to the increasing prevalence of antibiotic-resistant and disease-causing bacteria. Because antibiotic-resistant genes are the most commonly used type of selectable marker in genetic engineering and are rarely deleted from the resulting organism, incorporation of antibiotic-resistant markers into the genetic material of human pathogens could pose risks by increasing the prevalence of antibiotic-resistant and disease-causing bacteria. In 1989, dozens of people died and thousands others were crippled after consuming a batch of synthetic L-tryptophan produced using genetically-engineered bacteria. Although the exact cause of illness will never be known, it has been reported that a specific impurity stemming from the bacterial strain may be cause of the syndrome (Raphals 1990).

A fourth health concern is that genetically engineered crops may be able to transfer their foreign gene(s) to other unrelated microorganisms. Genetically engineered oilseed rape (canola), black mustard, thorn-apple, and sweet peas all contain an antibiotic-resistance gene and were grown together with fungus *Aspergillus niger*, or their leaves were added to the soil. The fungus was shown to have incorporated the antibiotic-resistance gene in all coculture experiments. (Hoffman et al. 1994).

Other health risks created through the use of biotechnology and genetic engineering include the ability of genetically engineered organisms to survive and harm nontarget organisms, the creation of new toxic organisms, or both (Goldburg and Tjaden 1990; Jepson et al. 1994). The risk of a new host being infected by a virus or recombining to form a more deadly, virulent virus must be investigated further (Greene and Allison 1994; Tepfer et al. 1994). For example, the risk of recombination between the engineered vaccine virus and other orthopox viruses endemic in wildlife, such as the cowpox virus, still needs to be investigated accurately (Boulanger 1995).

Social and Economic Impact

Proponents often cite biotechnology and genetic engineering as the way to help solve the world's food problems (i.e., improve agriculture in developing countries). However, some critics who are skeptical about of ability of biotechnology and genetic engineering to increase food production project there will only be about a 1% increase in crop yield on the basis of the anticipated contributions of biotechnology during the next two decades (Duvick 1989; Brown et al. 1990). In addition, no increase in crop yields is projected for Africa because no major advances in biotechnology are expected to be applied in Africa in the near future (Pimentel et al. 1992a).

Some persons further question why, after 20 years of research, biotechnologists have not produced a single high-yielding variety of wheat, rice, or corn. The answer, according to plant scientists, is that plant breeders using traditional techniques may have largely exploited the genetic potential for increasing the share of photosynthate that goes into the seed (Brown 1997). Others feel that the present channeling of funds to expensive biotechnology projects diverts scarce resources from other research that could focus on more practical solutions to pressing social problems such as hunger and food insecurity (Third World Network 1995).

Some authors (Krimsky and Wrubel 1996) have also observed that many of the crops being engineered for herbicide resistance belong to the group of key crops in Western agriculture. This circumstance may reflect the domination of the majority of the biotechnology industry by transnational companies (TNC) in the developed world whose business it is to generate profits (Mannion 1995). If sustaining the Third World is the target of genetically engineered crops, other vegetables and crops have to be considered. Also, helping these countries bypass expensive, high-input crop production and move their traditional agriculture toward low-input sustainable practices is desirable as well (Odum 1989).

TRANSGENIC ANIMALS

Benefits

Among the benefits of producing genetically engineered or transgenic animals are the ability to manufacture more cost-effective drugs (including vaccines), new animal models to study human disease, human tissue and organ harvesting, and improvements in the food supply (e.g., modifying the shape, size, or nutritional quality of animals). In the early 1980s, the first successful experiments creating transgenic vertebrates were reported (Krimsky and Wrubel 1996). Palmiter et al. (1982) created a transgenic mouse by transferring a growth hormone into the embryo of the mouse. In 1988, a patent was awarded for a mouse that was genetically engineered with human genes designed to be tumorigenic. This mouse, commonly referred to in the scientific literature as the "oncomouse" (oncology is the study of tumors), prompted the first patent issued for a transgenic animal (OTA 1989).

Transgenesis, or the transfer of genes across species lines, opened the way for animals to be used as an alternative to tissue-culture productions of human protein (Krimsky and Wrubel 1996). In 1987, a transgenic mouse was created that demonstated the viability of tissue-specific expression of foreign proteins (Gordon et al. 1987). As reported by Krimsky and Wrubel (1996, p. 192), "The mouse was genetically engineered to make the clot-dissolving factor tissue plasminogen activator (TPA), which is viewed as a highly promising drug for the treatment of coronary heart disease and a strong competitor of the widely acclaimed streptokinase."

Krimsky and Wrubel (1996) also note, "Finnish researchers have developed a genetically-modified cow that purportedly can produce milk containing large amounts of erythropoietin (red cell growth factor) used to treat anemia. If successful, this method will replace the costlier cell culture techniques. Other more remote applications of transgenic animals include human blood and organ production." (Krimsky and Wrubel 1996, p. 195).

Trangenic animals under development include swine with the human growth hormone gene, genetically engineered livestock designed to tolerate extreme climatic conditions, transgenic sheep that grow faster than normal sheep, engineered sheep that secrete insect repellent and produce moth-proof wool, and genetically engineered sheep and cows that produce milk consumable by lactose-intolerant individuals (Krimsky and Wrubel 1996). Many different species of fish are being genetically engineered to increase fish size and growth rate or improve survival in new environments. Genetic engineers have turned much of the attention toward finfish and shellfish (Warmbrodt 1993).

Many species of transgenic fish have been grown in the laboratory. Fast-growing Pacific salmon have been engineered by various groups of researchers worldwide. Scientists attached a switch to a growth hormone gene from coho salmon and injected the transgene into chinook salmon eggs. On average, the transgenic salmon grew to be elevenfold heavier than their age-mates (Devlin 1994). Fast-growing fish have also been produced outside the laboratory. In 1991, transgenic carp were tested in a high-security pond at Auburn University. These fish were fitted with a growth hormone gene from rainbow trout that enabled them to grow 40% faster than normal (Fishetti 1991). Researchers have also identified a protein in winter flounder, which has recently been transferred into Atlantic salmon, that prevents fish blood from freezing. Transgenic salmon with this trait could potentially be raised in sea pens farther north, where the species could not otherwise live (Fletcher 1992).

Risks

Environmental

Very real environmental risks may be associated with transgenic animals—particularly with certain applications, such as transgenic fish (Goldburg 1995; Regal 1994). For example, transgenic fish have the potential to disturb ecosystems seriously by gaining a competitive advantage in the wild ecosystem. A fast-growing transgenic fish could assume a higher than usual position in the food chain because of its greater size and ability to compete for food, which could harm native species, or a freeze-tolerant transgenic fish raised in the north could escape into a geographic area from which it was previously excluded, and cause competition that would harm native species

(Kapuscinski and Hallerman 1995). Fertile transgenic fish could also successfully invade ecosystems, thus exacerabating the present problem with exotic invaders in aquatic ecosystems (Carlton and Geller 1993).

Health

Using transgenic animals to harvest blood, tissues, or organs may create certain health risks. For example, take the case of deriving human hemoglobin from transgenic pigs. Human hemoglobin must be separated from that of the animal to ensure the purity of the product. Also, how humans will respond to such animal-derived human hemoglobin is still untested (Krimsky and Wrubel 1996).

Social and Ethical

Finally various social and ethical risks are associated with the production and use of transgenic animals. Certain animal rights groups (e.g., The Humane Society) question whether it is morally or ethically correct to turn animals such as mice, pigs, and sheep—which unlike plants and bacteria are sentient beings—into biomachines for the manufacture of proteins and other biological materials (Fox 1992). Some groups may also find it unethical to introduce human genes into livestock and plants. This application of genetic engineering could raise ethical concerns and seriously undermine the public's perceptions of biotechnology (Buttel 1988; Reiss and Straughan 1997). Religous groups may also have certain conflicts as to whether this new technology is compatible with each of their traditional norms. There may also be a conflict of interest within the value and belief systems of those groups for whom it is abhorrent to manipulate, control, experiment with, or consume animals of any type (Krimsey and Wrubel 1996).

GENERAL RISKS OF RELEASING GENETICALLY ENGINEERED ORGANISMS INTO THE ENVIRONMENT

Single-Gene Changes and Pathogenicity

Most single-gene changes are probably not likely to affect the pathogenicity and virulence of an organism in nature adversely (NAS 1987). However, some gene changes may have detrimental consequences. Certain genetic alterations in animal and plant pathogens, for example, have led to enhanced virulence and increased resistance to pesticides and antibiotics (Alexander 1985). For instance, some oat rust microbes, initially nonpest genotypes for a particular oat variety, became serious pest genotypes after a single gene change allowed the rust to overcome resistance in the oat genotype (Wellings and McIntosh 1990).

An important fungal disease of rice, rice blast, has been demonstrated to have genotypes with single-gene changes that cause the fungal organism to be potentially pathogenic to rice cultivars (Smith and Leong 1994). A similar phenomenon of single-gene changes resulting in pathogenicity has been documented with a related fungal pathogen that infects weeping love grass (Heath et al. 1990). This phenomenon has led plant pathologists to develop the "gene-for-gene" principle of parasite–host relationships in which a single mutation in a parasite overcomes single-gene resistance in the host (Person 1959). Furthermore, numerous instances have been documented in which insects, through a single-gene change, have overcome resistance in plant hosts or have evolved resistance to insecticides (Roush and McKenzie 1987). More than 500 species of arthropods have developed resistance to pesticides (Georghiou 1990).

Threats from Modified Native Species

Lindow (1983) has reported that there is little or no danger from the ice-minus strain of *Pseudomonas syringae* (Ps) because Ps is a native U.S. species that produces related phenotypes in nature. Other investigators have demonstrated that there are different genotypes of Ps, and some of these genotypes have genes for pathogenicity (Lindgren et al. 1988). Because some native species have the ability to alter their interactions within an ecosystem, the genetic modification and release of native species into the natural ecosystem may not always be safe. For example, from 60 to 80% of the major insect pests of U.S. and European crops, respectively, were once harmless native species in the United States and Europe (Pimentel 1993b). Many of the insects moved from benign feeding on natural vegetation to destructive feeding on introduced crops. For instance, the Colorado potato beetle moved from feeding on wild sandbar to feeding on the potato that was introduced from Peru and Bolivia (Elton 1958). This insect has become a serious pest of the potato in the United States and Europe.

Intentional Introduction of Crop Plants and Animals

Some proponents of biotechnology suggest that the intentional introduction of foregin plants and animals into the United States is a good model for predicting potential problems arising from biotechnology (NAS 1987). If so, there is reason for concern because several serious problems have resulted from the intentional introduction of what were believed to have been beneficial crops and animals. Genetic similarities between many of the crops and weeds are evident from the fact that 11 of the 18 most serious weeds of the world are crops in other regions of the world (Colwell et al. 1985). Of the several thousand crops that were intentionally introduced into the United States, 128 species of agricultural and ornamental plants have become serious pest weeds (Pimentel et al. 1989). Some of these introduced plant species, like Johnson grass, are among the most serious weed species in the United States—

especially in the southeastern United States. Johnson grass was introduced as a forage for livestock before it escaped and became a weed pest.

This pattern of native species and introduced plants species in the United States is not unique. For example, Florida has only 2525 indigenous plant species, but approximately 25,000 plant species have been introduced and are under cultivation there (Table 4) (Frank and McCoy 1995). An additional 925 species of exotic plants are established in nature in Florida, and several of these are currently displacing native plant species (Table 4). One exotic pest that is displacing native plant species is the melaleuca (*Melaleuca quinquenervia*) tree introduced from Australia. This clearly illustrates the threat to our natural ecosystem when so-called beneficial organisms are introduced and released into nature. The picture in Florida for insects is quite different than for plants (Table 4).

The great majority of insects species (11,512) are native, with nearly 1000 immigrant species. A relatively small number (42) of insect species are established in nature. One of these species that has established itself and is a serious pest in nature is the imported fire ant (Frank and McCoy 1995). Furthermore, 9 out of 20 introduced domestic animal species in the United States have displaced or destroyed native species (Pimentel et al. 1989). These introduced domestic animals, including donkeys, horses, and goats, have become serious ecological pests. A total of 10 other introduced animals, including mammals (e.g., mongoose and wild boars) and birds (e.g, English sparrow and mynah), have become pests (Pimentel et al. 1989).

Furthermore, at least 70 species of fish have been introduced and have become established in U.S. aquatic ecosystems (Countenay and Moyle 1992). These 70 species represent about 10% of all U.S. fish species. A total of 5 introduced fish species have become pests, displacing and reducing the number of native species and, in other cases, altering the habitat and making it uninhabitable for fish and other species (Courtenay and Moyle 1992). In addition to these intentionally introduced fish, there is concern about the introduction of transgenic fish and the potential ecological effects of these engineered fish in aquatic ecosystems in the United States (Kapuscinski and Hallerman 1991). This overall history suggests that the introduction of many types of foreign organisms in the ecosystem may have major negative impacts on many of the 500,000 beneficial plants and animals in the United States (Pimentel et al. 1992).

Are Ecological Niches Filled?

An estimated 1500 exotic insect species have been introduced and are established in the United States, and a few of these (17%) have become pests or have a negative impact on native species (Sailer 1983). This observation indicates that few of the niches in natural ecosystems are filled (Colwell et al. 1985; Herbold and Moyle 1986). There is ample opportunity, therefore, for many species to become established in the United States. Thus, although the argument that engineered organisms will not become established owing to competition with native species may apply in a few cases, most often this is not a valid argument (Pimentel et al. 1989).

USES OF BIOTECHNOLOGY AND GENETIC ENGINEERING FOR A SUSTAINABLE AGRICULTURAL SYSTEM

Acceptable and potentially sustainable options for agriculture that could be derived from biotechnology and genetic engineering have been put forward by several authors (Paoletti and Pimentel 1995, 1996). Some of the desirable areas of development for such technologies that have the potential to benefit agricultural sustainability, the integrity of the natural environment, and the health and safety of society are discussed in the following paragraphs.

Enhancing Crop Resistance to Pests

Approximately 500,000 kg of pesticides are applied each year in U.S. agriculture, and many nontarget species beneficial to the environment are negatively affected. Genetic engineering targeted for pest control could diminish the need for pesticides (Pimentel et al. 1992).

Resistance factors and toxins that exist in nature can be used for insect pest and plant pathogen control (Pimentel 1988). For example, more than 2000 plant species are known to posses some insecticidal activity (Crosby 1966), and approximately 700 natural substances in bacteria, fungi, and actinomycetes have fungicidal activity (Marrone et al. 1988). Traits for resistance to different insect pests and diseases already exist in many cultured crops, including corn, wheat, barley, soybeans, beans, apples, grapes, pears, tobacco, tomatoes, and potatoes (Russell 1978; Smith 1989).

Although some resistance characteristics have been reduced or eliminated in commercial crops, they still can be found in related wild varieties, which provide an enormous gene pool for the development of host–plant resistance (Boulter et al. 1990). For example a wild relative of tobacco that produces a single acetylated derivative of nicotine is reported to be 1000 times more toxic to the tobacco hornworm than cultivated tobacco (Jones et al. 1987).

Transferring this toxic gene to nonfood crops, such as ornamental shrubs and trees, may protect them from certain insect pests. In addition, thionins, proteases, lectins, and chitin-binding proteins that are often present in plants, especially in the seeds, help control some pathogens and pest insects in wild plants (Pimentel 1988; Garica-Olmedo et al. 1992; Boulter et al. 1990; Czapla and Lang 1990, Raikhel et al. 1993). For example, it has been shown that a cowpea protease inhibitor found in the cowpea *Vigna* spp. can now be engineered into tobacco (Gatehouse et al. 1993). Laboratory trials also indicate that the cowpea protease inhibitor provides protection against the cotton budworm (*Heliothis virescens*), which is a major pest of tobacco, cotton, and maize (Mannion 1995).

Development of Perennial Crops

At present, the major cereal crops of the world are annuals. The conversion of annual grains to perennial grains by genetic engineering will reduce tillage and erosion and conserve water and nutrients (Jackson 1991). Such crops will decrease labor costs, improve labor allocation, and, overall, improve the sustainability of agriculture. Energy efficiency in cultivation of perennial cereal crops will be greatly superior to that of annual crops (Jackson 1991).

Improved Botanical Pesticides

Only limited quantities of botanical pesticides, such as pyrethrums, are now used in developed countries in place of some synthetic pesticides. However, in some developing countries, including China and India, botanical pesticides such as neem are effectively used (NAS 1992; Vietmeyer 1992). Increasing the effectiveness of neem and other available botanical pesticides by genetic engineering would be an asset to farmers because these substances are relatively effective and safe.

Bioindication Needs for Sustainable Use of Genetically Engineered Plants

Bioindication is a strategy that adopts and assesses biological units, species, assemblages of species, and ecosystem models to determine the impact of a selected contaminant such as pesticide residues or fertilizers on the environment (Paoletti and Bressan 1996; Paoletti 1997). This strategy is aimed at using biological nontarget organisms, both in microcosm-modeled and in field arenas, to assess the environmental problems created by adopting certain new management techniques in agroecosystems, including genetically engineered organisms. As observed by several authors (Jepson et al. 1994; Paoletti and Pimentel 1995; Paoletti and Pimentel 1996), little work has been dedicated to assessing the true environmental impacts of genetically engineered crop plants (Rissler and Mellson 1995). For example, there are no data on pollinators of engineered plants modified with the BT δ-endotoxin. It would be rather useful to work with the nontarget species linked to the rural landscapes in which the genetically modified crops are expected to be introduced.

PUBLIC PERCEPTIONS OF BIOTECHNOLOGY

Current public opinion polls show that consumers find some applications of biotechnology and genetic engineering more acceptable than others (Hoban and Kendall 1992; Martin and Tait 1992). For example, in a national random telephone survey conducted in the United States by Hoban and Kendall (1992), 66% of respondents considered plant-to-plant gene transfers acceptable, 25% found animal-to-plant gene exchanges acceptable, 40% found animal-to-animal gene transfers acceptable, and only 10% approved of human-to-animal gene transfers.

A suvery of public attitudes toward biotechnology in the United Kingdom revealed that a large percentage of respondents would accept genetically manipulated foodstuffs under the condition they were confident about testing and that the food product(s) look and taste the same or better. One-half of respondents felt it would be a good thing to use genetic manipulations to solve the food problems in the Third World, whereas relatively few people felt it would be a good thing to use genetic manipulation to provide or improve the food supply in the Western world. A majority of respondents (79%) also felt that it would be a bad thing to use genetic manipulation for products they did not feel were needed (Martin and Tait 1992). Consumer surveys in the United States also suggest that the public is skeptical and cautious with regard to certain applications of biotechnology in agriculture and food production (Hoban and Kendall 1992; Wyse and Krivi 1987). Consumer apprehensions center around biotechnology's perceived unpredictability, risks to the environment, and moral and social questions (Bruhn 1992).

In Europe, public attitudes toward genetically engineered foods appear less favorable than in the United States (Krimsky and Wrubel 1996). Recently, the European Commission has approved mandatory labeling of all genetically modified organisms (GMOs). The European community states that the intent of the label is to serve as a source of information for consumers, not as a warning (Institute for Agriculture and Trade Policy 1997). Results from consumer surveys in the United States, Canada, and the United Kingdom also indicate that most consumers are in favor of labeling genetically engineered foods (Feder 1997; Slusher 1991; Optima Consultants 1994; Martin and Tait 1992).

USE OF BIOTECHNOLOGY AS A WAY
TO PRESERVE BIODIVERSITY

Benefits

Proponents of biotechnology claim that use of biotechnology and its subdiscipline of genetic engineering is needed to renew the momentum of plant and animal genetics and is one of the factors that has contributed to food production gains achieved thus far (Avery 1993). Advocates also claim that biotechology has the potential to enhance species preservation, increase the value of biodiversity, and promote ecosystem conservation through decreased environmental degradation.

Enhance Species Preservation

The International Board for Plant Genetics (IBPGR) as well as many others has taken up the job of collecting critically valuable germ plasm. The IBPGR gene banks now have more than a half a million plant "accessions" representing thousands of varieties and hundreds of different species. Biotechnology has encouraged the collection of genes through germ plasm secondary to its being viewed as a valuable and prosperous activity. Utilizing biotechnology for collecting germ plasm may help ensure the world's biodiversity is maintained by fostering preservation of species that are in danger of becoming displaced by new varieties (IBPGR 1992).

Increase the Value of Biodiversity

Biotechnology may increase the value that is placed on biodiversity through increased returns on investments in research and development of biotechnologies that may generate animal and crop breeds of potential value. The present economic benefits of biotechnology products are significant and are conservatively estimated to be between $2–3 billion/yr in the United States alone. Worldwide, current benefits are about $6.2 billion/yr. Nearly half of the current economic benefits relate to agriculture with significant benefits to the pharmaceutical industry (Kathuri et al. 1993). For example, biotechnology, in the form of bioassays, has reduced the time and cost of screening for pharmaceutical and other uses and has thus increased the value of the underlying genetic resources. For this reason, pharmaceutical companies are becoming more interested in the potential biochemical properties of tropical species and varieties for developing new drugs. On the basis of data from Costa Rica, Aylward (1993) estimates that the net private returns to pharmaceutical companies prospecting for biological resources are around $4.8 million per new drug per year. Over 50% of these royalty returns could realistically be allocated to biodiversity protection in Costa Rica.

Conserve Ecosystems Through Decreased Environmental Degradation

The improvement of crop and livestock productivity could, in principle, create an environmental advantage. Less land, particularly less marginal land, would need to be cultivated, thereby reducing problems such as soil erosion and desertification and promoting ecosystem conservation. However, this would require more substantial productivity gains than have been achieved thus far with crop cultivars that have been produced through biotechnology and genetic engineering. In addition to high-yield agriculture, biotechnology may produce high-yield forestry, which could not only speed up growth of tree crops such as rubber, pulpwood, and cocoa but would also help preserve rapidly depleting forests (Avery 1993). Although not yet achieved, biotechnology could also help promote environmental management by reducing the use of artificial fertilizers and chemicals and fossil-fuel consumption, which leads to environmental and biodiversity destruction.

Risks

On the other hand, biotechnology and genetic engineering may produce various environmental and social and ethical risks that could ultimately lead to further habitat destruction and depletion of genetic resources. These potential risks include further depletion of biodiversity, harm to nontarget organisms, and exploitation of farmers in developing countries by transnational companies.

Environmental

Further Depletion of Biodiversity

Use of biotechnology could actually lead to the transformation of more land and thus cause the removal of yet more natural ecosystems into agricultural land, which would further deplete biodiversity. For example, the ecosystems of the tropics, where land may not be suitable for agriculture of any kind, may be particularly vulnerable. Encouraging use of certain genetic resources in the production of pharmaceuticals may also lead to a further depletion of biodiversity. Biotechnology and genetic engineering may also favor monocultures and threaten the global centers of ordered biodiversity—the basis for ecological stability, which has already been seriously undermined primarily as a result of global industrialization, urbanization, and overexploitative agricultural practices (Third World Network 1995; Hopkins et al. 1991; Rissler and Mellon 1993).

Harm to Nontarget Organisms

Other environmental risks created through the use of biotechnology and genetic engineering include the ability of genetically engineered organisms to survive and harm nontarget organisms and the creation of new toxic organisms (Goldburg and Tjaden 1990; Jepson et al. 1994). The genetic engineering of viruses and bacteria could lead to the accidental production of toxic or environmentally harmful strains (Greene and Allision 1994). Not all ecological experiments are successful with their potential predictions, and such errors of judgment have had negative effects on the environment in the past.

Socioeconomic

Transnational Companies (TNCs) May Exploit Farmers in Developing Countries

One social concern of biotechnology is that TNCs are building monopolies over transgenic seed production and may eventually disadvantage poor farmers in developing countries (Shiva 1993; Marks et al. 1992). For example, the packaging of seeds with engineered herbicide resistance and the herbicides by agrochemical companies could be viewed in this light (Mannion 1995). The situation is made even more complex because the majority of the genetic resources, and thus biodiversity, on which biotechnology depends are found in developing countries. A second concern is that because genes extracted from ecosystems in developing nations can be engineered and turned into valuable assets it will be possible for the genes to be patented by developed nations, which will result in developing-world farmers paying for products that originated from their nation's own resources (Shand 1993).

Risk Assessment and Policy Recommendations

If biotechnology is to contribute to sustainable agricultural development, polices must be adopted to ensure that the profit generated by biotechnological research and development is invested in the conservation of the habitats that produced it and that prospecting ventures contribute economic benefits that build technological capacity in the country of origin (Reid 1993). An example that illustrates how this can be done is the partnership between Costa Rica's National Biodiversity Institute (INBio) and Merck and Company, Ltd., by which Merck provides INBio with a $1.1 million dollar budget and INBio provides Merck with plant and animal extracts from biologically diverse, undeveloped areas. In addition, INBio also receives a share of the royalties on any products that are ultimately developed. Mexico, Indonesia, and Kenya are establishing similar bilateral agreements (Reid 1993).

A second way that biotechnology can contribute to sustainable agricultural development is through adoption of an International Biosafety Protocol commissioned by article 19 of the Convention of Biological Diversity (CBD), which was introduced and opened for signature at the Earth Summit meeting in Rio de Janerio in 1992. Because genetically modified organisms (GMOs) are new and may pose novel risks, their future impact on the environment and health is uncertain. Many scientists and organizations, including the authors of this chapter, believe the international community would benefit from a legally binding protocol that sets basic standards for the release and export of GMOs and would prevent further damage to biodiveristy and the Earth's ecosystems (Mellon and Rissler 1995; Third World Network 1995; Goldburg 1995; Reid 1993).

Certain nongovernmental organizations (NGOs) have already begun to draft such a protocol. The drafted protocol would require exporters of biotechnology products to submit complete safety information on a case-by-case basis, establish an independent international body of experts to conduct risk assessments and make decisions on all transboundary trade of genetically modified organisms (GMOs), include public participation at every step of decision making, require mandatory labeling for genetically engineered food products, provide technical and analytical support for member countries, and establish liability standards. The protocol also calls for risk assessment to include social and cultural studies and states that genetic diversity "is dependent on the socioeconomic conditions of the peoples maintaining it" (Community Nutrition Institute 1996, 30). However, the United States would most likely only adopt a significantly modified position, that is one that is science-based and within a framework of risk assessment and management that has been proven adequate in the United States, Europe, and elsewhere" (Hoyle 1996).

Certain others in the scientific community do not support the adoption of the CBD and its legally binding International Biosafety Protocol and believe that regulation of biotechnology would serve as a hazard to the diffusion of biotechnology in the developing world; stifle the development of certain applications of biotechnology such as those that can assist in toxic waste removal, water purification, and displacement of agricultural chemicals; and would not likely meet the protocol's goal of safety enhancement cost-effectively (Miller 1996).

CONCLUSION

Techniques of biotechnology and genetic engineering, if applied responsibly, have the potential to increase productivity in crops and livestock, control pests, produce new food and fiber crops, and develop effective medicines (Paoletti and Pimentel 1996). Potential environmental and economic benefits from biotechnology include the reduction of fossil fuel in agriculture and forestry through improved nutrient availability in crops and livestock, use of fewer artificial inputs (e.g., synthetic nitrogen fertilizers, insecticides, and fungicides), and more cost-effective and environmentally friendly waste management practices such as bioremediation. If realized, these improvements will help protect ecological systems by reducing habitat degradation (Paoletti and Pimentel 1996). Although biotechnology and genetic engineering can be expected to provide major benefits to agriculture and the environment, risks with the use of this technology should also be recognized.

Environmental risks include: the potential to alter basic interactions in natural ecosystems; create plants that become new weeds; release pest control organisms that evolve resistance or harm nontarget organisms, or both; and deplete biodiversity further. Potential health risks include: the possibility to introduce new allergens into the food supply, increase levels of antibiotic residues in the food supply, increase the prevalence of antibiotic-resistant bacteria, and create new virulent strains of bacteria and viruses. Socioeconomic risks include: the ability of transnational companies (TNCs) to create monopolies and exploit farmers in developing countries. Ethical risks include issues related to the inhumane treatment of animals and conflicts arising in value and belief systems of certain individuals and organizations.

The public appears to find certain applications of biotechnology and genetic engineering acceptable, including plant-to-plant gene transfers and use of this technology to solve food problems in developing countries. However, other applications of biotechnology, such as animal-to-plant gene exchanges are reported to be less acceptable. The majority of consumers across international borders support mandatory labeling of genetically engineered foods. Because the goal of biotechnology is to reduce rather than increase risk in the food supply, a more effective international regulatory policy is needed. Adoption of an International Biosafety Protocol, as commissioned by article 19 of the Convention on Biological Diversity (CBD), would help reduce the new and novel risks associated with the use of biotechnology and genetic engineering.

REFERENCES

Abd Alla, M. H., and Omar, S. A. 1993. Herbicides effects on nodulation, growth, and nitrogen yield of faba bean introduced by indigenous *Rhizobium leguminosarum. Zentralblatt für Mikrobiologie* 148(8):593–597.

Al Khatib, K., Parker, R., and Fuerst, E. P. 1992. Sweet cherry response to simulated drift from selected herbicides. *Proc. Western Soc. Weed Soc.* 45:2–27.

Alexander, M. 1985. Ecological consequences of reducing the uncertainties. *Issues Sci. Technol.* 1(3): 57–68.

AMA, Council on Scientific Affairs, American Medical Association. 1991. Biotechnology and the American agricultural industry. *J. Am. Med. Assoc.* 265:1429–1436.

Aspelin, A. L., Grube, A. H., and Torla, R. 1992. *Pesticide industry sales and usage: 1990 and 1991 market estimates.* Washington, D.C.: Environmental Protection Agency, Office of Pesticide Programs, Biological and Economic Analysis Division, Economic Analysis Branch.

Avery, D. 1993. *Biodiversity: Saving species with biotechnology.* Indianapolis: Hudson Institute.

Aylward, B. A. 1993. The economic value of pharmaceutical prospecting and its role in biodiversity conservation. LEEC Discussion Paper 93-105. London: International Institute for Environment and Development.

Bartlett, A. C. 1995. Resistance of the pink bollworm to BT transgenic cotton. In *Proceedings beltwide cotton conference, National Cotton Council of America,* eds. D. A. Richter and J. Amour, 766. Memphis, Tennessee: National Cotton Council of America.

Bennett, A. B., DellaPenna, D., Fisher, R. L., Giovannoni, J., and Lincoln, J. E. Tomato fruit polygalacturonase: Gene regulation and enzyme function. 1990. In *Biotechnology and food quality,* eds. S.-D. Kung, D. D. Bills, and R. Quatrano, 167–180. Boston: Buttersworth Publishing.

BIO (Biotechnology Industry Organization). 1990. *Biotechnology at work.* Washington, D.C.: Biotechnology Industry Organisation.

Biotechnology Information Center. 1996. Commercialized biological-based products in the food and agriculture industries. Beltsville, Maryland: National Agricultural Library.

Bleecker, A. B. 1989. Prospects for the use of genetic engineering in the manipulation of ethylene biosynthesis and action in higher plants. In *Biotechnology and food quality,* eds. S.-D. Kung, D. D. Bills, and R. Quatrano, 143–158. Boston: Buttersworth Publishing.

Boudry, P., Broomberg, K., Saumitou-Laprade, P., Morchen, M., Cuegen, J., and Van Dijk, H. 1994. Gene escape in transgenic sugar beet: What can be learned from molecular studies of weed beet populations? *Proceedings of the 3d international symposium on the biosafety results of field tests of genetically-modified plants and microorganisms, Monterey, California, November 13–16, 1994.* Oakland, California: The University of California, Division of Agriculture and Natural Resources, 75–83.

Boulanger, D., Brochier, B., Crouch, A., Bennett, M., Gaskell, R. M., Baxby, D., and Pastoret, P. P. 1995. Comparison of the susceptibility of the red fox (*Vulpes vulpes*) to a vaccinia–rabies recombinant virus and to cowpox virus. *Vaccine* 13(2):215–219.

Boulter, D., Gatehouse, J. A., Gatehouse, A. M. R., and Hilder, V. A. 1990. Genetic engineering of plants for insect resistance. *Endeavour* (Oxford) 14:185–190.

Boutin, V., Pannenbecker, G., Ecke, W., Schewe, G., Saumitou-Laprade, P., Jean, R., Vernet, P., and Michealis, G. 1987. Cytoplasmic male sterility and nuclear restorer genes in a natural population of *Beta maritima*: Genetical and molecular aspects. *Theor. Appl. Genet.* 73:625–629.

Broglie, K., Broglie, L., Benhamou, N., and Chet, I. 1993. The role of cell wall degrading enzymes in fungal disease resistance. In *Biotechnology in plant disease control,* ed. I. Chet, 139–156. New York: John Wiley & Sons.

Broglie, K., Chet, I., Holliday, M., Cressman, R., Biddle, P., Knowlton, S., Mauvais, C. J., and Broglie, R. 1991. Transgenic plants with enhanced resistance to the fungal pathogen *Rhizoctonia solani. Science* 254:1194–1997.

Broom, D. M. 1995. Measuring the effects of management methods, systems high productivity efficiency and biotechnology on farm animal welfare. In *Issues in agricultural bioethics,* eds. T. B. Mepham, G. A. Tucker, and J. Wiseman, 319–334. Nottingham (UK): Nottingham University Press.

Brown, L. R. 1997. *The agriculture link: How environmental deterioration could disrupt economic progress,* Worldwatch Paper 136. Washington, D.C.: Worldwatch Institute.

Brown, J., and Brown, A. P. 1996. Gene transfer between canola (*Brassica napus* and *B. campestris*) and related weed species. *An. Appl. Biol.* 129(3):513–522.

Brown, L. R., Durning, A., Flavin, C., French, H., Jacobson, J., Lowe, M., Postel., S., Renner, M., Starke, L., and Young, J. 1990. *State of the world*. Washington, D.C.: Worldwatch Institute.

Bruhn, C. M. 1992. Consumer concerns and educational strategies: Focus on biotechnology. *Food Technol.* 46:80–102.

BSTID (Board on Science and Technology for International Development). 1996. *Lost crops of Africa.* Vol. 1, 1–383. Washington D.C.: National Academic Press.

Buhl, K. J., Hamilton, S. J., and Schmulbach, J. C., 1993. Chronic toxicity of bromoxynil formulation Buctril to *Daphnia magna* exposed continuously and intermittently. *Arch. Env. Cont. Toxicol.* 25(2): 152–159.

Buhler, T. A., Bruyere, T., Went, D. F., Stranzinger, G., and Burki, K. 1990. Rabbit b-cassein promoter directs secretion of human interleukin-2 into the milk of transgenic rabbits. *Bio/Technol.* 8:140–143.

Burnside, O.-C., Ahrens, W.-H., Holder, B.-J., Wiens, M.-J., Johnson, M.-M., and Ristau, E.-A. 1994. Efficacy and economics of various mechanical plus chemical weed control systems in dry beans *(Phaseolus vulgaris)*. *Weed Technol.* 8(2):238–244.

Burton, J. L., McBride, B. W., Block, E., Glimm, D. R., and Kennelly, J. J. 1994. A review of bovine growth hormone. *Can. J. Anim. Sci.* 74:164–201.

Buttel, F. 1988. *Social impacts of biotechnology on agriculture and rural America: Neglected issues and implications for agricultural research and extension policy*. Cornell Rural Sociology Bulletin Series No. 150. Ithaca, New York: Cornell University Press.

Carlton, J. T., and Geller, J. B. 1993. Ecological roulette: The global transport of nonindigenous marine organisms. *Science* 261:78–82.

Chen, Z., and Gu, H. 1993. Plant biotechnology in China. *Science* 262:377–378.

Clark, A. J., Bessos, H., Bishop, J. O., Brown, P., Harris, S., Lathe, R., McClenaghan, M., Prowse, C., Simons, J. P., Whitelaw, C. B. A., and Wilmut, I. 1989. Expression of human anti-hemophilic factor IX in the milk of transgenic sheep. *Bio/Technol.* 7:487–492.

Colwell, R. K., Norse, E. A., Pimentel, D., Sharples, F. E., and Simberloff, D. 1985. Letter to the editor on genetic engineering in Agriculture. *Science* 229:111–112.

Community Nutrition Institute. 1996. *Draft biosafety protocol to the convention on biological diversity*, 30. Washington, D.C.: Community Nutrition Institute.

Courtenay, W. R., and Moyel, P. B. 1992. Crimes against biodiversity: The lasting legacy of fish introductions. In *Transactions of the North American wildlife and natural resources conference 57: Crossroads of conservation: 500 Years after Columbus*, ed. R. E. McCabe, 365–372. Washington, D.C.: Wildlife Management Institute.

Cox, C. 1992. Sulfometuron methyl (oust). *J. Pest. Ref.* 13(4):30–35.

Cox, C. 1995a. Glyphosate, part 1: Toxicology. *J. Pest. Ref.* 15(3):14–20.

Cox, C. 1995b. Glyphosate, part 2: Toxicology. *J. Pest. Ref.* 15(4)14–20.

Cox, C. 1996. Herbicide factsheet: Glufosinate. *J. Pest. Ref.* 16(4):15–19.

Cramer, H. H. 1967. Plant protection and world crop production. *Pflanzenschutz Nachr. Bayer.* 20:2.

Crosby, D. G. 1966. *Natural pest control agents, a symposium*. Washington, D.C.: American University.

Crossley, D. A., Mueller, B. R., and Perdue, J. C. 1992. Biodiversity of microarthropods in agricultural soils: Relations to processes. *Agric. Ecosyst. Environ.* 40:37–46.

Cuguen, B., Wattier, R., Saumitou-Laprade, P., Forcioli, D., Morchen, M., Van Dilk, H., and Vernet, P. 1994. Gynodiecy and mitrochondrial DNA polymorphism in natural populations of *Beta vulgaris ssp. maritima. Genet. Sci. Evol.* 26:87–101.

Cummins, R. 1995. *Pure Food Campaign Report June 16–23, 1995*.

Cuozzo, M., O'Connell, K. M., Kaniewski, W., Fang, R.-X., Chua, N.-H, and Tumer, N. E. 1988. Viral protection in transgenic tobacco plants expressing the cucumber mosaic virus coat protein or its antisense RNA. *Bio/Technol.* 6:549–557.

Czapla, T. H., and Lang, B. A. 1990. Effect of plant lectins on larval development of European corn borer (Lepidoptera: *Pyralidae*) and southern corn rootworm (Coleoptera: *Chrysomilidae*). *J. Econ. Entogomol.* 86:2480–2485.

de Zoeten, G. A. 1991. Risk assessment: Do we let history repeat itself. *Phytopathol.* 81:585–586.

Dekker, J., and Comstock, G. 1992. Ethical and environmental considerations in the release of herbicide-resistant crops. *Agric. Human Values* 9(3):32–43.

Devlin, R. H. 1994. Extraordinary salmon growth. *Nature* 371:209–210.

Dong, J.-Z., Yang, M.-Z., Jia, S.-R., and Chua, N.-H. 1991. Transformation of melon (*Cucumis melo* L.) and expression from the cauliflower mosaic virus 35S promoter in transgenic melon plants. *Bio/ Technol.* 9:858–863.

Duvick, D. 1989. Personal communication. Des Moines:IA: Pioneer Hybrid Seed.

Ebert, K. M., Selgrath, J. P., DiTullio, P., Denman, J., Smith, T. E., Memon, M. A., Schindler, J. E., Monastersky, G. M., Vitale, J. A., and Gordon, K. 1991. Transgenic production of a variant of human tissue-type plasminogen activator in goat milk: Generation of transgenic goats and analysis of expression. *Bio/Technol.* 9:835–838.

Edwards, C. A., and Bohlen, P. J. 1992. The effects of toxic chemicals on earthworms. *Rev. Environ. Contam. Toxicol.* 125:23–99.

Elton, C. S. 1958. *The ecology of invasions by animals and plants.* London: Methuen.

EPA (Environmental Protection Agency). March 1991. Biodegradation of trichloroethylene by a genetically modified bacterium. *EPA* 540/2-91/007 (2):1–16.

Etherton, T. D. 1994. The impact of biotechnology on animal agriculture and the consumer. *Nutr. Today* 29(4):12–18.

FDA, BST Update: First year experience reports. 1995. FDA Press Release No. 95. 082. Washington, D.C.: Food and Drug Administration.

Feder, B. J. 1997. Biotech firms advocate labels on genetically altered products. *The New York Times,* 24 February 1997.

Fernandez-Cuartero, B., Burgyan, J., Aranda, M. A., Salanki, K., Moriones, E., and Garcia-Arenal, F. 1994. Increase of the relative fitness of a plant virus RNA associated with its recombinant nature. *Virol.* 203:373–377.

Fishetti, M. 1991. A feast of gene-splicing down on the fish farm. *Science* 253:512.

Fitch, M. M. M., Manshardt, R. M., Gonsalves, D., Slightom, J. L., and Sanford, J. C. 1992. Virus-resistant papaya plants derived from tissues bombarded with the coat protein gene of papaya ringspot virus. *Bio/Technol.* 10:1466–1472.

Fletcher, G. L., Davies, P. L., and Hew, C. L. 1992. *Genetic engineering and freeze-resistant Atlantic salmon, in transgenic fish,* eds. C. L. Hew and G. L. Fletcher. World Scientific Publishing Co.

Forschler, R. T., All, J. N., and Gardner, W. A. 1990. Steinermafeltiae activity and infectivity in response to herbicide exposure in aqueous and soil environments. *J. Inverteb. Pathol.* 55(3):375–379.

Fox, M. W. 1992. *Superpigs and wondercon.* New York: Lyons & Burford.

Fox, J. L. 1997. Farmers say Monsanto's engineered cotton drops bolls. *Nature Bio/Technol.* 15:1233.

Frank, J. H., and McCoy, E. D. 1995. Introduction to insect behavioral ecology: The good, the bad, and the beautiful: Non-indigenous species in Florida. *Florida Entomol.* 78(1):1–15.

GAO. Government Accounting Office of the U.S. Congress. 1992. *Recombinant bovine growth hormone: FDA approval should be withheld until mastitis issue is resolved. Report to Congressional requesters.* Washington, D.C.: Government Accounting Office.

Garcia-Olmedo, F., Carmona, M. J., Lopez-Fando, J. J., Fernandez, J. A., Castagnaro, A., Molina, A., Hernandez-Lucas, C., and Carbonero, P. 1992. Characterization and analysis of thionin genes. In *Genes involved in plant defense,* eds. T. Boller and F. Meins, 283–302. New York: Springer–Verlag.

Gene Exchange. 1997. Washington, D.C.: Union of Concerned Scientists, 6–8.

Georghiou, G. P. 1990. Overview of insecticide resistance. In *Managing resistance to agrochemicals: From fundamental research to practical strategies,* eds. M. B. Green, H. M. Le Baron, and W. K. Moberg, 18–41. Washington, D.C.: American Chemical Society.

Goldburg, R. 1995. Pause at the amber light. *Ceres* 153:21–26.

Goldburg, R., Rissler, J., Shand, H., and Hassebrook, C. 1990. *Biotechnology's bitter harvest.* Washington, D.C.: The biotechnology working group.

Goldburg, R. J., and Tjaden, G. 1990. Are B. T. K. plants really safe to eat? *Bio/Technol.* 8:1011–1015.

Gonsalves, D., Chee, P., Provvidenti, R., Seem, R., and Slightom, J. L. 1992. Comparison of coat protein-mediated and genetically-derived resistance in cucumbers to infection by cucumber mosaic virus under field conditions with natural challenge inoculations by vectors. *Bio/ Technol.* 10:1562–1570.

Gonsalves, D., Fuchs, M., Klas, F., and Tennant, P. 1994. Field assessment of risks when using trans-genic papayas, cucurbits, and tomatoes expressing viral coat protein genes. In *Proceedings of the 3d international symposium on the biosafety results of field tests of genetically-modified plants and microorganisms, Monterey, California, November 13–16, 1994.* Oakland, California: The University of California, Divsion of Agriculture and Natural Resources, 117–127.

Gordon, K., Lee. E., Vitale, J. A., Smith, A. E., Westphal, H., and Hennignausen, L. 1987. Production of human tissue plasminogen factor in transgenic mouse milk. *Bio/Technol.* 5:1183–1187.

Grasser, C. S., and Fraley, R. T. 1992. Transgenic crops. *Scientific Ame.* 266:34–39.

Greene, A. E., and Allison, R. F. 1994. Recombination between viral RNA and transgenic plant transcript. *Science* 263:1423–1425.

Gressel, J. 1991. The needs for new herbicide-resistant crops. In *Achievements and Developments in Combating Pesticide Resistance*, ed I. Denholm, A. L. Devonshire, and D. W. Hollomon. New York: Elsevier Science Publisher.

Gressel, J. 1992. Genetically-engineered herbicide resistant crop—A moral imperative for world food production. Agro-Food Industry. *Hi tech-Jonpaper* 15.

Hack, R., Ebert, E., Ehling, G., and Leist, K. H. 1994. Glufosinate ammonium: Some aspects of its mode of action in animals. *Food Chem. Toxicol.* 32(5):461–470.

Hammond et al. 1996. The feeding value of soybeans fed to rats, chickens, cat fish and dairy cattle is not altered by genetic incorporation of glyphosate tolerance. *J. Nutrit.* 126(3):717–727.

Harms, C. T. 1992. Engineering genetic disease resistance into crops: Biotechnological approaches to crop protection. *Crop Pot.* 11:291–306.

Hayakawa, T., Zhu, Y., Itoh, K., Kimura, Y., Izawa, T., Shimamoto, K, and Toriyama, S. 1992. Genetically-engineered rice resistant to rice stripe virus, an insect-transmitted virus. *Proc. Natl. Acad. Sci. U.S.A.* 89:9865–9869.

Hayenga, M. L., Thompson, C., Chase, C., and Kaaria, S. 1992. Economic and environmental implications of herbicide-tolerant corn and processing tomatoes. *J. Soil Water Conserv.* 47:411–417.

Heath, M. C., Valent, B., Howard, R. J., and Chumley, F. G. 1990. Correlations between cytological detected plant-fungal interactions and pathogenicity of *Magnaporthe grisea* toward weeping lovegrass. *Phytopathol.* 80(12):1382–1386.

Henry, C. J., Higgins, K. F., and Buhl, K. J. 1994. Acute toxicity and hazard assessment of Rodeo, Spreader, and Chem-Trol to aquatic invertebrates. *Act. Environ. Contam. Toxicol.* 27(3):392–399.

Herbold, B., and Moyle, P. B. 1986. Introduced species and vacant niches. *Am. Naturalist.* 128:751–760.

Hill, K. K. et al. 1991. The development of virus-resistant alfalfa, *Medicago sativa L. Bio/ Technology* 9:373–377.

Hoban, T. J., and Kendall, P. A. 1992. *Consumer attitudes about the use of biotechnology in agriculture and food production.* Raleigh, North Carolina: U.S. Department of Agriculture Extension Service and North Carolina State Universtiy.

Hodgson, J. 1992. Whole animals for wholesale protein production. *Bio/Technol.* 10:863–866.

Hoffman, T., Golz, C., and Schieder, O. 1994. Foreign DNA sequences by a wild-type strain of *Asperillus niger* after co-culture with transgenic higher plants. *Curr. Genet.* 27:70–76.

Holden, C. 1989. A tasty new tomato? *Science* 246:889.

Hopkins, D. D., Goldburg, R. J., and Hirsch, S. A. 1991. *A mutable feast: Assuring food safety in the era of genetic engineering.* New York: Environmental Defense Fund.

Hoyle, R. 1996. Biosafety protocol draft spooks U.S. biotechnology officials. *Nature Bio/Technol.* 14:803

Hull, R. 1990. The use and misuse of viruses in cloning and expression in plants. In *Recognition and response in plant-virus interactions.* NATO ASI, Vol. H41, ed. R. S. S. Frazer, 443–457. Berlin: Springer–Verlag.

Iannacone, R., Grieco, P. D., and Cellini, F. 1997. Specific sequence modifications of cry3B endotoxin gene result in high levels of expression in insect resistance. *Plant Mol. Biol.* 34:485–496.

IBPGR (International Board for Plant Genetic Resources). 1992. *Annual Report for 1992.* Rome: IBPGR.

Institute for Trade and Agricultural Policy. 1997. European commission approves compulsory labeling of genetically modified organisms (GMOs). *Labels: Linking Consumers and Producers* 1(1):3.

Jackson, W. 1991. Development of perennial grains. Paper presented at the Eighteenth International Conference on the Unity of Sciences, August 23–26, 1991; Seoul, Korea.

Jepson, P. C., Croft, B. C., and Pratt, G. E. 1994. Test systems to determine the ecological risks posed by toxin release from *Bacillus thuringiensis* genes in crop plants. *Mol. Ecol.* 3:81–89. *Gene Exchange* 1996. Vol. 6, No. 4. Washington, D.C.: Union of Concerned Scientists.

Joel, D., Kleifeld, M., Losner-Goshen, Y., Hezlinger, G., and Gessler, J. 1995. Transgenic crops against parasites. *Nature* 374:220–221.

Jones, D., Huesing, J., Zador, E., and Heim, C. 1987. The tobacco-insect model system for genetically-engineered plants for non-protein insect resistance factors. In: *UCLA symposium on molecular cell biology,* 469–478. New York: Alan R. Liss.

Kapuscinski, A. R., and Hallerman, E. M. 1995. *Benefits, environmental risks, social concerns, and policy implications of biotechnology in aquaculture.* Washington, D.C.: U.S. Office of Technology Assessment.

Kathuri, C. E., Polastro, E. T., and Mellor, N. 1993. Biotechnology in an uncommon market. *Bio/Technol.* 10:1545–1547.

Koyama, K., Koyama, K., and Goto, K. 1997. *Toxicol. Appl. Pharmacol.* 14(2):409–414.

Krimpenfort, P., Rademakers, A., Eyestone, W., van der Schans, A., van der Broek, S., Kooiman, P., Kootwijk, E., Platenburg, G., Pieper, F., Strijker, R., and de Boer, H. 1991. Generation of transgenic dairy cattle using 'in vitro' embryo production. *Bio/Technol.* 9:844–847.

Krimsky, S., and Wrubel, R. P. 1996. *Agricultural biotechnology and the environment.* Urbana and Chicago: University of Illinois Press.

LeBaron, H. M., and McFarland, J. E. 1990. Resistance to herbicides. *Chemtech.* 20:508-511.

Lee, L. 1996. Turfgrass biotechnology. *Plant Science (Limerick)* 115(1):1–8.

Lin, W., Anuratha, C. S., Datta, K., Potrykus, I., Muthukrishnan, S., and Datta, S. K. 1995. Genetic engineering of rice for resistance to sheath blight. *Bio/ Technol.* 13:686–691.

Lindgren, P. B., Panopoulos, N. J., Staskawicz, B. J., and Dahlbeck, D. 1988. Genes required for pathogenicity and hypersensitivity are conserved and interchangeable among pathovars of *Peudomanas syringae. Mol. Gen. Genet.* 211(3):499–506.

Lindow, S. E. 1983. Methods of preventing frost injury caused by epiphytic ice-nucleation-active bacteria. *Plant Dis.* 67:327–333.

Lindsay, K. F. 1992. 'Truly degradable' resins are now truly commercial. *Modern Plastics* 2:62–64.

Mannion, A. M. 1995. Agriculture, environment, and biotechnology. *Agric. Ecosyst. Environ.* 53:31–45.

Marks, L. A., Kerr, W. A., and Klein, K. K. 1992. Assessing the potential impact of agrobiotechnologies on Third World countries. *Sci. Technol. Develop.* 10:1–32.

Marrone, P. G., Stone, T. B., Sims, S. R., and Tran, M. T. 1988. Discovery of microbial natural products as source of insecticidal genes, novel synthetic chemistry, or fermentation products. In *Strategies for genetic engineering of fungal entomopathogens,* ed. R. Granados, 112–114. Ithaca, New York: Boyce Thompson Institute, Cornell University.

Martin, S., and Tait, J. 1992. *Release of genetically modified organisms: public attitudes and understanding.* Milton Keynes, UK: The Open University.

McCandless, L. 1996. Genetic engineering performs miracles with plants. *Cornell Focus* 6(1):20–24.

Mellon, M., and Rissler, J. 1995. Transgenic crops: USDA data on small-scale tests contribute little to commercial risk assessment. *Bio/Technol.* 13:96.

Mikkelsen, T. R., Anderson, B., and Jorgensen, R. B. 1996. The risk of crop transgene spread. *Nature* 380:31.

Milestone, E., Brunner, E., and White, I. 1994. Plagiarism or protecting public health? *Nature* (London) 371:647–648.

Miller, H. 1996. Biotechnology and the UN: New challenges, new failures. *Nature Bio/ Technol.* 14:831–834.

Miller, J. J., Hill, B. C., Chang, C., and Lindwall, C. W. 1995. *Can. J. Soil Sci.* 75(3):349–356.

Moffat, A. S. 1992. Plant biotechnology explored in Indianapolis. *Science* 255:25.

Mohamed, A.-I., Nair, G.-A., Kassem, H.-H., and Nuruzzaman, M-. 1995. Impacts of pesticides on the survival and body mass of the earthworm *Aporrectodea caliginosa (Annelida: Oligochaeta). Acta Zoologica Fennicica* 196:344–347.

NAS (National Academy of Sciences). 1987. *Regulating pesticides in food. The Delaney Paradox.* Washington, D.C.: National Academy of Sciences.

NAS (National Academy of Sciences). 1987. *Introduction of recombinant DNA-engineered organisms into the environment: Key issues.* Washington, D.C.: National Academy of Sciences.

NAS (National Academy of Sciences). 1992. *Neem, the tree that might help everyone.* Washington, D.C.: National Academy of Sciences Board on Science and Technology for International Development.

Nash, L. 1993. Water quality and health. In *Water in crisis: A guide to the world's fresh water resources.* ed. P. Gleick, 25–39. Oxford: Oxford University Press.

Nawrath, C., Poirier, Y., and Somerville, C. 1994. Targeting of the polyhydroxybutyrate biosynthetic pathway to the plastids of *Arabidopsis thaliana* results in high levels of polymer accumulation. *Proc. Natl. Acad. Sci. U.S.A.* 91:21760–21764.

Newton, L. H., and Dillingham, C. K. 1994. *Classic cases in environmental ethics.* Belmont: Wadsworth Publishing Co.

Nordlee, J. A., Taylor, S. L., Townsend, J. A., Thomas, L. A., and Bush, R. K. 1996. Identification of a Brazil nut allergen in transgenic soybeans. *New Eng. J. Med.* 334:688–692.

Odum, E. P. 1989. *Ecology and our endangered life-support system.* Sunderland, Massachusetts: Sinauer.

Optima Consultants. 1994. *Understanding the consumer interest in the new biotechnology industry.* Ottawa, Canada: Communications Branch, Industry Canada.

OTA (Office of Technology Assessment), U.S. Congress. 1989. *New developments in biotechnology: Patenting life*, OTA-BA-360. Washington, D.C.: Government Printing Office.

Palmiter, R. D., Brinster, R. L., Hammer, R. E., Trumbauer, M. E., Rosenfeld, M. G., Brinberg, N. C., and Evans, R. M. 1982. Dramatic growth of mice that develop from eggs microinjected with metallothionein-growth hormone fusion genes. *Nature* 300:611–615.

Palukaitis, P. 1991. Virus-mediated genetic transfer in plants. In *Risk assessment in genetic engineering*, eds. M. Levin and H. Strauss, 140–162. New York: McGraw-Hill.

Paoletti, M. G., and Pimentel, D. 1995. The environmental and economic costs of herbicide resistance and host–plant resistance to plant pathogens and insects. *Technologic. Forecast. Soc. Change* 50:9–23.

Paoletti, M. G. 1997. Elsevier book, in press.

Paoletti, M. G., and Bressen, M. 1996. Soil invertebrates as bioindicators of human disturbance. *Crit. Rev. Plant Sci.* 15(1):21–62.

Paoletti, M. G., Favretto, M. R., Stinner, B. R., Purrington, F. F., and Bater, J. E. 1991. Invertebrates as bioindicators of soil use. *Agric. Ecosyst. Environ.* 34:341–362.

Paoletti, M. G., Iovane, E., and Cortese, M. 1988. Pedofauna bioindicators and heavy metals in five agro-systems in northeastern Italy. *Revue D'Ecologie Biologie du Sol.* 25:33–58.

Paoletti, M. G., and Pimentel, D. 1996. Genetic engineering in agriculture and the environment. *Biosci.* 46:665–673.

Pertucci, R., and Scarponi, L. 1994. Effects on the herbicide imazethapyr on soil microbial biomass and various soil enzyme activities. *Biol. Fertil. Soils* 17(3):237–240.

Perlak. F. J., Stone T. B., Muskopf Y. M., Petersen L. J., Parker G. B., McPerson, S. A., Wymam, J., Love, S., Reed D., Biever, D., and Fischoff, D. A. 1993. Genetically improved potatoes: Protection from damage by Colorado potato beetles. *Plant Mol. Biol.* 22:313–321.

Person, C. 1959. Gene-for-gene relationships in host–parasite systems. *Can. J. Bot.* 37:1101–1130.

Pimentel, D. 1988. Herbivore population feeding pressure on plant host: Feedback evolution and host conservation. *Oikes* 53:289–302.

Pimentel, D., McLaughlin, L., Zepp, A., Lakitan, B., Kraus, T., Kleinman, P., Vancini, F., Roach, W. J., Graap, E., Heeton, W. S., and Selig, G. 1993. Environmental and economic effects of reducing pesticide use in agriculture. *Agric. Ecosyst. Environ.* 46:273–288.

Pimentel, D. 1993b. Habitat factors in new pest invasions. In *Evolution of insects—Patterns of variation*, eds. K. C. Kim and B. A. McPheron. New York: John Wiley & Sons.

Pimentel, D., Brown, N., Vecchio, F., La Capra, V., Hausman, S., Lee, O., Diaz, A., Williams, J., Cooper, S., and Newburger, E. 1992. Ethical issues concerning global climate change on food production. *J. Agric. Environ. Ethics.* 5(2):113–146.

Pimentel, D., Hunter, M. S., LaGro, J. A., Efroymson, R. A., Landers, J. C., Mervis, F. T., McCarthy, C. A., Boyd, A. E. 1989. Benefits and risks of genetic engineering in agriculture. *BioScience* 39:606–614.

Pimentel, D., Stachow, U., Takacs, D. A., Brubaker, H. W., Dumas, A. R., Meaney, J. J., O'Neil, J., Onsi, D. E., and Corzilius, D. B. 1992. Conserving biological diversity in agricultural/forestry systems. *Bio-Science* 42:354–362.

Pimentel, P., Wilson, C., McCullum, C., Huang, R., Dwen, P., Flack, J., Tran, Q., Saltman, T., and Cliff, B. 1997. Economic and environmental benefits of biodiversity. *BioScience* 47(11):747–757.

Poirier, Y., Dennis., D. E., Klomparens, K., and Somerville, C. 1992. Polyhydroxybutyrate, a biodegradable thermoplastic, produced in transgenic plants. *Science* 256:520–523.

Poirier, Y., Nawrath, C., and Somerville C. 1995. Production of polyhydroxyalkanoates, a family of biodegradable plastics and elastomers, in bacteria and plants. *Bio/Technol.* 13:142–150.

Raikhel, N. V., Lee, H. I., and Broekaert, W. F. 1993. Structure and function of chitin-binding proteins. *Annu. Rev. Plant Physiol. Plant Mol. Biol.* 44:591–615.

Raphals, P. 1990. Disease puzzles near solution. *Science* 249:619.

Rebanova, V., Tuma, V., and Ondokova, K. 1996. Effect of the herbicide Roundup on earthworms of the family Lumbricidae in the mountainous meadow ecosystems. *Zootechnicka Rada, Ceske Budejovice* 13(2):63–70.

Regal, P. 1994. Scientific principles for ecologically-based risk assessment of transgenic organisms. *Mol. Ecol.* 3:5–13.

Reichad, S. L., Sulc, R. M., and Rhodes, L. H. 1997. Growth and reproduction of Sclerotinia trifoliorum as influenced by herbicides. *Mycologia.* 89(1):82–88.

Reid, W. V. 1993. Bioprospering: A force for sustainable development. *Environ. Sci. Technol.* 27(9):1730–1732.

Reiss, M. J., and Straughan, R. 1996. *Improving nature?* Cambridge (UK): Cambridge University Press.

Richards, F. F. 1993. An approach to reducing arthropod vector competence: Dispersion of insect-borne diseases could be modified by genetically-altered symbionts. *Am. Soc. Microbiol. News* 59:509–514.

Rissler, J., and Mellon, M. 1993. *Perils amidst the promise: Ecological risks of transgenic crops in a global market.* Cambridge, Massachusetts: Union for Concerned Scientists.

Rogers, H. J., and Parkes, H. C. 1995. Transgenic plants and the environment. *J. Exp. Bot.* 46:467–488.

Rohlfing, C. 1991. Longevity's latest drugs: Milks, carrots, bread, & oj. *Food Technol.* 44:55.

Roush, R. T., and McKenzie, J. A. 1987. Ecological genetics of insecticide and acaride resistance. *Annu. Rev. Entomol.* 32:361–380.

Russell, G. E. 1978. *Plant breeding for pest disease resistance.* New York: Butterworth.

Sailer, R. I. 1983. History of insect introductions. In *Exotic plant pests and North American agriculture,* eds. C. L. Wilson and C. L. Graham, 15–38. New York: Academic Press.

Salt, D. E., Blaylock, M., Kumar, N., Dushenkov, V., Ensley, B. D., Chet, I., and Raskin, I. 1995. Phytoremediation: A novel strategy for the removal of toxic metals from the environment using plants. *Bio/Technol.* 13(5):468–474.

Samsoe-Peterson, L. 1995. Effects of 67 herbicides and plant growth regulators on the rove beetle *Aleochara bilineata* (Col. Staphylinidae) in the laboratory. *Entomophaga* 40:95–104.

Sandermann, H., and Wellman, E. 1988. *Biologische Sicherheit* 1:285–292.

Santoni, S., and Berville, A. 1992. Characterization of the nuclear ribosomal DNA units and phylogeny of *Beta* L. wild forms and cultivated beets. *Theor. Appl. Genet.* 83:533–542.

Schroth, M. N., and McCain, A. H. 1991. The nature, mode of action, and toxicity of bactericides. In *CRC handbook of pest management,* eds. D. Pimentel and A. A. Hensen, 497–505. Boca Raton: CRC Press.

Shand, H. 1993. Agbio and Third World development. *Bio/Technol.* 11:513.

Shimamoto, K., Terada, R., Izawa, T., and Fujimoto, H. 1989. Fertile transgenic rice plants regenerated from transformed protoplasts. *Nature* (London) 338:274–276.

Shiva, V. 1993. *Monocultures of the mind.* London and Penang, Malaysia: Zed Books Ltd.

Slusher, B. J. 1991. Consumer acceptance of food production innovations: An empirical focus on biotechnology and BST. In *Enhancing consumer choice,* ed. R. N. Mayer, 105–106. Columbia, Missouri: American Council on Consumer Interest.

Smith, C. M. 1989. *Plant resistance to insects: A fundamental approach.* New York: John Wiley & sons.

Smith, J. R., and Leong, S. A. 1994. Mapping of a magnaporthe grisea locus affecting rice *(Orysa sativa)* cultivar specificity. *Theor. Appl. Genet.* 88(8):901–908.

Springett, J. A., and Gray, R. A. J. 1992. Effect of repeated low doses of biocides on the earthworm *Apporectodea caliginosa* in laboratory culture. *Soil Biol. Biochem.* 24(12):1739–1744.

Steffey K. L. 1995. Crops, genetic technology, and insect management: Let's talk. *Am. Entomol.* 41(4): 205–206.

Stein, R. S. 1992. Polymer recycling: Opportunities and limitations. *Proc. Natl. Acad. Sci. U.S.A.* 89:835–838.

Strizhov, N., Keller, M., Mathur, J., Konz-Lalman, Z., Bosch, D., Prudovsky, E., Schell, J., Sneh, B., Koncz, C., and Zilberstein, A. 1996. *Proc. Acad. Sci. U.S.A.* 93(26):15012–15017.

Tabashnik, B. E. 1992. Resistance risk management: Realized heritability of resistance to *Bacillus thuringiensis* in diamond back moth (Lepidoptera: *Pluetellidae*), tobacco budworm (Lepidoptera: *Noctuidae*), and Colorado potato beetle (Coleoptera: *Chrysomelidae*). *J. Econ. Entomol.* 85:1551–1559.

Tada, Y., Shimada, H., and Fujimura, T. 1994. Reduction of allergenic protein in rice grain. *Proceedings of the 3d International Symposium on the Biosafety Results of Field Tests of Genetically-Modified Plants and Microorganisms, Monterey, California, November 13–16, 1994.* Oakland, California: The University of California, Division of Agriculture and Natural Resources, 290.

Tennant, P. F., Gonsalves, C., Ling, K. S., Fitch, M., Manshardt, R., Slightom, J. L., and Gonsalves, D. 1994. Differential protection against papaya ringspot virus isolates in coat protein gene transgenic papaya and classically cross-protected papaya. *Phytopathol.* 84:1359–1366.

Tepfer, M. 1993. Viral genes and transgenic plants. *Bio/Technol.* 11:1125–1129.

Third World Network. 1995. *The need for greater regulation and control of genetic engineering. A statement by scientists concerned about the trends in the new biotechnology.* Penang, Malaysia: Third World Network.

Thwaites, T. 1993. Modified clover could create super sheep. *New Scientist* 138:19.

Truve, E. et al. 1993. Transgenic potato plants expressing mammalian 2'–5' oligoadenylate synthetase are protected from potato virus X infection under field conditions. *Bio/Technol.* 11:1048–1052.

USGS (United States Geological Survey). 1995. USGS toxic substances hydrology program. Washington, D.C.: U.S. Geological Survey, http://h20.usgs.gov/public/wid/html/ toxic.htm.

Van den Elzen, P. J. M., Jongedijk, E., Malchers, L. S., and Cornelissen, B. J. C. 1993. Virus and fungal resistance: From laboratory to field. *Phil. Trans. Roy. Soc. (London)* B342:271–278.

Van Straalen, N. M. 1997. Community structure of soil arthropods as bioindicator of soil health. In *Biological indicators of soil health*, eds. C. Pankhurst, B. M. Doube, and V. V. S. R. Gupta, 235–263. London: CAB International.

van Weerden, E. J., and Verstegen, M.W.A. 1989. Effect of pST on environmental N pollution. In *Biotechnology for control of growth and product quality in swine. Implications and acceptability*, eds. P. van der Wal, G. J. Nieuwhof, and R. D. Politick, 273–243. Wageningen, the Netherlands: Pudoc Wageningen.

Vietmeyer, N. D. 1992. *Neem, a tree for solving global problems*. Washington, D.C.: National Academy Press.

Waite, D. T., Grover, R., Westcott, N. D., Irvine, D. G., Kerr, L. A., and Sommerstad, H. 1995. Atmospheric deposition of pesticides in a small southern Saskatchewan watershed. *Environ. Toxicol. Chem.* 14(7):1171–1175.

Warmbrodt, R. D., and Stone, V. 1993. *Transgenic fish research: A bibliography*. Bibliographies and Literature of Agriculture, No. 117. Beltsville, Maryland: National Agricultural Library.

Watanabe, T., and Iwase, T. 1996. Developmental and dysmorphongenic effects of glufosinate ammonium on mouse embryos in culture. *Teratogen. Carcinogen. Mutagen.* 16(6):287–299.

Wellings, C. R., and McIntosh, R. A. 1990. *Puccina straiiformis F sp tririci* in Australia: Pathogenic changes during first 10 years. *Plant Pathol. (London)* 39(2):316–325.

WHO (World Health Organization). 1994. *Glyphosate*. Geneva, Switzerland: World Health Organization.

Wilkinson, C. F. 1990. Introduction and overview. In *the effects of pesticides on human health, proceedings of a workshop May 9–11, 1988, Keystone, Colorado*, eds. S. R. Baker and C. F. Wilkinson, 5-33. Princeton: Princeton Scientific Publishing.

Woodruff, D. S., and Gall, G. A. 1992. Genetics and conservation. *Agric. Ecosyst. Environ.* 42:53–73.

Wrage, K. 1995. Plastic producing plants. *Biotech. Reporter* 12:1–4.

Wrubel, R. P., and Gressel, J. 1994. Are herbicide mixtures useful for delaying the rapid evolution of resistance? A case study. *Weed Technol.* 8:635–648.

Wyse, R., and Krivi, G. C. 1987. Strategies for educating the public concerning biotechnology. In *Public perceptions of biotechnology*, eds. L. R., Batra, and W. Klassen. Bethesda. Maryland: Agricultural Research Institute.

Xue, B., Gonsalves, C., Provvidenti, R., Slightom, J. L., Fuchs, M., and Gosalves, D. 1994. Development of transgenic tomato expressing a high level of resistance to cucumber mosaic virus strains of subgroup I and II. *Plant Dis.* 78:1038–1041.

Yahiro, Y., Kimura, Y., and Hayakawa, T. 1994. Biosafety results of transgenic rice plants expressing rice stripe virus–coat protein gene. In *Proceedings of the 3d international symposium on the biosafety results of field tests of genetically-modified plants and microorganisms, Monterey, California, November 13–16, 1994*. Oakland, California: The University of California, Division of Agriculture and Natural Resources, 23–29.

Zinnen, T., and Voichick, J. 1994. *Biotechnology and food*. North Central Regional Publication No. 569. Madison, Wisconsin: Cooperative Extension Publications, University of Wisconsin-Extension.

9

Biotechnology and Risk

Henry I. Miller

*Hoover Institution and Institute for International Studies, Stanford University,
Stanford, California*

SCIENTIFIC CONSENSUS ABOUT BIOTECH RISK

For two decades, international scientific organizations and professional groups have grappled with the correct assumptions about risk on which to base regulation of the new biotechnology. To the pivotal question of whether the use of the new techniques is singular or widespread enough to require a new regulatory paradigm, their answer has been remarkably consistent: *Oversight should focus on the risk-related characteristics of products* (whether a living organism or an inert derivative), *not on the use of certain techniques for genetic manipulation.* It is worth describing the consensus in some detail because of the centrality of the issue and the striking congruence in the conclusions and recommendations.

The United States National Academy of Sciences (NAS) published a 1987 white paper on the planned introduction of genetically modified organisms into the environment (NAS 1987). This document has had wide-ranging impacts in the United States and internationally. Its most significant conclusions and recommendations include the following:

- Recombinant DNA techniques constitute a powerful and safe new means for the modification of organisms.
- Genetically modified organisms will contribute substantially to improved health care, agricultural efficiency, and the amelioration of many pressing environmental problems that have resulted from the extensive reliance on chemicals in agriculture and industry.
- There is no evidence of unique hazards either in the use of rDNA techniques or in the movement of genes between unrelated organisms.
- The risks associated with the introduction of rDNA-engineered organisms are the same in kind as those associated with the introduction of unmodified organisms and organisms modified by other methods.
- The assessment of the risks associated with introducing recombinant DNA organisms into the environment should be based on the nature of the organism and of the

environment into which the organism is to be introduced and be independent of the method of engineering per se.

In a 1989 extension of that analysis, the United States National Research Council (NRC), the research arm of the NAS, concluded that "no conceptual distinction exists between genetic modification of plants and microorganisms by classical methods or by molecular techniques that modify DNA and transfer genes" whether in the laboratory, in the field, or in large-scale environmental introductions (NRC 1987). The NRC report enlarged upon this statement by offering extensive observations of past experience with plant breeding, introduction of genetically modified plants, and introduction of genetically modified microorganisms as follows:

- The committees [of experts commissioned by the NRC] were guided by the conclusion (NAS 1987) that the *product* of genetic modification and selection should be the primary focus for making decisions about the environmental introduction of a plant or microorganism and not the *process* by which the products were obtained (NRC 1987, 14).
- Information about the process used to produce a genetically modified organism is important in understanding the characteristics of the product. However, the nature of the process is not a useful criterion for determining whether the product requires less or more oversight (NRC 1987, 14).
- The same physical and biological laws govern the response of organisms modified by modern molecular and cellular methods and those produced by classical methods (NRC 1987, 15).
- Recombinant DNA methodology makes it possible to introduce pieces of DNA, consisting of either single or multiple genes, that can be defined in function and even in nucleotide sequence. With classical techniques of gene transfer, a variable number of genes can be transferred, the number depending on the mechanism of transfer; but predicting the precise number or the traits that have been transferred is difficult, and we cannot always predict the phenotypic expression that will result. With organisms modified by molecular methods, we are in a better, if not perfect, position to predict the phenotypic expression (NRC 1987, 13).
- With classical methods of mutagenesis, chemical mutagens such as alkylating agents modify DNA in essentially random ways; it is not possible to direct a mutation to specific genes, much less to specific sites within a gene. Indeed, one common alkylating agent alters a number of different genes simultaneously. These mutations can go unnoticed unless they produce phenotypic changes that make them detectable in their environments. Many mutations go undetected until the organisms are grown under conditions that support expression of the mutation (NRC 1987, 14).
- Crops modified by molecular and cellular methods should pose risks no different from those modified by classical genetic methods for similar traits. As the molecular methods are more specific, users of these methods will be more certain about the traits they introduce into the plants (NRC 1987, 3).
- The types of modifications that have been seen or anticipated with molecular techniques are similar to those that have been produced with classical techniques. No new or inherently different hazards are associated with the molecular techniques. Therefore, any oversight of field tests should be based on the plant's phenotype and genotype and not on how it was produced (NRC 1987, 70).
- Established confinement options are as applicable to field introductions of plants modified by molecular and cellular methods as they are for plants modified by classical genetic methods (NRC 1987, 69).

The NRC proposed that the evaluation of experimental field testing be based on three considerations: familiarity (i.e., the sum total of knowledge about the traits of the organism and the test environment), the ability to confine or control the spread of the organism, and the likelihood of harmful effects if the organism should escape control or confinement.

The same principles were emphasized in the comprehensive 1992 report by the United States National Biotechnology Policy Board (on which I served as a charter member), which was established by the Congress and composed of representatives from the public and private sectors. The report concluded that

> the risks associated with biotechnology are not unique, and tend to be associated with particular products and their applications, not with the production process or the technology per se. In fact biotechnology processes tend to reduce risks because they are more precise and predictable. The health and environmental risks of not pursuing biotechnology-based solutions to the nation's problems are likely to be greater than the risks of going forward (NIH 1992).

These findings resonate with the observations and recommendations of the United Kingdom's (U.K.'s) House of Lords Select Committee on Regulation of the U.K. Biotechnology Industry and Global Competitiveness, which urged that, "as a matter of principle, GMO-derived products should be regulated according to the same criteria as any other product." This committee went on to excoriate the process-based approach to oversight of the United Kingdom and the European Union (EU), calling it, "excessively precautionary, obsolescent and unscientific" (HL 1993).

The Paris-based Organization for Economic Cooperation and Development (OECD), in approaching the question from the perspective of the safety of new varieties of foods, came to similar conclusions in a 1993 report titled *Concepts and Principles Underpinning Safety Evaluation of Foods Derived by Modern Biotechnology* (OECD 1993). The OECD report described several concepts related to food safety that are consistent with the NRC's and related findings as follows:

- In principle, food has been presumed to be safe unless a significant hazard was identified (OECD 1993, 13).
- Modern biotechnology broadens the scope of the genetic changes that can be made in food organisms and broadens the scope of possible sources of foods. This does not inherently lead to foods that are less safe than those developed by conventional techniques (OECD 1993, 13).
- Therefore, evaluation of foods and food components obtained from organisms developed by the application of the newer techniques does not necessitate a fundamental change in established principles, nor does it require a different standard of safety (OECD 1993, 13).
- For foods and food components from organisms developed by the application of modern biotechnology, the most practical approach to the determination of safety is to consider whether they are *substantially equivalent* to analogous conventional food product(s), if such exist (emphasis in original) (OECD 1993, 14).

Various other national and international groups have repeatedly echoed or extended these conclusions; their observations are described briefly below.

- In a joint statement from the International Council of Scientific Unions' (ICSU) Scientific Committee on Problems of the Environment (SCOPE) and the Committee on Genetic Experimentation it was noted that

> the properties of the introduced organisms and its target environment are the key features in the assessment of risk. Such factors as the demographic characterization of the introduced organisms; genetic stability, including the potential for horizontal transfer or outcrossing with weedy species; and the fit of the species to the physical and biological environment . . . apply equally to both modified or unmodified organisms; and, in the case of modified organisms, they apply independently of the techniques used to achieve modification. That is, it is the organism itself, and not how it was constructed, that is important (ICSU 1987).

- The report of a NATO Advanced Research Workshop (Rome 1987) contains the following observation:

> In principle, the outcomes associated with the introduction into the environment of organisms modified by rDNA techniques are likely to be the same in kind as those associated with introduction of organisms modified by other methods. Therefore, identification and assessment of the risk of possible adverse outcomes should be based on the nature of the organism and of the environment into which it is introduced, and not on the method (if any) of genetic modification (Fiksel and Covello 1988).

- The report of the UNIDO/WHO/UNEP Working Group on Biotechnology Safety (Paris 1987) states that

> the level of risk assessment selected for particular organisms should be based on the nature of the organism and the environment into which it is introduced (UNIDO/WHO/UNEP 1987).

RISK-ASSESSMENT EXPERIMENTS

The statements by scientific academies and professional groups, based largely on theoretical principles and the conviction that a continuum exists between "conventional" and "new" biotechnology, were not the only available "evidence" on which to base conclusions. In the late 1970s, when the regulation of recombinant DNA experimentation was at its most restrictive, it was decided that a bona fide risk assessment experiment to test one worrisome scenario would allay concerns that had been expressed. This scenario was that cloning the DNA of an animal tumor virus into a bacterium that normally resides in the human gut could produce human cancer—with the potential of an epidemic should the bacterium escape from the laboratory. Consequently, elaborate experiments were carried out to test whether a disabled strain of recombinant *Escherichia coli* containing the entire genome of the polyoma virus within its own DNA could transmit the virus to a permissive (murine) host while growing within the animal's intestine. The result was negative; that is, no viruses (or tumors) were found (Israel et al. 1990).

Although this experiment has often been cited as "evidence" of the safety of recombinant DNA experimentation, arguably it was not a particularly good risk-assessment experiment: a "positive" result (that is, a deleterious effect on a mouse) required a highly improbable sequence of molecular events. This experiment has a more recent analog: the Silwood study, a complex and elaborate experiment by Crawley and colleagues (Crawley et al. 1993) intended to assess the "invasiveness" of transgenic oilseed rape (canola), is subject to the same criticism, among others. It was a well executed but poorly designed risk assessment experiment.

In three climatically distinct sites and four habitats, the experiment compared the "invasiveness" of three variants of oilseed rape plants (two varieties modified with recombinant DNA techniques and one "unmodified" variety) over three growing seasons. The results indicated no important differences in invasiveness among the three varieties.

Although the methodology was elegant and scientifically sound, the negative result was predictable. It added little to risk assessment research to conclude that transgenic plants with genes for resistance to an antibiotic or to an herbicide are no more invasive than their unmodified parent in the absence of selection pressure from either the antibiotic or the herbicide in the test environment.

The stated purpose of the experiment, to ascertain how invasiveness "is affected by genetic engineering," is problematic when one considers the numerous analyses, including those of the NAS and NRC cited above and Crawley's own observation that "the ecology of genetically engineered organisms is exactly the same as the ecology of any other living thing" (Crawley 1990). Arguably, more useful information would have been derived from exploring the question of how invasiveness is affected by the introduction of certain traits of interest. Even Crawley, the study's director, appears to agree, admitting that if more appropriate and " 'risky' constructs like drought-tolerant perennial grasses or insect-resistant weeds had been available, then we would certainly have applied for permission to release them" (Crawley 1993). That better experimental subjects were not available was hardly a suitable rationale for proceeding.

Moreover, the result is not generalizable. The conclusion drawn from the Silwood study, that the particular variants of oilseed rape did not differ in their invasiveness under the conditions tested, enables one to infer nothing about plants in general or about genetically engineered plants in particular.

Finally, at the risk of splitting hairs, Crawley et al. refer to the recombinant DNA-modified oilseed rape as "transgenic," a term that has come to signify an organism in which genes have been introduced from across species lines by molecular techniques. However, there is evidence that the unmodified *Brassica napus*, the species of oilseed rape used in the Silwood study, originated in nature by hybridization between different species or genera. According to Goodman and colleagues (Goodman et al. 1987), the ancestor of *B. napus* was itself a hybrid between *B. oleracea* and *B. campestris*. Thus, the distinction between "wild type" (that is, unmodified) and "genetically engineered" or "transgenic" is not clear-cut, and it is open to question whether the three variants of *B. napus* were sufficiently different to provide a rationale for such a large and expensive ecological study (especially in view of the minimal scientific interest of the "transgenics" that were tested).

The point of this somewhat lengthy digression on the Silwood study is the importance of scientifically valid assumptions about risk. Without them, we design the wrong risk-assessment experiments, establish flawed regulatory paradigms, and regulate individual experiments in such a way that degree of oversight is disconnected from degree of risk.

Other investigators have examined issues in biotechnology risk assessment. In oilseed rape, J. W. Crawford and his collaborators investigated the likelihood of glufosinate-tolerant transgenes first moving into and then influencing the survival of feral populations of the crop. Monitoring airborne pollen densities at various distances from isolated oilseed rape fields over 3 years, the investigators concluded that significant quantities of pollen travel over hundreds of meters and that pollen viability is unaffected by dispersal. In view of the low frequency of herbicide application to control feral populations of oilseed rape and the absence of evidence of glufosinate application in any commercial formulation of herbicide, the investigators concluded that, "in the area surveyed, the possession of glufosinate-tolerance is unlikely to affect the survival of feral populations significantly" (Timmons et al. 1996).

Given these hardly unexpected or alarming results and conclusions, the authors' subsequent call for a "case-by-case approach to the risk-assessment of genetically modified organisms" is difficult to credit. Like Crawley et al., Crawford and his collaborators do not appear to give adequate consideration to the vast experience of plant biologists, plant breeders, and farmers with transgenic-but-nonrecombinant-DNA-manipulated plants. For example, according to Goodman et al. (1987)

> there is evidence that some of our modern crop species, such as rapeseed (*Brassica napus*), tobacco (*Nicotiana tabacum*), and wheat, originated in nature by hybridization between different species or genera. The available evidence indicates, for example, that the ancestor of *B. napus* was a hybrid between *B. oleracea* and *B. campestris*. The creation of new plant species has been mimicked in the modern era by the intentional hybridization of species from the genera *Secale* (rye) and *Triticum* (wheat) to create a new cereal crop, *Triticosecale* (triticale).

Gene flow is well-documented and certainly not unique to recombinant DNA-manipulated plants. For example, neutral or favorable cultivar genes have been found to move progressively from cultivated sunflowers to nearby wild sunflowers as a function of time. Similarly, in a population of wild strawberries growing within 50 m of a field of cultivated strawberries, more than 50% of the wild plants contained marker genes from the cultivated strawberries (King 1996).

Therefore, Crawford's apparent dismissal of the fundamental similarity (with respect to environmental or food safety risk) between his experimental molecular transgenic cultivar of oilseed rape and what I have called "transgenic-but-nonrecombinant-DNA-manipulated" cultivars of oilseed rape and triticale is puzzling. Why should recombinant cultivars require case-by-case review whereas other meagerly characterized cultivars containing *trans*-species or *trans*-genera genetic material receive no oversight whatever?

If one sets the Silwood and Crawford experiments aside, it is worth considering how one might better go about understanding the risks of recombinant DNA-modified plants. Two alternative general approaches are discussed in the next section.

THE SEARCH FOR A VALID REGULATORY PARADIGM

An understanding of the risks of recombinant DNA-modified organisms may be acquired in one of two ways. The first is by performing well-designed risk assessment experiments, to accumulate data on all the relevant issues. For plants, for example, such studies may attempt to quantify experimentally the likelihood of transforming a benign, nontoxic, noninvasive plant into one that possesses an undesirable trait(s), or the ability to induce any plant with a certain newly acquired trait to transfer that trait by means of outcrossing. An example of the latter would be herbicide resistance under the positive selection pressure of the herbicide in the test environment—an experiment similar to Crawley's but with glufosinate application to provide selective pressure.

However, many such experiments would have a very low probability of a positive result unless they were carefully designed to maximize the occurrence of a "positive" event and to detect rare events. This approach is cumbersome and often provides data of only limited usefulness. Moreover, as discussed below, it is largely unnecessary.

Alternatively, one can exploit the consensus view discussed in the previous section that there is no conceptual distinction between "conventional" (that is, pre-rDNA) genetically altered organisms and those modified with molecular techniques. On the basis of the appropriate risk assumptions that contributed to that consensus view, the United States National Science Foundation has concluded (Covello and Fiskel 1985) that risk assessment paradigms and regulatory methods in use before the introduction of the federal Coordinated Framework (EOP 1986) in 1986 provided both a useful foundation and "a systematic means of organizing a variety of relevant knowledge" for the assessment and management of recombinant DNA-modified products. Thus, the rational approach to risk-assessment when risk is not readily demonstrable (a situation sometimes referred to as "very low risk") will entail using established scientific principles and will identify significant gaps in understanding that can be addressed by conducting properly designed experiments.

This is a "vertical" approach to understanding risk that draws on appropriate precedents. It relies heavily on existing knowledge about the behavior of genetic variants produced by nature or human intervention. Thus, a tomato breeder or a government regulator of polio vaccines assessing the potential risks of a new recombinant DNA-derived tomato or vaccine, respectively, is likely to rely more heavily on background information about tomatoes and poliovirus manipulated via traditional techniques than on information about recombinant DNA-manipulated pigs or bacteria.

Important scientific questions relevant to the behavior and performance of organisms in field trials can and should be addressed systematically and in ways that are consistent with recognized scientific principles and procedures. Scientific experience and common sense can suggest approaches that avoid unnecessary and costly experiments performed in the name of risk assessment. For example, after a field trial of an obviously innocuous recombinant DNA-modified strain of *Bradyrhizobium japonicum*, Louisiana regulators "felt that continued monitoring of the field [was] needed," and a Louisiana State University professor proposed that this constituted "an opportunity to obtain valuable scientific information from the careful long-term study of the impact of this release on the environment" (USDA 1991).

The vertical approach to risk can also guide effective governmental regulation by drawing attention to the nature of the product (such as identifying a need for a high level of quality control for potentially high-hazard products such as airplanes, artificial hearts, and nuclear reactors), to circumstances related to risk (such as widespread exposure of the workplace to hazardous products or processes), and, by extension, to existing applicable risk management systems. In other words, effective risk analysis (of which risk assessment is but one component) constitutes a vertical approach to risk.

A vertical approach to biotechnology risk would take into account that governmental agencies in the United States, the European Union, Japan, and certain other countries are experienced regulators. For decades they have overseen the safety of food plants and animals, pharmaceuticals, pesticides, and other products that now can be produced using new biotechnology methods. A vertical approach to risk would emphasize the fundamental similarities among products with similar characteristics and the ways that regulators have grouped products or processes in the past, in order to oversee them effectively. A relevant example is the CDC/NIH handbook, *Biosafety in Microbiological and Biomedical Laboratories*, which specifies risk categories for a variety of pathogenic viruses, fungi, bacteria, and other agents along with the physical containment and handling procedures appropriate for each category (HHS 1988). This is a genuinely risk-based, vertical approach to risk and oversight in which any mention of whether organisms are wild-type, mutated, or recombinant DNA-manipulated is conspicuously absent. However, factors that *can* alter the risk category are taken into consideration; they include using high-concentration cultures or conditions that produce aerosols. (It is noteworthy, as well, that these guidelines for the oversight of dangerous pathogens are voluntary.)

A variation on the theme of the vertical, historical approach to risk would apply when recombinant DNA-engineered organisms are introduced into quite different environments or geographical locations from where the (parental) organism evolved: in such cases, the accumulated experience with introduced species (that is, exotics) is most appropriate for risk analysis.

A distorted alternative to the *vertical* approach to regulation and risk assessment is what I have referred to as the "horizontal" approach (Miller and Gunary 1993). In the context of biotechnology, the horizontal approach is predicated on the notion that there is something systematically similar and functionally important about the set of organisms whose only common characteristic is genetic manipulation using the techniques of the new biotechnology. The horizontal approach is focused on an artifactual "category"; namely, products made or manipulated with these techniques.

As scientific consensus dictates and the NIH polyoma virus and Silwood risk-assessment experiments illustrate, the horizontal approach is not a fruitful way to formulate or organize policy toward recombinant DNA-modified organisms, which simply do not constitute a category amenable to generalizations about safety or risk. Nonetheless, the horizontal approach is evidenced in a variety of risk assessment experiments and has been the theme of conferences and symposia. At best, the horizontal approach to recombinant DNA technology has been a wasteful nuisance; at worst, it has contributed to specious generalizations and flawed assumptions in policy making (Miller and Gunary 1993; Huttner 1995).

A horizontal approach to regulatory oversight seldom makes sense, but least so as a basis for crafting governmental regulation. Consider the hypothetical but analogous example of widely disparate regulatory requirements for peaches or tomatoes, depending on whether they were picked mechanically or by hand. In the absence of extenuating circumstances—mechanical pickers spraying toxic oil on the fruit or workers defecating in the fields during harvesting, for example—dissimilar regulation of the fruit would be hard to justify. The same can be said about dissimilar regulation of phenotypically similar or identical organisms subjected to very different regulatory strictures merely because different genetic techniques were used to craft them.

A PARADIGM FOR BIOTECHNOLOGY REGULATION

A paradigm for regulating products of the new genetic engineering may be summarized in a syllogism:

- Industry, government, and the public already possess considerable experience with the planned introduction and uses of traditional genetically modified organisms: plants for food and fiber, microorganisms live attenuated vaccines as and other uses such as sewage treatment and mining.
- Existing regulatory mechanisms have generally protected human health and the environment effectively without stifling industrial innovation.
- There is no evidence that unique hazards exist either in the use of recombinant DNA techniques or in the movement of genes between unrelated organisms.
- *Therefore*, for recombinant DNA-manipulated organisms, there is no demonstrated need for additional regulatory mechanisms to be superimposed on existing regulation.

The syllogism assumes, of course, that leaving aside the new biotechnology and its products, adequate governmental control exists over the testing and use of living organisms and their products. This assumption is certainly open to question, particularly where known dangerous pathogens, chemicals, and similar products are largely unregulated or where regulations are widely ignored. Nevertheless, throughout our pre–recombinant-DNA history of scientific research, in the United States and elsewhere, scientists have had a high degree of unencumbered freedom of experimentation with pathogens as well as nonpathogens, indoors and out. The resulting harmful incidents have been few, and the benefits, both intellectual and commercial, have far exceeded any detrimental effects.

The syllogism can be illustrated graphically. In figure 1(A), the large triangle represents the entire universe of field trials. The horizontal lines divide the universe into classes according to the safety category of the experimental organism (some examples of organisms are given on the right side of the figure). These categories can take into account the effect of a genetic change—whether it is a consequence of spontaneous mutation or the use of conventional or new techniques of genetic

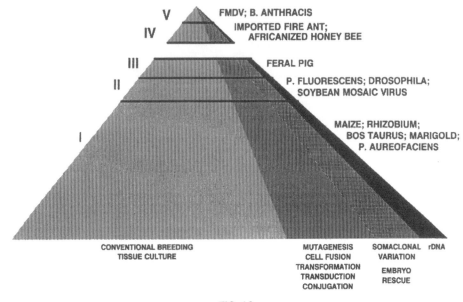

FIG. 1A.

PLANNED INTRODUCTIONS INTO THE ENVIRONMENT

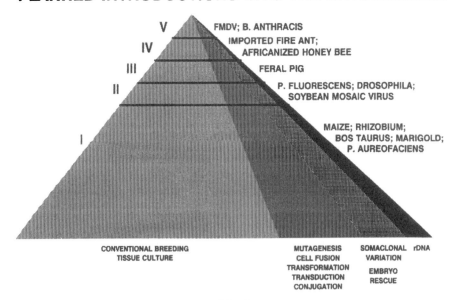

FIG. 1B.

manipulation. Such genetic changes can cause the organism to be shifted from one safety category to another. For example, if one were to grow mutagenized cultures of *Neisseria gonorrhea* (which causes gonorrhea) or *Legionella pneumophila* (which causes Legionnaire's disease) in the presence of increasing concentrations of antibiotics to select for antibiotic-resistant mutants, the classification might change from, say, Class III to Class IV.

Conversely, deletion of the entire botulin gene from *Clostridium botulinum* (the organism that causes botulism) could shift the organism from Class III to Class II or even Class I. The oblique lines divide the universe according to the use of various techniques (with techniques generally moving from left to right as they become more recent).

The oblique lines in the figure divide the universe of field trials according to the use of various genetic manipulation techniques. It is evident schematically that the use of various techniques does not itself confer safety or risk (except insofar as a genetic change wrought with recombinant DNA techniques is likely to be more precise and more fully characterized). Rather, risk is primarily a function of the *phenotype* of an organism (whether the wild type or with traits newly introduced or enhanced in some way), which, in turn, is determined by its genomic information: some organisms are destined primarily to exist symbiotically and innocuously on the roots of legumes (like the bacterium *Rhizobium*) and others to infect and kill mammals at low-inoculum concentration (Lassa fever virus).

In a risk-based oversight scheme the use of genetic manipulation techniques generally or of certain techniques in particular should not, per se, dictate the degree of oversight required. In the words of the 1989 report of the National Research Council, "Information about the process used to produce a genetically modified organism is important in understanding the characteristics of the product. However, the nature of the process is not a useful criterion for determining whether the product requires less or more oversight" (NRC 1989). Thus, the trigger for oversight or for enhanced scrutiny should be a function of the characteristics of an organism; there is no coherent scientific rationale for defining the scope of a regulatory net as, say, the "recombinant DNA slice" in Figure 1, though that has often been adopted or proposed.

Figure 1(B) differs from 1(A) by having categories IV and V lifted off to illustrate that this high-risk set of organisms is more appropriate for heightened scrutiny than the "recombinant DNA slice"—particularly if one were designing a regulatory scheme de novo. This suggestion is merely illustrative rather than prescriptive, however; depending on various considerations such as the amount of risk inherent in the test organism and the resources available for regulatory activities, the degree of regulation could be graduated in a way that oversight is commensurate with risk. As an example of such a regulatory regime, category IV and V organisms might be circumscribed for case-by-case, "every-case" reviews by governmental agencies before field trials; category III organisms, for only a notification to regulators; and category I and II organisms considered exempt.

Such a risk-based approach reflects the rational consideration of factors relevant to cost-effective protection from risks, while rejecting discredited process- or technique-based oversight. This approach also provides maximum flexibility because it can be adapted to make regulatory regimes either more risk averse (leaning toward more categories requiring case-by-case review and notification, and therefore fewer exempt) or less risk averse (leaning toward fewer categories requiring case-by-case review and notification, and more exempt). Most important, these kinds of choices about the extent and kind of government agency involvement in risk analysis would be made within a rational context.

BIOTECHNOLOGICAL FOODS AND RISK

New biotechnology, says the scientific consensus, lowers even further the already minimal risk associated with introducing new plant varieties into the food supply. The use of the latest biotechnology techniques makes the final product even safer, for it is now possible to introduce pieces of DNA that contain one or a few well-characterized genes in contrast to older genetic techniques that transfer or modify a variable number of genes haphazardly. This means users of the new techniques can be more certain about the traits they introduce into the plants. Thousands of products from plant varieties crafted with the older techniques have entered the marketplace in the last three or four decades. Only three of them (two squash varieties and one potato type) had unsafe levels of toxins, and one celery variety caused allergic skin reactions in some farm and supermarket workers, but the use of recombinant DNA techniques mitigates against any repetition of such problems.

Nevertheless, antibiotechnology activists have argued that the new molecular techniques represent a technological disjunction and therefore deserve extraordinary testing and scrutiny. They have even argued—to no avail, fortunately—the need for *clinical trials* of foods produced with new biotechnology. Thwarted in their desire to get regulators to require clinical trials or case-by-case evaluation of biotechnology-derived foods, activists have retreated to demanding labeling that would inform consumers when the techniques of new biotechnology were used in a food's manufacture. The ostensible rationale for such a requirement is that information is power and that consumers can never know too much about the products they buy—especially for foods; the more information, the better. But this is not necessarily true. A message can mislead and confuse consumers if it is irrelevant, unintelligible, or crafted to tell only part of the truth. Moreover, a requirement for labeling carries added production expenses and raises costs to both producers and consumers to a level that can constitute a barrier to the development of, and access to, new products.

To serve the consumer best, regulation should focus on genuine risks and should require only information about a food's origin or use that is relevant to safety and that supports informed choice. Mandatory labeling of all biotechnology-derived foods would achieve none of this.

Consider, for example, the long-standing commonsense approach to food labeling of the Food and Drug Administration (FDA). The FDA has required that label information be both accurate and "material." The FDA does not require a "product of biotechnology" or "genetically engineered" label for foods from plants or animals that have been improved with recombinant DNA techniques. In a 1992 food policy statement, the FDA said that labeling is required "if a food derived from a new plant variety differs from its traditional counterpart such that the common or usual name no longer applies to the new food, or if a safety or usage issue exists to which consumers must be alerted" (FDA 1992).

The 1992 policy statement emphasized that, as for other foods derived from new plant varieties, no premarket review or approval is required unless the characteristics of the new biotechnology products raise explicit safety issues. The policy emphasized that these concerns could be raised about food from new plant varieties irrespective of how they were created. The safety issues identified by FDA include the introduction of a substance that is new to the food supply (and, hence, lacks a history of safe use), increased levels of a natural toxicant (elevated alkaloids in a new variety of potato), changes in the levels of a major dietary nutrient (citrus lacking vitamin C), and transfer of an allergen to a milieu in which a consumer would not expect to find it (say, peanut protein transferred to a potato). This is the essence of a risk-based approach to food regulation.

The policy statement emphasized that any new food raising any of these safety issues requires consultation with the FDA and could be subject to regulations for premarket testing or product labeling or could be removed from the marketplace. The FDA cited the example of new allergens in a food as a possible material fact whose omission could make a label misleading. The agency reiterated that the genetic method used in the development of a new plant variety is *not* considered to be material information because there is no evidence that new biotechnology-derived foods are different from other foods in ways related to safety. Therefore, the FDA said that product labeling will not be required to disclose the method through which a new plant variety was developed. Biotechnology-derived foods would not be required to be labeled, as such.

The 1992 FDA policy statement has already been tested. A scientific article in the March 1996 issue of the *New England Journal of Medicine* reported that allergenicity common to Brazil nut proteins was transferred into soybeans by recombinant DNA manipulation and was readily identified by routine procedures (Nordler et al. 1996). In effect, this report validates at least part of the FDA policy. The plant breeder, Pioneer Hi-Bred International, was required to, and did consult with the FDA during product development. During the course of consultation and subsequent analysis, the allergenicity was identified. Confronted with the dual prospects of potential product liability and the costs of labeling all products derived from the new plant variety, the company abandoned all plans for using the new soybeans in consumer products. Not a single consumer was exposed to or injured by the newly allergenic soybeans. In what might be considered a "positive control," the system worked.

The approach the FDA took in its 1992 policy statement is consistent with scientific consensus that the risks associated with new biotechnology-derived products are fundamentally the same as for other products. Dozens of new plant varieties modified with traditional genetic techniques (such as hybridization and mutagenesis) enter the marketplace every year without premarketing regulatory review or special labeling. Moreover, many of these products are from "wide crosses" in which genes have been moved across natural breeding barriers (and without recombinant DNA techniques).

Not one of these plants exists in nature or requires or receives a premarket review by a government agency. (Safety tests by plant breeders primarily involve taste and appearance and, in the case of plants with high levels of known intrinsic toxicants—such as tomato and potato—the measurement of levels of certain alkaloids.) Nonetheless, wheat, corn, rice, oats, black currants, pumpkins, tomatoes, and potatoes have become an integral, familiar, and safe part of our diet.

The FDA's approach is consistent with that of OECD, which came to similar conclusions in a 1993 report, "Concepts and Principles Underpinning Safety Evaluation of Foods Derived by Modern Biotechnology" (OECD 1993). The OECD report described several concepts related to food safety that are consistent with the FDA's policy as well as the consensus scientific view of new biotechnology as follows:

- In principle, food has been presumed to be safe unless a significant hazard was identified (OECD 1993, 13).
- Modern biotechnology broadens the scope of the genetic changes that can be made in food organisms and broadens the scope of possible sources of foods. This does not inherently lead to foods that are less safe than those developed by conventional techniques (OECD 1993, 13).
- Therefore, evaluation of foods and food components obtained from organisms developed by the application of the newer techniques does not necessitate a fundamental change in established principles, nor does it require a different standard of safety (OECD 1993, 13).
- For foods and food components from organisms developed by the application of modern biotechnology, the most practical approach to the determination of safety is to consider whether they are *substantially equivalent* to analogous conventional food product(s), if such exist (emphasis in original) (OECD 1993, 14).

The FDA's and OECD's analyses have gone unheeded by a few antitechnology advocacy groups that have pushed for labels disclosing the use of certain genetic engineering techniques. The activists have found allies in the European Parliament, in spite of compelling reasons why special regulations and labeling requirements are often not in the best interest of consumers.

The now mandatory European labels will add significantly to the costs of processed foods made from fresh fruits and vegetables. The precise costs will vary according to the product. But a company using a gene-spliced, higher-solids, less-watery tomato (more favorable for processing), for example, would have the additional costs of segregating the product at all phases of production (planting, harvesting, processing and distribution), which adds costs and eliminates economies of scale. Labels would likely have to appear on minestrone soup indicating the presence of any amount of gene-spliced tomato, potato, or other products. The added production costs are a particular disadvantage to products in this competitive, low-profit-margin market segment and will likely relegate many gene-spliced products to the status of expensive "boutique" foods that will be out of the reach of less affluent consumers.

The EU compromise was reached after 5 years of negotiations by a joint committee of the European Parliament and the EU Council of Ministers (which represents the 15 states). The Council and the European Commission had preferred labeling only when the new food or ingredient was "significantly different" from its predecessors, but the European Parliament had its way, and labeling will be required for "live" genetically modified products—those that could, in theory, grow if put in soil—such as tomatoes or potatoes (Butler 1996). One wonders whether the label is intended to discourage the purchasers of fresh vegetables from planting them or perhaps to suggest that they should do so only after notifying the local constabulary.

The compromise has not mollified the radicals. A new "technical amendment" to the regulation adopted by the European Commission in April 1997 would require the labeling of seed products that will give rise to transgenic plants.

The European Novel Food Regulation, with or without the amendment, is irrelevant to public health. A label that proclaims that a food is derived from "genetically modified" plants provides no useful or material information to consumers but merely imposes significant cost.

It is worth noting that nongovernmental mechanisms have proved adequate to provide labeling in situations where consumers want—and are willing to pay for—certain kinds of information. For example, private-sector groups oversee the labeling of foods that are "organic" or that conform to rigorous Jewish dietary laws.

Quite apart from gene-splicing considerations, other parts of the regulation also fail to take into account scientific principles and precedents. For example, new varieties of wheat improved by the introduction of genes from hardy grasses (a common plant-improvement strategy) might be deemed "different in comparison with a conventional food or food ingredient." Under such circumstances, says the regulation, the varieties are "no longer equivalent" to preexisting foods or ingredients and would require special—and costly—labeling. (Consider, also, that through the use of sophisticated analytical techniques, the chemical composition of English potatoes can easily be distinguished from those grown in Italy; under the regulation, the two varieties—even if the same species and cultivar—may be considered nonequivalent and therefore would need to be distinguished by labeling.)

As often happens with political compromises by politicians who neither know nor care about the underlying rationality of their actions, European consumers are compromised by an outcome that makes neither scientific nor economic sense. Superfluous labeling requirements for new biotechnology-derived products constitute, in effect, an unwarranted, punitive tax on the use of a new, superior technology. The requirement exacts excess costs and reduces profits to plant breeders, farmers, food processors, grocers, and others in the distribution pathway. The power of regulatory disincentives is such that this burden could obstruct the use of new biotechnology tools from food research, development, and production.

This latter point is illustrated by a 1994 analysis of the economic impacts of a labeling requirement for new biotechnology-derived foods by the California Department of Consumer Affairs (CDCA) (Taylor 1994). The report predicted that the additional costs would be "substantial" and that "while the American food processing industry is large, it is doubtful that it would be either willing or able to absorb most of the additional costs associated with labeling biotech foods." The analysis concluded that "there is cause for concern that consumers will be unwilling to pay even the increased price for biotech foods necessary to cover biotechnology research and development, much less the additional price increases necessary to cover the costs associated with labeling biotech food."

The CDCA assessment implies another outcome of unwarranted but compulsory labeling. Overregulated and, therefore, overpriced biotechnology-derived products would be limited to upscale, higher-income markets. Wealthier consumers will be able to pay more for the improved products, whereas the less affluent will simply do without them. (The implications of such policies for developing countries are obvious—and ominous.)

Consumers, whose prices will be raised and choices diminished by a regulatory tax in the form of compulsory labeling for biotechnology-derived foods would be better served by industry's spending its resources on research and development to create better, safer, more economical products.

The Quest for Risk-Based Regulation

The uses of the new biotechnology in "contained" laboratories, pilot plants, greenhouses and production facilities have engendered relatively little controversy. The NIH Guidelines for Research Involving Recombinant DNA Research now exempt from oversight more than 99% of laboratory experiments, and this has allowed organisms of low risk to be handled under modest containment conditions. These conditions permit large numbers of living organisms to be present in the workplace and even to be released routinely from laboratories (Lincoln et al. 1983). Despite more than two decades of extensive work in thousands of laboratories throughout the United States with millions of individual genetic clones, there has been no report of these incidental releases causing a human illness or any injury to the environment.

The regulatory picture is entirely different for the testing of recombinant DNA-modified organisms outside "containment," however. As discussed above, the central issue has been the scope of regulation—in other words, what experiments and products should fall into the regulatory net. Many negligible-risk experiments have been subjected to extreme regulatory scrutiny and lengthy delays solely because recombinant DNA techniques were employed—even when the genetic change was completely characterized and benign and the organism demonstrably innocuous. The impacts have been substantial. Investigators have shied away from areas of research that require field trials of recombinant organisms (Ratner 1990); companies have avoided the newest, most precise and powerful techniques in order to manage research and development costs (Naj 1989); and investors have avoided companies whose recombinant DNA-derived products became caught up in the public controversy and new regulation (Miller 1990).

U.S. government agencies have variously regulated new biotechnology products using either previously existing regimes (FDA) or by instituting new ones (EPA, USDA, and NIH). Whether regulatory strategies are new or old, certain cardinal principles should apply. First, triggers to regulation—the criteria that specify which products and experiments warrant regulation—must be scientifically defensible. Second, the degree of oversight and compliance burdens must be commensurate with scientifically measurable risk. Some commentators have contended that this may be obvious in theory but difficult to achieve in practice. Skeptics of risk-based oversight contend that if we knew a priori which experiments and products are risky, agencies could just perform "armchair" risk assessment and exempt those proposals that pose negligible risk. Both assertions are weak. The United States and other nations have devised other regulatory nets based on assumptions about the magnitude or the distribution of risk. For example, we require permits for field trials with certain organisms known or considered to be plant pests, whereas we exempt similar organisms based on a knowledgeable assessment of predicted risk. The validity of these assumptions determines the integrity of the regulatory scheme; without them, we might as well flip a coin or exempt field trials proposed on certain days of the week.

Consistent with this regulatory philosophy, the federal government's 1986 Coordinated Framework attempted, at least on paper, to focus oversight and regulatory triggers on the *characteristics of products* and on their intended use rather than on *processes* used for genetic manipulation (EOP 1986).

In spite of the Coordinated Framework's stated goals, during the past decade the USDA and EPA have created oversight regimes for tests in the environment that conflicted with them. The agencies' policies have discriminated against organisms manipulated with recombinant DNA techniques. This discrepancy is particulary egregious given the scientific guidance offered to government regulators by the National Academy of Sciences and the National Research Council, whose reports have been described.

The Bush administration tried to adopt the recommendations of the scientific community and to guide agencies toward more scientific and risk-based policies via a 1992 policy statement on the appropriate "scope" of biotechnology regulation. Borrowing language from the NRC report, the policy stated that "the same physical and biological laws govern the response of organisms modified by modern molecular and cellular methods and those produced by classical methods . . . [Therefore] no conceptual distinction exists between genetic modification of plants and microorganisms by classical methods or by molecular techniques that modify DNA and transfer genes" (EOP 1992). In short, the regulation of biotechnology products, whether in agriculture, medicine, pharmaceuticals, manufacturing, or bioremediation should be based on any inherent risk in the product, not on the process by which it is made.

Regulators, who are often more intent on acquiring additional mandates, larger budgets, and more expansive bureaucratic empires than on serving the public interest, have expressed disinterest in the scientific consensus. Many American, foreign, and international regulatory proposals continue to be based on process; that is, on the use of certain genetic techniques. Sometimes, the regulations forthrightly circumscribe all recombinant DNA-manipulated organisms for review. On occasion, the case-by-case review requirement is more convoluted—for example, limited to organisms manifesting phenotypes that "do not exist in nature" according to the rationale that such organisms are "unfamiliar" and, by extension, pose potentially high risk. These proposals focus implicitly or explicitly on a process-determined definition of "familiar," an approach that seems to be derived from the prosaic meaning of the word; namely, "accepted, accustomed, well-known." They equate "familiarity" inappropriately with safety. Demonstrative of the circularity and inappropriateness of this approach is the practice of considering organisms "familiar" solely because they are "natural" or have been created by older, more "familiar" genetic manipulation techniques. No matter how pathogenic, invasive, or otherwise hazardous these "familiar" or "natural" organisms may be, often they are exempted from the regulatory net. At the same time, obviously innocuous recombinant DNA-manipulated organisms require case-by-case reviews.

A Risk-Based Protocol for Field Trials

As has been discussed, the regulation of recombinant DNA-manipulated organisms can be approached in the same manner as other organisms; that is, according to intended use (such as vaccines, pesticides, or food additives) or according to intrinsic risk (a function of characteristics such as pathogenicity, toxigenicity, and invasiveness). It is ironic that while various regulatory agencies in many nations were struggling to make *technique*-based regulatory schemes plausible, *risk*-based alternatives were widely available and in operation.

Field trials of any organisms that pose significant risk warrant biosafety oversight and appropriate precautions. In 1995, several colleagues and I described a biosafety protocol for field trials that is scientifically defensible and risk-based (Miller et al. 1995). The basis of the protocol is the tabulation of organisms into risk categories. It accommodates any organism, whether naturally occurring or genetically modified by old or new methods. It can provide the foundation for a cost-effective oversight system. It is adaptable to the resources and needs of different forms of oversight and regulatory mechanisms, whether they are implemented by governments or by other institutions.

According to the protocol, a researcher intending to perform a field trial would determine the biosafety level for the test organism and locale on the basis of lists that stratify or categorize organisms according to risk. This tabulation would be based on scientific knowledge and experience as compiled by experts. Several factors determine this overall level of safety concern, including ability to colonize, ecological relationships, effects on humans, potential for genetic change, and potential for monitoring and control.

Thus, the lists would provide an indication of the intrinsic level of risk of the organism ranging from, say, Level 1 (lowest safety concern) to Level 5 (greatest safety concern). An important factor in stratifying plants according to potential risk, for example, would be the presence in a geographic area of cross-hybridizing relatives of the plant to be tested (NRC 1989). The proximity of a relative does not, however, alone confirm a risk. For example, there is limited gene flow from maize to nearby teosinte (and vice versa). Even when such gene flow occurs, it appears neither to be detrimental to the teosintes nor to change their basic nature as distinctive wild races and species. Thus, the presence of teosinte near a field trial of maize does not alter the assessment of maize as posing negligible risk (Category 1). By contrast, distinct varieties of oilseed rape (*Brassica napus*, or canola) with widely differing concentrations of toxic erucic acid (and intended for different applications) should be kept segregated to avoid outcrossing between varieties. For example, high-erucic-acid canola may be classified as Category 1 in regions where that variety of the plant is grown but perhaps Category 4 where low-erucic-acid canola is grown.

This approach is analogous to that used for categorizing microorganisms by the U.S. Centers for Disease Control (CDC) and the National Institutes of Health (NIH) to establish laboratory safety standards for the handling of pathogens (HHS 1988) and to one proposed (but never implemented) by the U.S. Department of Agriculture 1991). Foreign countries' regulatory approaches that employ inclusive lists of regulated articles such as plant pests or animal pathogens operate within similar principles (Frommer et al. 1989).

As a practical matter, it would be difficult for experts to stratify every organism in every geographic region according to risk. However, categorizing, a few dozen of the major crop plants that are the likeliest candidates for field trials would be a feasible and useful beginning. Additional panels of experts could then address subsets of additional plants, fish, terrestrial animals, microorganisms, insects and other groups.

This process has, in fact, begun. A working group was assembled in January 1997 at the International Rice Research Institute in the Philippines as part of a project undertaken by the author and several collaborators at Stanford University to evaluate and refine this model and actually to begin to stratify organisms according to risk (Barton et al. 1997). The success of this exercise indicates that other groups of organisms can be approached in a similar way.

Using the Protocol to Determine Degree of Oversight

For regulators, especially those in the developing world, a crucial "first cut" is the designation of organisms that are considered to be of negligible risk (Level 1) or low risk (Level 2). This is important because, arguably, field trials in these lowest-risk categories can be exempt from case-by-case review and managed using standard research practices appropriate to the test organism and test site.

By contrast, field trials with organisms in the highest-risk categories may automatically require biosafety evaluation. (It is worth noting that this graduated approach to regulated and nonregulated field trials is both more scientific and more risk-based than the biotechnology regulatory regimes of the EPA, USDA, EU and UN).

In theory, the degree of oversight of proposed field trials can vary widely between *exempt* (that is, subject to only the usual "good agricultural practices" with the test organism) and *prior approval required* (that is, by a national, regional or international agency), with various levels in between. However, in the 1995 paper we proposed only three levels of oversight: exempt, notification to a local or international agency, or prior approval required.

In this approach, which initially estimates the risk of the field trial, other factors that are part of the process of "risk analysis" can be invoked to influence the degree of oversight required. These factors may include the available regulatory resources and the financial and manpower burden that regulation imposes on researchers and the government. As noted above, within the constraints of the model, a national or other policymaking authority could choose to apply regulatory strictures relatively more stringently (tending toward more prior approval and fewer exempt categories) or less stringently (tending toward more risk categories being exempt or requiring only notification). Of paramount importance, however, is that *the overall approach always operates within a scientifically sound context and the degree of oversight is commensurate with risk.*

For example, regulatory authorities may require case-by-case review for organisms in Categories 4 and 5, exempt experiments in Categories 1 and 2, and require a simple notification (describing the organism to be tested, the site, the identity of the investigators, and so forth) for Category 3. Within this internally consistent scheme, other permutations are possible that would be chosen to meet regional preferences and needs. Most important, this flexibility in deciding the appropriate match between risk categories and risk-management regimes would be exercised within a scientific and logical framework.

This protocol is very flexible and applicable to any organism. It meets the basic requirements of a biosafety regime: it is risk-based, scientifically defensible, and focused on the characteristics of the test organisms and the environment of the field trial. The protocol is highly adaptable: it can be incorporated into existing regulatory regimes in industrialized countries or be used by nations that currently lack such mechanisms. Moreover, it can offer adequate safety precautions to protect the public health and the environment from significant risk coupled with the cost-effectiveness demanded by limited government resources.

CONCLUSIONS

The saga of the new biotechnology and risk assessment is not a happy one. Scientific considerations have frequently fallen victim to politics, cupidity, and the self-interest of government regulators. Too often, scientists have found themselves doing paperwork instead of experiments. A poorly served public has understandably been confused by the extraordinary scrutiny applied to the testing and use of the products of the new biotechnology. As the president of a national consumer advocacy group observed cogently, "For obvious reasons, the consumer views the technologies that are *most* regulated to be the *least* safe ones. Heavy involvement by government, no matter how well intended, inevitably sends the wrong signals. Rather than ensuring confidence, it raises suspicion and doubt" (NIH 1982).

Note Added in Proof

Ongoing research on risk assessment and risk management continues to reinforce the view that recombinant DNA-modified transgenic plants do not pose unique problems and that the power of the techniques can, in fact, be used to reduce agronomic or environmental risks. Chévre et al. reported on experiments using an intergeneric model of gene flow from transgenic oilseed rape (*Brassica napus* L., or canola) containing a single copy of the *bar* gene, which confers resistance to the herbicide Basta (glufosinate ammonium), to wild radish (*Raphanus raphanistrum* L.), a widely distributed weed (Chévre, A-M, Eber, F., Baranger, A., and Renard, M. 1997. Gene flow from transgenic crops. *Nature* (London) 389:924). They concluded that intergeneric gene flow appears to occur mainly by transgene introgression within the genome of the weeds, but that this occurs slowly and at a low probability under "natural optimal conditions." Four generations were needed to provide herbicide-resistant plants (resulting from gene transfer) with a chromosome number and morphology close to that of the weed. The authors conclude that "it is likely that under normal agricultural conditions this event is rare when the wild radish is the female parent."

Daniell et al. reported a development the has important implications for the risk assessment and risk management of recombinant DNA-modified crops (Daniell, H., Datta, R., Varma, S., Gray, S., and Lee, S-B. 1998. Containment of herbicide resistance through genetic engineering of the chloroplast genome. *Nature Biotechnology* 16:345–348). Recombinant DNA techniques are used most often to modify plants by inserting genes into DNA in the nucleus and the genes can therefore spread to other plants or to wild relatives by the movement of pollen. By engineering tolerance to the herbicide glyphosate into the tobacco *chloroplast* genome, the researchers have not only obtained high levels of transgene expression but, because chloroplasts are inherited maternally in many species, the construction prevents transmission of the gene by pollen. The effect of this chloroplast-localized construction is to close a potential escape route for transgenes into the environment. This development is of particular importance for species in which pollen dispersal is extensive and male-sterility systems are absent or unreliable.

For risk analysis, as for many things, science lights the way—but only if policy makers and regulators wish to see.

REFERENCES

Barton, J., Crandon, J., Kennedy, D., and Miller, H. A model protocol to assess the risks of agricultural introductions. *Biotechnology*, 1997, in press.

Butler, D. Europe agrees: A compromise. 1996. *Nature* (London) 384: 502–503.

Covello, V. T., and J. R. Fiskel, eds. 1985. *The suitability and applicability of risk assessment methods for environmental applications of biotechnology*. Washington, D.C.: U.S. National Science Foundation.

Crawley, M. J., Hails, R. S., Rees, M., Kohn, D., et al. 1993. *Nature* 363:620–623.

Crawley, M. J. 1990. The ecology of genetically engineered organisms: Assessing the environmental risks. In *Introduction of genetically modified organisms into the environment*, eds. Mooney and G. Bernardi. New York: John Wiley & Sons.

Crawley, M. J. 1993. Arm-chair risk assessment. *Bio/Technol.* 11:1496.

EOP (Executive Office of the President). 1986. Coordinated framework for regulation of biotechnology. *Fed. Reg.* 51:23302–23347.

EOP (Executive Office of the President). 1992. Exercise of federal oversight within scope of statutory authority: Planned introductions of biotechnology products into the environment. *Fed. Reg.* 57:6753–6762.

FDA (Food and Drug Administration). 1992. Statement of policy: Foods derived from new plant varieties. *Fed. Reg.* 57:22984–23005.

Fiksel, J., and Covello, V. T. eds., 1988. Safety assurance for environmental introductions of genetically-engineered organisms. NATO ASI Series. Berlin: Springer–Verlag.

Frommer, W., Ager, B., Archer, L., et al. 1989. Safe biotechnology: III. Safety precautions for handling microorganisms of different risk classes. *Appl. Microbiol. Biotech.* 30:541.

Goodman, R. M., et al. 1987. Gene transfer in crop improvement. *Science* 236:48–54.

HHS (Department of Health and Human Services). 1988. *Biosafety in microbiological and biomedical laboratories.* Washington, D.C.: U.S. Department of Health and Human Services, HHS Publication No. (NIH) 88-8395.

HL (House of Lords). 1993. Report on regulation of the United Kingdom biotechnology industry and global competitiveness. HL Paper 80. London: Her Majesty's Stationery Office.

Huttner, S. L. 1995. Government, researchers and activists: The crucial public policy interface. In *Biotechnology.* ed. D. Brauer, 459–494. Weinheim, Germany: VCH.

ICSU (International Council of Scientific Unions). 1987. *Joint statement from the International Council of Scientific Unions' (ICSU) Scientific Committee on Problems of the Environment (SCOPE) and the Committee on Genetic Experimentation (COGENE).* Bellagio, Italy.

Israel, M. A., Chan, H. W., Rowe, W. P., et al. 1980. Molecular cloning of polyoma virus DNA in *Escherichia coli*: Plasmid vector system. *Science* 203:883–887.

Kling, J. 1996. *Science* 274:180–181.

Lincoln, D. R., Fisher, E. S., Lambert, D., et al. 1983. Release and containment of microorganisms from applied genetics activities. Report submitted in fulfillment of EPA Grant No. R-808317-01.

Miller, H. I. 1990. Governmental regulation of the products of the new biotechnology: A U.S. perspective. In *Proceedings of trends in biotechnology: An international conference organized by the Swedish council for forestry and agricultural research and the Swedish recombinant DNA advisory committee.* Stockholm: AB Boktryck HBG.

Miller, H. I. 1997. Chap. 3 of *Policy controversy in biotechnology: An insider's view.* New York: Academic Press and RG Landes, Inc..

Miller, H. I., and Gunary, D. 1993. Serious flaws in the horizontal approach to biotechnology risk. *Science* 262:1500–1501.

Miller, H. I., Altman, D. W., and Barton, J. H., et al. 1995. *Biotechnology oversight in developing countries: A risk-based algorithm. Bio/technol.* 13:955–959.

Naj, A. K. 1989. Clouds gather over the biotechnology industry. *The Wall Street Journal*, 30 Jan. 1989, 11.

NAS (National Academy of Sciences). 1987. *Introduction of recombinant DNA-engineered organisms into the environment: Key issues.* Washington, D.C.: National Academy Press.

NIH (National Institutes of Health). 1992. *National biotechnology policy board report.* Bethesda, Maryland: National Institutes of Health, Office of the Director.

NRC (National Research Council). 1989. Field testing genetically modified organisms: Framework for decisions. Washington D.C.: U.S. National Research Council/National Academy Press.

NSF (National Science Foundation). 1986. Coordinated framework for regulation of biotechnology. *Fed. Reg.* 51:23302–23347.

Nordler, J. A., Taylor, S. L., Townsend, J. A., et al. 1996. Identification of a Brazil nut allergen in transgenic soybeans. *New Eng. J. Med.* 334:688–699.

OECD (Organization for Economic Cooperation and Development). 1993. Safety evaluation of foods derived by modern biotechnology: Concepts and principles. Paris: Organization for Economic Cooperation and Development.

OECD (Organization for Economic Cooperation and Development). 1993. Safety evaluation of foods derived by modern biotechnology: Concepts and principles. 1993. Paris: Organization for Economic Cooperation and Development.

Ratner, M. 1990. BSCC addresses scope of oversight. *Bio/Technol.* 8:196–198.

Taylor, V. J. 1994. Memorandum to the labeling subcommittee, California interagency biotechnology task force.

Timmons, A. M., et al. 1996. *Nature* (London) 380:487.

UNIDO/WHO/UNEP. 1987. *Report of the UNIDO/WHO/UNEP working group on biotechnology safety.* 1987. Paris.

USDA (United States Department of Agriculture). 1991. Proposed guidelines for research involving the planned introduction into the environment of organisms with deliberately modified traits. *Fed. Reg.* 56:4134.

USDA (U.S. Department of Agriculture). 1991. *Minutes of the agricultural biotechnology research advisory committee.* Document N. 91-01. Washington, D.C.: Office of Agricultural Biotechnology, U.S. Department of Agriculture: 25–30.

Biotechnology and Safety Assessment, 2nd ed.
Edited by John A. Thomas
Copyright © 1998 Taylor & Francis

10

Safety Assessment of Insect-Protected Corn

Patricia R. Sanders, Thomas C. Lee, Mark E. Groth,
Jim D. Astwood, and Roy L. Fuchs

Monsanto Company, Saint Louis, Missouri

The success of the genetic modification of plant species for insect and virus resistance and herbicide tolerance has led to the rapid development of improved commercial varieties for many crops. There are over 40 different genetically modified plant products that have been approved in at least one country, and about 30 million acres of genetically modified plant varieties were planted in 1997 (James 1997). One of these products, the insect-protected corn event MON 810 (YieldGard™) expresses the *Bacillus thuringiensis* protein Cry1Ab. YieldGard™ corn is protected from feeding damage normally caused by the European corn borer (ECB, *Ostrinia nubilalis*), southwestern corn borer (SWCB, *Diatraea grandiosella*), and pink borer (*Sesamia cretica*). A critical step in the commercial development of these products is gaining regulatory approvals for both production and import of the genetically modified plant products. Food, feed, and environmental approvals, and, in some countries, variety registration for new plant varieties, are necessary for genetically modified plants. This chapter is an overview of the information developed and provided to the regulatory agencies to obtain food, feed, and environmental approvals for YieldGard™ corn. Variety registration is the same for varieties developed through traditional breeding or biotechnology and therefore is not discussed in this chapter.

The European corn borer (ECB, *Ostrinia nubilalis*) (Hubner) is an economically important pest of corn that has spread throughout the major corn-growing regions of the United States and Europe (Showers et al. 1989). Yield losses due to ECB damage are estimated to be 3 to 7% per borer per plant (Lynch 1980; Gay 1993), which causes annual losses from $37 to $172/ha of corn (Rice 1994). This insect typically has one to three generations per year. Insect damage to the plant includes leaf feeding by the first generation; stalk tunneling by the first and second generations; and leaf sheath, collar feeding, and ear damage by the second and third generations. Chemical control is ineffective once the borers have tunneled into the plant because the insect pest is then inaccessible to the chemical.

The benefits of planting insect-protected corn include: 1) a reliable means to control these corn pests; 2) control of target insects while maintaining beneficial species;

3) reduced use of chemical insecticides; 4) reduced applicator exposure to chemical pesticides; 5) fit with integrated pest management (IPM) and sustainable agricultural systems; 6) reduced fumonisin (class of mycotoxins) levels in maize kernels (Munkvold et al. 1997); and 7) no additional labor or machinery is required, allowing both large and small growers to maximize hybrid yields. The average yield advantage measured during the first year of commercial planting of YieldGard^{TM} corn in the US was approximately 15%.

The development of corn transformation methodology (Fromm et al. 1990) created the opportunity to protect corn plants from insect-feeding damage using genes isolated from the bacterium *Bacillus thuringiensis*. The *cryIAb* gene (Höfte and Whiteley 1989) was isolated from the *Bacillus thuringiensis kurstaki (B.t.k.)* HD-1 strain used in DiPel® (Abbott Laboratories, Chicago, Illinois), the leading microbial insecticide in agricultural use. Several laboratories have produced transgenic corn plants expressing proteins of the Cry1A class in corn (Hill et al. 1995; Armstrong et al. 1995). In 1997, insect-protected corn varieties derived from the corn event MON 810 (YieldGard^{TM}), which expresses the Cry1Ab protein, were commercialized in the United States.

The food, feed, and environmental safety assessment data generated for corn line MON 810 has been used to obtain the regulatory approvals from the United States Environmental Protection Agency (EPA) and the United States Department of Agriculture (USDA) and to complete the consultation process with the Food and Drug Administration (FDA). Food, feed, and environmental approvals have also been obtained from the appropriate agencies in Canada and Japan. Approvals in Western Europe are nearing completion. Regulatory approvals continue to be a critical step in the expanding commercial launches of YieldGard^{TM} corn in global markets.

Food, feed, and environmental safety assessments include product characterization consisting of molecular analysis of the inserted DNA, protein characterization, and protein expression levels; protein safety evaluation; compositional analysis of food components to establish substantial equivalence to commercial varieties; and environmental assessment to ensure that there will be no deleterious effect on the environment.

MOLECULAR CHARACTERIZATION

The insect-protected corn line MON 810 was produced by microprojectile bombardment of embryogenic tissue (Armstrong et al. 1991) with plasmid PV-ZMBK07 (Fig. 1). The PV-ZMBK07 plasmid contains the *cryIAb* gene, which produces the Cry1Ab protein that has insecticidal activity. The *cryIAb* gene from *Bacillus thuringiensis* subspecies HD-1 (Fischhoff et al. 1987) was modified to increase the levels of expression in plants. The enhanced cauliflower mosaic virus (CaMV) 35S promoter (Kay et al. 1985; Odell et al. 1985) and hsp70 maize intron (Rochester et al. 1986) regulate the expression of the Cry1Ab protein. The 3' nontranslated region of the nopaline synthase (NOS) gene, isolated from the Ti plasmid of *Agrobacterium tumefaciens*, terminates transcription and directs polyadenylation of the mRNA (Fraley

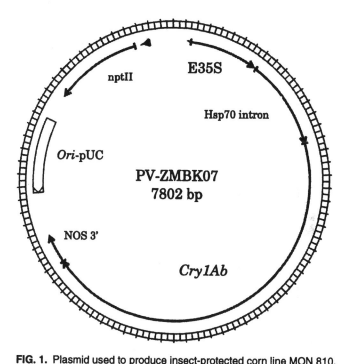

FIG. 1. Plasmid used to produce insect-protected corn line MON 810.

et al. 1983). The plasmid contains the bacterial-selectable marker gene neomy-cin phosphotransferase (*nptII*) (Beck et al. 1982), which confers resistance to amino-glycoside antibiotics (e.g., kanamycin).

Southern blot analysis of corn line MON 810 demonstrated that a single copy of the *cry1Ab* gene was integrated into the corn genome. The *nptII* gene was not integrated during transformation. The *cry1Ab* gene is inherited in the expected Mendelian pattern and is transmitted through pollen, which demonstrates stable integration into the nuclear genome. The integrity of the insert has been maintained during extensive breeding into commercial corn hybrids.

Cry1Ab PROTEIN LEVELS IN THE CORN PLANT

The Cry1Ab protein expression levels were determined in order to: define the level of active ingredient for the EPA product label, calculate expected exposure levels, support the effective dose–insect-resistance management strategy, and demonstrate the stability of the product during breeding. The Cry1Ab protein expression levels were measured on samples from four different field trials: 1994 and 1995 trials in the United States and 1995 and 1996 trials in Europe. A direct double antibody sand-wich enzyme-linked immunosorbent assay (ELISA) was developed and validated to

TABLE 1. *Levels of Cry1Ab protein in plant samples of MON 810 plants*

Plant tissue	Cry1Ab protein (mean; μg/g fresh weight of tissue)			
	1994 U.S.[a] 6 sites	1995 U.S. 5 sites	1995 E.U.[a] 4 sites	1996 E.U. 3 sites
Leaf[b]	9.3	8.9	8.6	12.2
Forage[c]	4.1	3.3	4.8	4.9
Grain[b]	0.3	0.5	0.5	0.4
Overseason Leaf[d]	(1st) 9.7 (2d) 8.4 (3d) 4.9			

[a]U.S. is United States; E.U. is European Union.
[b]The mean was calculated from the analyses of plant samples from each field site.
[c]The mean was determined from the analysis of two plants from one site in the United States and all sites in the European Union.
[d]Leaf samples collected at 2-week intervals from V4 stage until pollination.

quantify the levels of Cry1Ab protein in various plant tissues. The Cry1Ab protein levels in tissues collected from plants of line MON 810 have been consistent across several years of evaluation in the United States and Europe (Table 1). The consistency of Cry1Ab expression through years of breeding supports the stability of the insert, which is an important component of product performance. The Cry1Ab protein levels in leaf samples collected during the season and whole plant samples are sufficient to provide effective protection from both first and second generation ECB feeding damage throughout the growing season.

SAFETY ASSESSMENT OF THE Cry1Ab PROTEIN

Safety assessment of the Cry1Ab protein includes protein characterization, digestion in simulated gastric and intestinal fluids, acute oral toxicity evaluation in mice, and amino acid sequence comparison to known toxins and allergens. The use of *Escherichia coli*-produced Cry1Ab protein for safety assessment studies was justified by demonstrating the equivalence of the *E. coli*-produced and MON 810-produced protein (Lee et al. 1995).

Cry1Ab Protein Mode of Action and Specificity

The mode of action of *B. thuringiensis* delta-endotoxins such as the Cry1Ab protein has been studied extensively and reviewed (Gill et al. 1992; English and Slatin 1992; Dean et al. 1996; Yamamoto and Powell 1993; Knowles 1994). The Cry1Ab protein must be ingested by the susceptible insect to produce an insecticidal effect (Huber and Lüthy 1981). Following ingestion, Cry1-type toxins are solubilized and proteolytically processed to the active toxic core protein. After traversing the insect midgut peritrophic membrane, Cry1-type toxins selectively bind to specific receptors localized on the brush border midgut epithelium (Hofmann et al. 1988a and b). Cation-specific pores are formed that disrupt midgut ion flow and thereby cause paralysis and death of the susceptible insect.

The Cry1Ab protein is insecticidal to only lepidopteran insects. Seven of the eighteen insects screened were sensitive to Cry1Ab and Cry1Ac proteins (MacIntosh et al. 1990) and all seven insects were, as expected, lepidopterans. This specificity is directly attributable to the presence of Cry1A specific receptors in the target insects (Van Rie et al. 1990; Hofmann et al. 1988b).

There are no receptors for the protein delta-endotoxins of *B. thuringiensis* subspecies on the surface of mammalian intestinal cells; therefore, humans are not susceptible to these proteins (Hofmann et al. 1988a; Noteborn et al. 1995; Sacchi et al. 1986). In addition to the lack of receptors for the *Bacillus thuringiensis kurstaki (B.t.k.)* proteins, the absence of adverse effects in humans is further supported by numerous reviews on the safety of the *B.t.* proteins and a long history of safe use of microbial *B.t.* products (Ignoffo 1973; Shadduck 1983; Siegel and Shadduck 1989; McClintock et al. 1995).

Digestion of Cry1Ab Protein in Simulated Gastric and Intestinal Fluids

The trypsin-resistant core of the Cry1Ab protein was used in the simulated digestion study because this is the insecticidally active form of the Cry1Ab protein (Huber and Lüthy 1981). In gastric fluid, the Cry1Ab protein degraded rapidly; more than 90% of the initially added Cry1Ab protein degraded within 30 s of incubation in simulated gastric fluids, as assessed by Western blot analysis. The Cry1Ab protein bioactivity, as measured by insect bioassay, also dissipated readily; 74 to 90% of the added Cry1Ab activity dissipated within 2 min during incubation in simulated gastric fluids, which was the earliest time point measured. To put the rapid degradation of the Cry1Ab protein in the simulated gastric system into perspective, approximately 50% of solid food has been estimated to empty from the human stomach within 2 h, whereas liquid empties in approximately 25 min (Sleisenger and Fordtran 1989). In intestinal fluid, as expected, the Cry1Ab trypsin-resistant core protein did not degrade substantially after a 19.5-h incubation, as assessed by both Western blot analysis and insect bioassay. The tryptic core of this and other *B. thuringiensis* insecticidal proteins are widely known to be relatively resistant to digestion by serine proteases like trypsin, a major protease in intestinal fluid (Bietlot et al. 1989).

Acute Mouse Gavage Study with Cry1Ab Protein

An acute mouse gavage study was performed to directly assess any potential toxicity associated with the Cry1Ab protein. An acute study was considered appropriate because toxic proteins are only known to exert acute effects (Sjoblad et al. 1992).

The Cry1Ab protein was administered by gavage to 3 groups of 10 male and female mice. The targeted doses of Cry1Ab protein administered to mice were 0, 400, 1000, and 4000 mg/kg. The highest dose embodied the maximum hazard dose concept of the Environmental Protection Agency. Bovine serum albumin (BSA) was gavaged at 4000 mg/kg to the control group. At the time of sacrifice, 7 days after dosing, there were no statistically significant differences in mortality, body weights, cumulative body weight, or total food consumption between the BSA control groups and the Cry1Ab protein-treated groups. Results from this study demonstrate that the Cry1Ab protein is, as expected, not acutely toxic to mammals.

Lack of Homology of Cry1Ab Protein to Known Protein Toxins

One method for the assessment of potential toxic effects of proteins introduced into plants is to compare the amino acid sequence of the protein to known toxic proteins. Homologous proteins derived from a common ancestor are likely to share function. Therefore, it is undesirable to introduce DNA that encodes for proteins homologous to toxins. Homology is determined by comparing the degree of amino acid similarity between proteins using published criteria (Doolittle 1990). The Cry1Ab protein does not show meaningful amino acid sequence similarity when compared with known protein toxins present in the PIR, EMBL, SwissProt, and GenBank protein databases with the exception of other *B.t.* proteins.

Lack of Homology of Cry1Ab Protein to Known Allergens

The most important factor to consider in assessing allergenic potential is whether the source of the gene being introduced into plants is allergenic (FDA 1992; Metcalfe et al. 1996). *B. thuringiensis*, the source of the *cry1Ab* gene, has no history of causing allergy. In over 30 years of commercial use, there have been no reports of allergenicity to *B. thuringiensis*, including occupational allergy associated with manufacture of products containing *B. thuringiensis*.

In addition, the biochemical profile of the Cry1Ab protein provides a basis for allergenic assessment when compared with known protein allergens. Protein allergens must be stable to the peptic and tryptic digestion and the acid conditions of the digestive system if they are to pass through the intestinal mucosa to elicit an allergic response. Another significant factor contributing to the allergenicity of proteins is their high concentration in foods that elicit an allergic response (Taylor 1992; Taylor et al. 1987; Taylor et al. 1992; Metcalfe et al. 1996). The physiochemical properties of the Cry1Ab protein are clearly distinct from these characteristics of known allergens.

A comparison of the amino acid sequence of an introduced protein with the amino acid sequences of known allergens is also a useful indicator of allergenic potential. (Metcalfe et al. 1996) defined an immunologically relevant sequence comparison test for similarity between the amino acid sequence of the introduced protein and known allergens as a match of at least eight contiguous identical amino acids.

The amino acid sequences of the 219 allergens present in public domain genetic databases (GenBank, EMBL, PIR, and SwissProt) have been searched for similarity to the amino acid sequence of the Cry1Ab protein using the FASTA computer program (Pearson and Lipman 1988). No allergen homologies (Doolittle 1990) or immunologically significant sequences were identified in the Cry1Ab protein (Metcalfe et al. 1996). The two conclusions derived from this research were that (1) the *cry1Ab* gene introduced into corn does not encode a known allergen, and (2) the introduced protein does not share any immunologically significant amino acid sequences with known allergens.

In summary, the Cry1Ab protein shows no amino acid sequence similarity to known protein toxins, other than other *B.t.* proteins, and the Cry1Ab protein is rapidly degraded and its insecticidal activity lost under conditions that simulate mammalian digestion. There were no indications of toxicity as measured by treatment-related adverse effects in mice to which Cry1Ab protein was administered by oral gavage. The *cry1Ab* gene was not derived from an allergenic source, does not possess immunologically relevant sequence similarity with known allergens, and does not possess the characteristics of known protein allergens. These studies support the safety of Cry1Ab protein, and are fully consistent with the extensive history of safe use for the Cry1Ab protein, which has high selectivity for insects with no deleterious effects on other types of organisms such as mammals, fish, birds, or invertebrates (EPA 1988; McClintock et al. 1995).

COMPOSITIONAL ANALYSIS OF CORN GRAIN AND FORAGE

The design of a food and feed safety assessment program for a genetically engineered crop requires detailed understanding of the uses of the crop and crop products in animal and human nutrition. Approximately 80% of the total quantity of corn grain is fed to livestock and poultry in the United States. Food and industrial uses account for the balance. Corn is palatable, readily digested by humans and by monogastric and ruminant animals, and is one of the best sources of metabolizable energy among the grains (Wright 1988). Corn is a major food and feed source worldwide; therefore, extensive data were generated to demonstrate the food and feed safety of the insect-protected corn line MON 810.

New corn varieties developed by traditional breeding are typically selected for yield potential, and measurements of nutritional parameters are not routine. Corn feed is supplemented with protein, minerals, and vitamins to meet the nutrient requirements of animals. Food safety can be demonstrated by confirming that the new food is substantially equivalent (i.e., as safe as) the conventional food. The establishment of substantial equivalence is an important component of a food safety assessment (WHO 1991; OECD 1993; WHO 1995; FAO/WHO 1996). Completed compositional analyses demonstrate that corn line MON 810 is substantially equivalent to current corn varieties. Compositional analyses were performed on grain harvested from the 1994 U.S. field trials (Figs. 2–5), forage (Fig. 6), and grain (data not shown) harvested

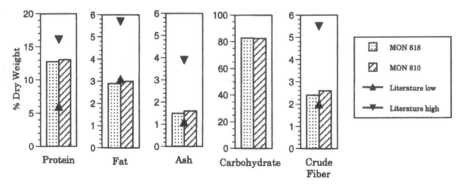

FIG. 2. Proximate analysis of corn grain harvested from six U.S. field sites in 1994. Carbohydrate values were not reported in the literature. MON 818 was the nontransgenic control; MON 810 was the transgenic corn line.

from the 1995 European Union field trials. Grain and forage of corn line MON 810 were compared with those of the control line as well as published literature values. The compositional parameters measured on grain samples from the 1994 U.S. field trials included proximate analyses (protein, fat, ash, crude fiber, and moisture), amino acid composition, fatty acid profile, calcium, and phosphorus. Carbohydrates were determined by calculation. Proximate analysis was performed on forage samples from the 1995 field trials in Europe.

The results of the proximate analysis of corn grain are summarized in Figure 2. There were no statistically significant differences between the values for the control, MON 818, and MON 810 for proximate analyses of the grain samples analyzed from the 1994 U.S. field trials. The values for the MON 818 and MON 810 lines were also comparable with the published literature (Watson 1987; Jugenheimer 1976).

Amino acid composition was completed on these corn grain samples, and the results are presented in Figure 3. In the grain samples analyzed from the 1994 U.S. field trials, no statistically significant differences were observed in 10 of the 18 amino acids. Statistically significant increases were observed for eight amino acids (cystine, tryptophan, histidine, phenylalanine, alanine, proline, serine, and tyrosine). In the samples from the 1995 European field trial, no statistically significant differences for 16 of the 18 amino acids (data not shown) were observed. Methionine and tryptophan levels were lower in MON 810 grain from Europe compared with the control grain. The values for cystine, histidine, and glutamic acid were slightly higher than the published literature range (Watson 1982) but similar to the nonmodified control. The few differences measured are minor, not consistent across multiple-year data, and are within either the published literature ranges or the values measured for the control. Therefore, these differences are not considered biologically meaningful. The observed variance could be due to environmental conditions, differences in analytical methodology, or differences from the older hybrids included in the published literature (Watson 1982).

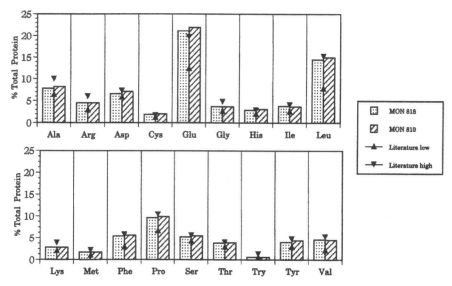

FIG. 3. Amino acid composition of corn grain harvested from six U.S. field sites in 1994. MON 818 was the nontransgenic control; MON 810 was the transgenic corn line.

The fatty acid composition was determined for the grain of corn line MON 810 and the control line, and the results are summarized in Figure 4. Ten fatty acids for which the measured values were near or below the limit of detection of the assay (arachidic, arachidonic, behenic, caprylic, capric, eicosadienoic, eicosatrienoic, eicosenoic, heptadecenoic, lauric, myristic, myristoleic, palmitoleic, and pentadecanoic) were excluded from the figure. No statistically significant differences were observed for the five fatty acids in the grain samples analyzed from the 1994 U.S. field trials. There were no statistically significant differences for four of the five fatty acids in grain

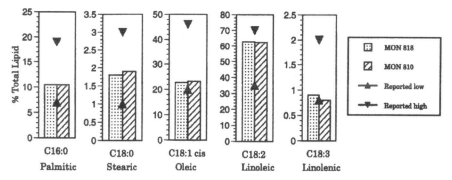

FIG. 4. Fatty acid analysis of corn grain harvested from six U.S. field sites in 1994. MON 818 was the nontransgenic control; MON 810 was the transgenic corn line.

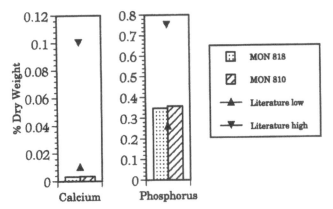

FIG. 5. Calcium and phosphorus analysis of corn grain harvested from six U.S. field sites in 1994. MON 818 was the nontransgenic control; MON 810 was the transgenic corn line.

from the 1995 European trials (data not shown). Palmitic acid was slightly higher in MON 810 grain compared with the control grain. The higher palmitic acid value was not consistent across multiple-year data and was within published ranges; therefore, it was not considered biologically meaningful.

The values for the mineral components, calcium and phosphorus, are reported in Figure 5. Statistically, the calcium level in MON 810 was significantly higher than the control line, MON 818, but this small difference was not considered biologically meaningful. The calcium values for both MON 810 and MON 818 are below the published literature range. The observed difference from literature values could be due to variances in hybrids included in the analyses, differences in analytical methodology, or both. No statistically significant differences in the phosphorus values between the MON 810 line and the control line MON 818 were noted.

The results of proximate analyses of forage collected from the 1995 European field trials are presented in Figure 6. There were no statistically significant differences in the

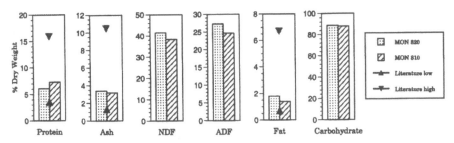

FIG. 6. Analysis of forage from four field sites in France in 1995. MON 820 was the nontransgenic control; MON 810 was the transgenic corn line. Carbohydrates, NDF (neutral detergent fiber), and ADF (acid detergent fiber) were not reported in the literature (Watson 1982).

values for fat, ash, neutral detergent fiber, acid detergent fiber, carbohydrate, and dry matter content between the MON 810 line and the control line MON 820. Statistically, the protein level was significantly increased in forage of MON 810 compared with the control protein value, but this was not considered biologically important. The observed ranges are within the published literature ranges (Watson 1982).

The grain and forage compositional data confirmed that corn line MON 810 is substantially equivalent to the parental hybrid as well as traditional corn hybrids. Processing is unlikely to alter the compositional components of corn; therefore, products derived from corn grain will also be substantially equivalent and as safe as current corn-hybrid-derived products. Similar data have been published to confirm the substantial equivalence of several genetically modified crops: Roundup Ready™ soybean (Padgette et al. 1996); New Leaf™ potato (Lavrik et al. 1995), Bollgard™ cotton (Berberich et al. 1996), Roundup Ready™ cotton (Nida et al. 1996); and Bt-tomatoes (Noteborn et al. 1995).

ENVIRONMENTAL IMPACT OF THE PRODUCT

There is extensive information on the lack of nontarget effects from microbial preparations of *B.t.k.* strains containing the Cry1Ab protein (Melin and Cozzi 1990). The full-length Cry1Ab protein encoded by the *cry1Ab* gene used to produce the insect-protected corn plants and the insecticidally active core protein produced in the insect gut following ingestion are identical to the respective full-length and trypsin-resistant core Cry1Ab proteins contained in microbial formulations that have been used safely for over 30 years. The *B.t.k.* Cry1A proteins are extremely selective for the lepidopteran insects (MacIntosh et al. 1990; Klausner 1984; Aronson et al. 1986; Dulmage 1981; Whiteley and Schnepf 1986), bind specifically to receptors on the midgut of lepidopteran insects (Wolfersberger et al. 1986; Hofmann et al. 1988a; Hofmann et al. 1988b; Van Rie et al. 1989; Van Rie et al. 1990), and have no deleterious effect on beneficial nontarget insects, including predators and parasitoids of lepidopteran insect pests or honeybee (*Apis mellifera*) (Flexner et al. 1986; Krieg and Langenbruch 1981; Cantwell et al. 1972; EPA 1988; Vinson 1989; Melin and Cozzi 1990).

To confirm and expand on the preceding results produced for the microbial products containing the same Cry1A protein as produced in YieldGard™ corn, the potential impact of the Cry1Ab protein on nontarget organisms was assessed on several representative organisms. These studies were conducted with the trypsin-resistant core of the Cry1Ab protein because this is the insecticidally active portion of the protein. The nontarget insect species included larvae and adult honey bee (*Apis mellifera* L.), a beneficial insect pollinator; green lacewing larvae (*Chrysopa carnea*), a beneficial predatory insect; hymenoptera (*Brachymeria intermedia*), a beneficial parasite of the housefly; the ladybird beetle (*Hippodamia convergens*), a beneficial predaceous insect; and earthworms (*Eisenia fetida*). Leaf material of MON 810 plants was used for the Collembola (*Folsomia candida*) nontarget soil organism study. Because of

TABLE 2. *Summary of ecological effects studies*

Organism	NOEL
Larval honey bee	$\geq$20 ppm
Adult honey bee	$\geq$20 ppm
Green lacewing larvae	$\geq$16.7 ppm
Parasitic hymenopteran	$\geq$20 ppm
Ladybird beetle	$\geq$20 ppm
Earthworms	$\geq$200 mg/kg dry soil
Collembola	50.6 μg/g leaf tissue
Daphnia magna	$\geq$100 mg pollen/liter

the potential exposure of aquatic invertebrates to corn pollen containing the Cry1Ab protein, a toxicity test was performed with *Daphnia magna*.

The mortality of nonlepidopteran insect species and three other representative organisms exposed to the Cry1Ab protein did not differ significantly from control mortality (Table 2). The no-observed-effect level (NOEL) (Table 2) was the highest concentration tested.

In addition to these studies, no acute detrimental effects were observed for three predator species (*Coleomegilla maculata, Orius insidiosus*, and *Chrysoperla carnea*) exposed to pollen of plants expressing the Cry1Ab protein (Pilcher et al. 1997). No effects were observed when *Folsomia candida* and *Oppia nitens* were fed cotton leaf material containing the Cry1Ac and Cry1Ab proteins (Yu et al. 1997). These results demonstrate the safety of the Cry1Ab protein to nontarget organisms.

In addition to assessing the potential effect of the Cry1Ab protein on nontarget organisms, a study was conducted to confirm the expected rapid degradation of the Cry1Ab protein in soil. Corn plant tissues remaining after harvest may be tilled into the soil or remain on the soil surface (no till), depending upon agricultural practices following harvest. The degradation rate of the Cry1Ab protein was assessed by measuring the decrease in insecticidal activity of transgenic corn tissue incubated in soil. The Cry1Ab protein, as a component of corn tissue, had an estimated DT50 (time to 50% reduction of bioactivity) and DT90 (time to 90% reduction of bioactivity) of 1.6 and 15 days, respectively (Sims and Holden 1996). This measured rate of degradation in soil is comparable to that reported for the Cry1Ab and Cry1Ac proteins in genetically modified cotton (Palm et al. 1994) and to the degradation rate reported for microbial *B.t.* products (West et al. 1984; West 1984; and Pruett et al. 1980). This rapid degradation further supports the lack of deleterious effects on nontarget soil organisms.

Entomologists have observed that insect populations adapt to insecticides if those insecticides are not managed correctly. Integrated pest management (IPM) was developed as a result of industry experiences with chemical insecticides. The components of the insect resistance management strategy implemented to maximize the sustainability of YieldGard™ corn include (1) expanding the knowledge of insect biology and ecology, (2) using an effective dose product that kills nearly all of the resistant heterozygote insects, (3) creating refuges to support populations of Cry1Ab-susceptible insects, (4) monitoring for any incidents of pesticide resistance and implementing a containment plan, (5) employing IPM practices that encourage ecosystem diversity and multiple tactics for insect control, (6) educating growers to ensure implementation of these strategies, and (7) developing products with alternative modes of action.

SUMMARY

The MON 810 plants have demonstrated effective control of the targeted insect pests in corn field trials since 1993. The *cry1Ab* gene has been crossed into commercial corn inbreds to produce hybrids of superior agronomic performance with resistance to lepidopteran insects. Detailed food, feed, and environmental safety assessments have confirmed the safety of this product and supported the regulatory approval of insect-protected corn line MON 810. The analyses included (1) detailed molecular characterization of the DNA introduced, (2) safety assessment of the expressed Cry1Ab protein, (3) compositional analysis of corn grain and forage, and (4) environmental impact assessment of the corn plants. These studies demonstrated that the Cry1Ab protein is safe to nontarget organisms, including humans, animals, and beneficial insects. Additionally, MON 810 corn plants and grain were shown to be substantially equivalent to (i.e., as safe as) conventional corn varieties. These data continue to be used to obtain regulatory approvals from the appropriate regulatory agencies around the world. This safety assessment approach has been used for over 40 different genetically modified plant products.

ACKNOWLEDGMENTS

We thank the many scientists who have been involved in the development of the data presented: Bruce Hammond, Janice Kania, Pam Keck, Kim Magin, Bibiana Ledesma, Elaine Levine, Michael McKee, Joel Ream, Glen Rogan, Steve Sims, and Mark Naylor. We thank Diana Kester for generating the tables and figures.

REFERENCES

Armstrong, C. L., Green, C. E., and Phillips, R. L. 1991. Development and availability of germplasm with high type II culture formation response. *Maize Genet. Cooperation NewsLetter* 65:92–93.

Armstrong, C. L., Parker, G. B., Pershing, J. C., Brown, S. M., Sanders, P. R., Duncan, D. R., Stone, T., Dean, D. A., DeBoer, D. L., Hart, J., Howe, A. R., Morrish, F. M., Pajeau, M. E., Petersen, W. L., Reich, B. J., Rodriguez, R., Santino, C. G., Sato, S. J., Schuler, W., Sims, S. R., Stehling, S., Tarochione, L. J., and Fromm, M. E. 1995. Field evaluation of European corn borer control in progeny of 173 transgenic corn events expressing an insecticidal protein from *Bacillus thuringiensis*. *Crop Science* 35(2):550–557.

Aronson, A. I., Backman, W., and Dunn, P. 1986. *Bacillus thuringiensis* and related insect pathogens. *Microbiol. Rev.* 50:1–24.

Beck, E., Ludwig, G., Auerswald, E. A., Reiss, B., and Schaller, H. 1982. Nucleotide sequence and exact localization of the neomycin phosphotransferase gene from transposon Tn5. *Gene* 19:327–336.

Berberich, S. A., Ream, J. E., Jackson, T. L., Wood, R., Stipanovic, R., Harvey, P., Patzer, S., and Fuchs, R. L. 1996. The composition of insect-protected cottonseed is equivalent to that of conventional cottonseed. *J. Agri. Food Chem.* 44(1):365–371.

Bietlot, H., Carey, P. R., Choma, C., Kaplan, H., Lessard, T., and Pozsgay, Z. 1989. Facile preparation and characterization of the toxin from *Bacillus thuringiensis* var. *kurstaki. Biochem J.* 260:87–91.

Cantwell, G. E., Lehnert, T., and Fowler, J. 1972. Are biological insecticides harmful to the honey bees? *Am. Bee J.* 112:294–296.

Dean, D. H., Rajamohan, F., Lee, M. K., Wu, S. J., Chen, X. J., Alcantara, E., and Hussain, S. R. 1996. Probing the mechanism of action of *Bacillus thuringiensis* insecticidal proteins by site-directed mutagenesis— A minireview. *Gene* 179:111–117.

Doolittle, R. F. 1990. Searching through sequence databases. *Meth. Enzymol.* 183:110.

Dulmage, H. T. 1981. *Microbial control of pests and plant diseases 1970–1980*, ed. H. D. Burges, 193–222. London: Academic Press.

English, L., and Slatin, S. L. 1992. Mini-Review. Mode of action of delta-endotoxins from *Bacillus Thuringiensis*: A comparison with other bacterial toxins. *Insect Biochem. Mol. Biol.* 22(1):1–7.

EPA. 1988. Guidance for the reregistration of pesticide products containing *Bacillus thuringiensis* as the active ingredient. NTIS PB 89-164198. Washington, D.C.: Environmental Protection Agency.

FAO/WHO. 1996. Biotechnology and food safety. Report of a Joint JAO/WHO Consultation, Rome, Italy, 30 September–4 October 1996. FAO, Food and Nutrition Paper 61. Geneva: World Health Organization.

FDA. 1992. Statement of policy: Foods derived from new plant varieties. *Fed. Regist.* 57:22984–23005.

Fischhoff, D. A., Bowdish, K. S., Perlak, F. J., Marrone, P. G., McCormick, S. M., Niedermeyer, J. G., Dean, D. A., Kusano-Kretzmer, K., Mayer, E. J., Rochester, D. E., Rogers, S. G., and Fraley, R. T. 1987. Insect-tolerant transgenic tomato plants. *Biotechnol* 5:807–813.

Flexner, J. L., Lighthart, B., and Croft, B. A. 1986. The effects of microbial pesticides on non-target beneficial arthropods. *Agric. Ecosys. Environ.* 16:203–254.

Fraley, R. T., Rogers, S. G., Horsch, R. B., Sanders, P. R., Flick, J. S., Adams, S. P., Bittner, M. L., Brand, L. A., Fink, C. L., Fry, J. S., Galluppi, G. R., Goldberg, S. B., Hoffmann, N. L., and Woo, S. C. 1983. Expression of bacterial genes in plant cells. *Proc. Natl. Acad. Sci. U.S.A.* 80:4801–4807.

Fromm, M. E., Morrish, F., Armstrong, C. L., Williams, R., Thomas, J., and Klein, T. M. 1990. Inheritance and expression of chimeric genes in the progeny of transgenic maize plants. *Bio/Technol.* 8:833–839.

Gay, P. 1993. Semences et Biotechnologies, éléments pour une stratégie. *Phytoma* 451:16–19.

Gill, S. S., Cowles, E. A., and Pietrantonio, P. V. 1992. The mode of action of *Bacillus thuringiensis* endotoxins. *Annu. Rev. Entomol.* 37:615–636.

Hill, M., Launis, K., Bowman, C., McPherson, K., Dawson, J., Watkins, J., Koziel, M., and Wright, M. S. 1995. Biolistic introduction of a synthetic *Bt* gene into elite maize. *Euphytica* 85:119–123.

Hofmann, C., Lüthy, P., Hutter, R., and Pliska, V. 1988a. Binding of the delta endotoxin from *Bacillus thuringiensis* to brush-border membrane vesicles of the cabbage butterfly (*Pieris brassicae*). *Eur. J. Biochem.* 173:85–91.

Hofmann, C., Vanderbruggen, H. V., Höfte, H., Van Rie, J., Jansens, S., and Van Mellaert, H. 1988b. Specificity of *B. thuringiensis* delta-endotoxins is correlated with the presence of high affinity binding sites in the brush border membrane of target insect midguts. *Proc. Natl. Acad. Sci. U.S.A.* 85:7844–7848.

Höfte, H., and Whiteley, H. R. 1989. Insecticidal crystal proteins of *Bacillus thuringiensis. Microbiol. Rev.* 53:242–255.

Huber, H. E., and Lüthy, P. 1981. *Bacillus thuringiensis* delta-endotoxin: Composition and activation. In *Pathogenesis of invertebrate microbial diseases*, ed. E. W. Davidson, 209–234. Montclair, New Jersey: Allanheld, Osmun Publishers.

Ignoffo, C. M. 1973. Effects of entomopathogens on vertebrates. *Ann. N.Y. Acad. Sci.* 217:141–172.

James, C. 1997. Global status of transgenic crops in 1997. ISAAA Briefs No. 5. Ithaca, New York: ISAAA pp. 31.

Jugenheimer, R. W. 1976. Corns for special purposes and uses. In *Corn: Improvement, seed production, and uses*, 227. New York: John Wiley & Sons.

Kay, R., Chan, A., Daly, M., and McPherson, J. 1985. Duplication of CaMV 35S promoter sequences creates a strong enhancer for Plant Genes. *Science* 236:1299–1302.

Klausner, A. 1984. Microbial insect control. *Bio-Technol.* 2:408–419.

Knowles, B. H. 1994. Mechanism of action of *Bacillus thuringiensis* delta-endotoxins. *Adv. Insect Physiol.* 24:275–308.

Krieg, A., and Langenbruch, G. A. 1981. Susceptibility of arthropod species to *Bacillus thuringiensis*. In *Microbial control of pests and plant diseases*, ed. H. D. Burges, 837–896. London: Academic Press.

Lavrik, P. B., Bartnicki, D. E., Feldman, J., Hammond, B. G., Keck, P. J., Love, S. L., Naylor, M. W., Rogan, G. J., Sims, S. R., and Fuchs, R. L. 1995. Safety assessment of potatoes resistant to Colorado potato beetle. In *Genetically Modified Foods, safety aspects*, eds. K. H. Engel, G. R. Takeoka, and R. Teranishi, 148–158. Washington, D.C.: American Chemical Society.

Lee, T. C., Zeng, J., Bailey, M., Sims, S. R., Sanders, P. R., and Fuchs, R. L. 1995. Assessment of equivalence of insect protected corn- and *E. coli*-produced *B.t.k.* HD-1 protein. *Plant Physiol. Suppl.* 108:151.

Lynch, R. E. 1980. European corn borer: Yield losses in relation to hybrid and stage of corn development. *J. Econ. Entomol.* 73:159–164.

MacIntosh, S. C., Stone, T. B., Sims, S. R., Hunst, P., Greenplate, J. T., Marrone, P. G., Perlak, F. J., Fischhoff, D. A., and Fuchs. R. L. 1990. Specificity and efficacy of purified *Bacillus thuringiensis* proteins against agronomically important insects. *J. Insect Path.* 56:258–266.

McClintock, J. T., Schaffer, C. R., and Sjoblad, R. D. 1995. A comparative review of the mammalian toxicity of *Bacillus thuringiensis*-based pesticides. *Pestic. Sci.* 45:95–105.

Melin, B. E., and Cozzi, E. M. 1990. Safety to nontarget invertebrates of lepidopteran strains of *Bacillus thuringiensis* and their β-exotoxins. In *Safety of microbial insecticides*, eds. M. Laird, L. A. Lacey, and E.W. Davidson, 150–167. Boca Raton, Florida: CRC Press.

Metcalfe, D. D., Astwood, J. D., Townsend, R., Sampson, H. A., Taylor, S. L., and Fuchs, R. L. 1996. Assessment of the allergenic potential of foods derived from genetically engineered crop plants. *Crit. Rev. Food Sci. Nutri.* 36(s):S165–S186.

Munkvold, G. P., Hellmich, R. K., and Showers, W. B. 1997. Reduced fusarium ear rot and symptomless infection in kernels of maize genetically engineered for European corn borer resistance. *Phytopathol.* 87(10):1071–1077.

Nida, D. L., Patzer, S., Harvey, P., Stipanovic, R., Wood, R., and Fuchs, R. L. 1996. Glyphosate-tolerant cotton: The composition of the cottonseed is equivalent to that of conventional cottonseed. *J. Agri. Food Chem.* 44(7):1967–1974.

Noteborn, P. J. M., Bienenmann-Ploum, M. E., van den Berg, J. H. J., Alink, G. M., Zolla, L., Reynaerts, A., Pensa, M., and Kuiper, H. A. 1995. Safety assessment of the *Bacillus thuringiensis* insecticidal crystal protein CRYIA(b) Expressed in transgenic tomatoes. In *Genetically modified foods, safety aspects*, eds. K. H. Engel, G. R. Takeoka, and R. Teranishi, 134–147. Washington, D.C.: American Chemical Society.

Odell, J. T., Mag, F., and Chua, H.-H. 1985. Identification of DNA sequences required for activity of the cauliflower mosaic virus 35S promoter. *Nature* (London) 313:810–812.

OECD (Organization for Economic Cooperation and Development). 1993. *Safety evaluation of foods produced by modern biotechnology: Concepts and principles*. Paris: OECD.

Padgette, S. R., Taylor, N. B., Nida, D. L., Bailey, M. R., MacDonald, J., Holden, L. R., and Fuchs, R. L. 1996. The composition of glyphosate-tolerant soybean seeds is equivalent to that of conventional soybeans. *J. Nutri.* 126:702–716.

Palm, C. J., Donnegan, K., Harris, D., and Seidler, R. 1994. Quantification in soil of *Bacillus thuringiensis* var *kurstaki* delta-endotoxin from transgenic plants. *Mol. Ecol.* 3:145–151.

Pearson, W., and Lipman, D. 1988. Improved tools for biological sequence comparison. *Proc. Natl. Acad. Sci. U.S.A.* 85:2444–2448.

Pilcher, C. D., Obrycki, J. J., Rice, M. E., and Lewis, L. C. 1997. Preimaginal development, survival, and field abundance of insect predators on transgenic *Bacillus thuringiensis* corn. *Environ. Entomol.* 26(2):446–454.

Pruett, C. J. H., Burges, H. D., and Wyborn, C. H. 1980. Effect of exposure to soil on potency and spore viability of *Bacillus thuringiensis*. *J. Invertebr. Pathol.* 35:168–174.

Rice, M. E. 1994. Aerial application of insecticides for control of second generation European corn borer: 1991–1993. *Arthropod Manage. Tests* 19:204–206.

Rochester, D. E., Winter, J. A., and Shah, D. M. 1986. The structure and expression of maize genes encoding the major heat shock protein, hsp70. *EMBO J.* 5:451–458.

Sacchi, V. F., Parenti, P., Hanozet, G. M., Giordana, B., Luthy, P., and Wolfersberger, M. G. 1986. *Bacillus thuringiensis* toxin inhibits K^+-gradient-dependent amino acid transport across the brush border membrane of *Pieris brassicae* midgut cells. *FEBS Lett.* 204:213–218.

Shadduck, J. A. 1983. Some observations on the safety evaluation of nonviral microbial pesticides. *Bull. W.H.O.* 61:117–128.

Showers, W. B., Witkowski, J. F., Mason, C. E., Calvin, D. D., Higgins, R. A., and Dively, G. P. 1989. European corn borer development and management. North Central Regional Publication 327. Ames, Iowa: Iowa State University.

Siegel, J. P., and Shadduck, J. A. 1989. Safety of microbial insecticides to vertebrates humans. In *Safety of microbial insecticides*, eds. M. Laird, L. A. Lacey, and E. W. Davidson, 102–113. Boca Raton, Florida: CRC Press.

Sims, S. R., and Holden, L. R. 1996. Insect bioassay for determining soil degradation of *Bacillus thuringiensis* subsp. *kurstaki* CryIA(b) protein in corn tissues. *Physiol. Chem. Ecol.* 25(3):659–664.

Sjoblad, R. D., McClintock, J. T., and Engler, R. 1992. Toxicological considerations for protein components of biological pesticide products. *Regulat. Toxicol. Pharmacol.* 15:3–9.

Sleisenger, M. H., and Fordtran, J. S. 1989. *Gastrointestinal disease*. Vol. 1, *Pathophysiology diagnosis management*. 4th ed., 685–689, Toronto: W. B. Saunders Co.

Taylor, S. L. 1992. Chemistry and detection of food allergens. *Food Technol.* 39:146–152.

Taylor, S. L., Lemanske, R. F., Jr., Bush, R. K., and Busse, W. W. 1987. Food allergens: Structure and immunologic properties. *Ann. Allergy* 59:93–99.

Taylor, S. L., Nordlee, J. A., and. Bush, R. K. 1992. Food allergies. In *Food safety assessment*, ACS Symposium Series 484, eds. J. W. Finley, S. F. Robinson, and D. J. Armstrong. Washington, D.C.: American Chemical Society.

Van Rie, J., Jansens, S., Höfte, H., Degheele, D., and Van Mellaert, H. 1989. Specificity of *Bacillus thuringiensis* δ-endotoxins, importance of specific receptors on the brush border membrane of the midgut of target insects. *Eur. J. Biochem.* 186:239–247.

Van Rie, J., Jansens, S., Höfte, H., Deghelle, D., and Van Mellaert, H. 1990. Receptors on the brush border membrane of the insect midgut as determinants of the specificity of *Bacillus thuringiensis* delta-endotoxins. *Appl. Environ. Microbiol.* 56:1378–1385.

Vinson, S. B. 1989. Potential impact of microbial insecticides on beneficial arthropods in the terrestrial environment. In *Safety of microbial insecticides*, eds. M. Laird, L. A. Lacey, and E. W. Davidson, 43–64. Boca Raton, Florida: CRC Press.

Watson, S. A. 1982. Corn: Amazing maize. General properties. In *CRC handbook of processing and utilization in agriculture*, Volume II: Part 1 *Plant products*, ed. I. A. Wolff, 3–29. Boca Raton, Florida: CRC Press.

Watson, S. A. 1987. Structure and composition. In *Corn: Chemistry and technology*, eds. S. A. Watson and P. E. Ransted, 53–82. St. Paul, Minnesota: American Association of Cereal Chemists, Inc.

West, A. W., Burges, H. D., White, R. J., and Wyborn, C. H. 1984. Persistence of *Bacillus thuringiensis* parasporal crystal insecticidal activity in soil. *J. Invertebr. Pathol.* 44:128–133.

West, A. W. 1984. Fate of the insecticidal, proteinaceous parasporal crystal of *Bacillus thuringiensis* in soil. *Soil Biol. Biochem.* 16:357–360.

Whiteley, H. R., and Schnepf, H. E. 1986. The molecular biology of parasporal crystal body formation in *Bacillus thuringiensis*. *Ann. Rev. Microbiol.* 40:549–576.

WHO. 1991. *Strategies for assessing the safety of foods produced by biotechnology*. Report of a Joint FAO/WHO Consultation. Geneva: World Health Organization.

WHO. 1995. *Application of the principles of substantial equivalence to the safety evaluation of foods and food components from plants derived by modern biotechnology*. Geneva: World Health Organization, Food Safety Unit.

Wolfersberger, M. G., Hofmann, C., and Lüthy, P. 1986. Interaction of *Bacillus thuringiensis* delta-endotoxin with membrane vesicles isolated from lepidopteran larval midgut. In *Bacterial protein toxins*, eds. P. Falmagne, F. J. Fehrenbech, J. Jeljaszewics, and M. Thelestam. 237–238, Port Jervis, New York: Lubrecht & Cramer, Ltd.

Wright, K. N. 1988. Nutritional properties and feeding value of corn and its by-products. In *Corn: Chemistry and technology*, eds. S. A. Watson and P. E. Ramstad, 447–478. St. Paul, Minnesota: American Association of Cereal Chemists.

Yamamoto, T., and Powell, G. K. 1993. *Bacillus thuringiensis* crystal proteins: Recent advances in understanding its insecticidal activity. In: *Advanced engineered pesticides*, ed. L. Kim, 3–4. New York: Marcel Dekker, Inc.

Yu, L., Berry, R. R., and Croft, B. A. 1997. Effects of *Bacillus thuringiensis* toxins in transgenic cotton and potato on *Folsomia candida* (Collembola: Isotomidae) and *Oppia nitens* (Acari: Orbatidae). *Ecotoxicol.* 90(1):113–118.

Biotechnology and Safety Assessment, 2nd ed.
Edited by John A. Thomas
Copyright © 1998 Taylor & Francis

11

Pharmacology of Recombinant Proteins

James E. Talmadge

University of Nebraska Medical Center, Omaha, Nebraska

Cytokines, growth factors, and other recombinant proteins are some of the most rapidly growing areas of biotechnology. The development of these bioengineered drugs is occurring at an astonishing pace with rapid preclinical and clinical development and licensing by regulatory agencies. In addition, the availability of the gene sequences and rational drug design technologies has resulted in the development of engineered genes, proteins, and peptidomimetics. In contrast to traditional pharmacophores, which are developed based on the identification of the maximum tolerated dose (MTD), most recombinant proteins have a very high binding affinity, unique biodistributions, and unusual pharmacodynamic and pharmacokinetic attributes. In this chapter, representative cytokines, including IFN-α, IFN-γ, and interleukin-2 (IL-2), are used to demonstrate the importance of cytokine pharmacology in optimal cytokine administration for biologic activity and development. This includes the conceptual need for chronic immunoaugmentation to foster optimal therapeutic activity, the need to consider the pharmacokinetics of administration to optimize drug delivery, and the nonlinear dose–response relationship, which can result in a bell-shaped dose response.

Because these therapeutic agents have maximal potential in an adjuvant protocol, their development in concert with high-dose chemotherapy and stem cell rescue is discussed. The strategies for combination chemotherapy and immunotherapy, although they hold great promise, require close attention to the pharmacodynamics of protein administration to enhance the likelihood for failure-free and overall survival. This chapter discusses the pharmacodynamics of recombinant proteins that are currently either licensed or in active clinical trials in the United States, and these biopharmaceuticals are presented in the context of the strategies used for their development.

RECOMBINANT CYTOKINE

Cytokines and growth factors are relatively low-molecular-weight proteins that are secreted in minute quantities. They act either in an autocrine fashion on the cell from which they are secreted or in a paracrine manner on adjacent cells. The isolation of

TABLE 1. Approved biotechnology drugs and vaccines

Name	Company	Indication	Date
Actimmune, interferon gamma-1b	Genentech, Inc.	Management of chronic granulomatous disease	Dec 1990
Alferon N, interferon alfa-n3	Interferon Sciences, Inc.	Genital warts	Oct 1989
Betaseron, R interferon beta-1b	Berte Laboratories, Chiron Therapeutics	Relapsing, remitting multiple sclerosis	July 1993
Engerix-B Hepatitis B vaccine R	SmithKline Beecham	Hepatitis B	Sep 1989
Epogen, Epoetin alfa (EPO)	Amgen, Inc.	• Treatment of anemia associated with chronic renal failure, including patients on dialysis and not on dialysis; anemia in Retrovir-treated HIV-infected patients; treatment of anemia caused by chemotherapy in patients with nonmyeloid malignancies	Jun 1989; Apr 1993
Procrit, Epoetin (rEPO))	Ortho Biotech, Inc.	• Treatment of anemia associated with chronic renal failure, including patients on dialysis and not on dialysis; anemia in Retrovir-treated HIV-infected patients; treatment of anemia caused by chemotherapy in patients with nonmyeloid malignancies. • Procrit was approved for marketing under Amgen's Epoetin alfa PLA. Amgen manufacturers the product for Ortho Biotech. Amgen licensed the U.S. rights to Epoein alfa to Ortho Pharmaceutical for indications excluding dialysis and diagnostics.	Dec 1990; Apr 1993
Intron A, interferon alfa-2b (R)	Schering-Plough	Hairy cell leukemia; genital warts; AIDS-related Kaposi's sarcoma; hepatitis C; hepatitis B, melanoma	July 1992
Leukine sargramostim (yeast-derived GM-CSF)	Immunex Corporation	Autologous bone marrow transplantation	Mar 1991

Product	Company	Indication	Date
Neupogen, filgrastim (rG-CSF)	Amgen, Inc.	Chemotherapy-induced neutropenia; autologous or allogeneic bone marrow transplantation; chronic severe neutropenia	Feb 1991; Jun 1994; Dec 1994
Proleukin, aldesleukin (Interleukin-2)	Chiron Therapeutics	Renal cell carcinoma, melenoma	May 1992
Recombinate HB, hepatitis B vaccine	Merck	Hepatitis B prevention	July 1986
Roferon-A, interferon alfa-2a	Hoffmann—LaRoche	Hairy cell leukemia; AIDS-related Kaposi's sarcoma	Jun 1986; Nov 1988
IL-11 Neumega (GM-CSF)	Genetics Institute, Immunex	Thrombocytopenia prophylaxis and treatment of chemotherapy-induced neutropenia; prophylaxis of chemotherapy-induced neutropenia in acute myelogenous leukemia (AML); reduction of postoperative infections, neonatal sepsis; adjuvant to AIDS therapy, HIV infection	1997 Application Submitted Phase III Phase II

cDNA clones for cytokines and growth factors has permitted their production in large and reproducible quantities, which in turn has also accelerated the preclinical and clinical study of their functions and therapeutic attributes. The ability to cut and rejoin DNA at any desired site or to introduce point mutations at directed sites has resulted in the development and clinical use of mutant as well as chimeric therapeutic proteins. Thus, we can now utilize proteins that are either exact or mutated forms of the naturally occurring ones or design proteins that are composed of various polypeptide structures derived from different sequences (e.g., the humanized monoclonal antibodies). The introduction of point mutations has resulted in drugs with decreased toxicity, better production capabilities, or higher expression levels. These variants have included mutant IL-2, IL-3, or CSF-g molecules (Kuga et al. 1989; Lu et al. 1989; Rosenberg et al. 1994) in which, for example, serines are substituted for cysteines to reduce the development of aberrant tertiary structures. Furthermore, chimeric cytokines have been developed with properties associated with multiple parent structures such as PIXY, which is a single biologically active drug composed of IL-3 and CSF-g (Bhalla et al. 1992; Bruno et al. 1992). More recently, the active binding site of IL-3 has been combined with a variety of growth functions, including thrombopoietin (TPO) (MacVittie et al. 1996; Monroy et al. 1987).

Therapeutic proteins have emerged as an important class of drugs for the treatment of cancer, myelodysplasia, and infectious diseases (Oldham 1982). However, their development has been slowed by poorly predictive models and our superficial knowledge of the pharmacology and mechanism of action. To facilitate the development of these immunoregulatory proteins, additional information is needed on their pharmacology (Mihich 1986; Talmadge and Herberman 1986). One approach for the development of cytokines is to propose a clinical hypothesis based on the preclinical identification of therapeutic surrogates. Notably, this strategy was recently accepted formally by the Food and Drug Administration (FDA, 1993). A surrogate for clinical efficacy may be a phenotypic, biochemical, enzymatic, functional (immunologic, molecular, or hematologic), or quality-of-life measurement believed to be associated with therapeutic activity. Phase I clinical trials can then be designed to identify the optimal and maximum tolerated dose or treatment schedule that most enhances the augmentation of the surrogate end point(s). Subsequent phase II and III trials can then be established to determine if the changes in the surrogate levels correlate with therapeutic activity. Table 1 lists the immunologically and hematologically active proteins that have been approved for general use. The cytokines that are currently in development within clinical trials are listed in Table 2 along with the indications under investigation.

In contrast to strategies based on the identification of surrogates for efficacy, many approaches to the development of recombinant proteins have been predicted on the basis of practices used for conventional low-molecular-weight drugs. However, the latter approach of more is better is not advantageous for the development of proteins. The unique pharmacological attribute of proteins requires selective or targeted delivery to the desired site (i.e., the bone marrow, spleen, or tumor) and consideration of their pharmacokinetics (Tomlinson 1989; Tomlinson 1991). To administer proteins

TABLE 2. *Biotechnology drugs in development*

Approved drugs		
Product type	Abbreviated indication	U.S. status
Colony Stimulating Factors		
CSF-GM	Adjuvant to chemotherapy	Phase I/II
CSF-GM	Low blood cell counts	Submitted
Sargramostim (CSF-GM)	Allogeneic bone marrow transplantations, chemotherapy adjuvant	Phase III
	Adjuvant to AIDS therapy	Phase II
CSF-M	Fungal disease, hypercholesterolemia	Phase I
Filgrastim (R-CSF-G)	AIDS, leukemia aplastic anemia	Submitted
Sargramostim (CSF-GM)	Neutropenia to secondary chemotherapy	Phase III
PIXY 321	Neutropenia,/thrombocytopenia	Phase I/II
Stem-Cell Factor (SCF)	Neutropenia/thrombocytopenia	Phase II
TPO	Thrombocytopenia	Pahse I
Flt-3 ligand	Neutropenia/thrombocytopenia, vaccines	Phase I
Erythropoietins		
Epoetin beta	Anemia secondary to kidney disease	Submitted
	Autologous transfusion	Phase II/III
Epoetin alpha	Anemia of cancer and chemotherapy	Submitted
	Anemia of surgical blood loss, autologous transfusion	Phase III
Interferons		
Interferon gamma-1b	Small-cell lung cancer, atopic dermatitis	Phase III
	Trauma-related infections, renal cell carcinoma asthma and allergies	Phase II
Interferon alfa-n3	ARC, AIDS	Phase I/II
Interferon beta	Cancer	Phase I/II
Interferon gamma	Rheumatoid arthritis	Phase II/III
	Venereal warts	Phase II
Interferon consensus	Cancer, infectious disease	Phase II/III
Interferon gamma	Cancer, infectious disease	Phase II
Interferon alfa 2b	Superficial bladder cancer, basal cell carcinoma, chronic hepatitis B, delta hepatitis	Submitted
	Acute hepatitis B, delta hepatitis, acute chronic myelogenous leukemia	Phase III
	HIV (with Retrovir)	Phase I
Interferon beta	Unresponsive malignant diseases	Phase I
Interferon alfa-2a	Colorectal cancer (with 5-fluorouracil) chronic, acute hepatitis b; non-A, non-B	Phase II
	Hepatitis, chronic myelogenous leukemia, HIV positive, ARC, AIDS (with Retrovir)	
Interleukins		
PEG IL-2	AIDS (with Retrovir)	Phase I
Aldesleukin (IL-2)	Cancer	Phase II/III
	Kaposi's sarcoma (with Retrovir)	Phase I
Human IL-1 alpha	Bone marrow suppression (chemo-, radiotherapy)	Phase I/II
Human IL-1 beta	Bone marrow suppression, melanoma, immunotherapy	
	Wound healing	Phase II
Human IL-2	Cancer immunotherapy	Phase III
Human IL-2	Cancer immunotherapy (with Roferon-A)	In clinical trials

Continued.

TABLE 2. *Continued.*

Approved drugs		
Product type	Abbreviated indication	U.S. status
Human IL-3	Bone marrow failure, platelet deficiencies, autologous marrow transplant,	Phase II/III
	Chemotherapy adjuvant	Phase II
	Peripheral stem-cell transplant	
Human IL-4	Immunodeficient disease, cancer therapy, vaccine adjuvant, immunization	Phase I/II
Human IL-4	Cancer immunomodulator	Phase II
Human IL-6	Platelet deficiencies	Phase II
Human IL-9	Thrombocytopenia	Phase I/II
Human IL-11	Reduction of postoperative infections, Neonatal sepsis, adjuvant in AIDS therapy	Phase II/III
Human IL-12	Neoplasia (renal cell)	Phase I
Tumor Necrosis Factors		
TNF	Cancer	Phase II
TNF-β	Cancer	Phase II
Others		
Anakinra (IL-1 receptor antagonist)	AML, CML, inflammatory bowel disease, rheumatoid arthritis, sepsis, septic shock	Phase II

optimally as drugs and ensure their targeting are the primary challenges for their development. An additional difficulty in the development of a recombinant protein is that in many instances there is little relationship between the dose administered and the biological effect. Indeed, in some instances the nonlinear dose relationship has been described as bell-shaped (Talmadge et al. 1987). This dose–response relationship, or

TABLE 3. *Hematopoietic growth factor characteristics*

GF	Other	Chrom	Homo	Source	Molecular weight	N Link	S-S
IL-3	MCGF	5q23-31	29%	T, NK, mast cells	15–17	2	1
EPO		7pter-q22	80%	Kidney, liver	36	3	2
G-CSF		17q21-22	73%	M, Str, Endo, Fib	21	0	2
GM-CSF		5q21-q32	56%	T, Mac, Fib, Endo	22	2	2
M-CSF	CSF-1	5q33.1	82%	L, Mac, Fib, Epit, Endo, Osteo, Myo	45–90	3	7–9
SCF	SF, cKitL,		81%	Str, Brain, liver, kidney, placenta, fib, testes, oocytes	36	5	2
TPO	cMPL L		84%	Liver, kidney, Mus, aplastic	60	6	2
FLT3	FLK2		86%	β, BM cells, placenta, brain, gonads	135–155	4	

lack thereof, may be due to a nonlinear dispersion in the body, a poor ability to enter into a saturable receptor-mediated transport process, chemical instability, a sequence of administration with other agents, or an inappropriate location or response of the target cells. Further, a bell-shaped dose–response curve can be associated with the tachyphylaxis of receptor expression or a signal transduction mechanism whereby the cells become refractory to subsequent receptor-mediated augmentation. Because the regulation of biological control can, and likely will, lead to physiologically unwanted events, it is important that the protocol of administration for recombinant proteins be tailored to ensure the desired physiological activity.

Several paradigms distinguish the therapeutic activity of proteins from classical low-molecular-weight drugs. These differences are predominantly associated with the pharmacologic attributes of the proteins. Thus, it is critical to understand the pharmacology of these drugs to optimize their therapeutic activity or, as is more generally the case, to identify their therapeutic activity. These paradigms include the following:

1. The short half-life of proteins and the requirement for subcutaneous or continuous infusion delivery for maximal activity.
2. The apparent bell-shaped dose response curve.
3. The need for chronic administration, which is associated with the perceived mechanism of action by these agents.
4. The optimal activity of these agents is as an adjuvant therapeutic administered in conjunction with chemotherapy, radiotherapy, or both.
5. The maximum adjuvant immunotherapy is found in patients with minimal residual disease.

This chapter discusses these paradigms using individual cytokines as examples with those agents that have been approved or are currently in clinical trials.

Cytokine Pharmacokinetics and Immunoregulatory Properties

RIL-2 and Pharmacokinetics

Over the past decade, recombinant IL-2 (rIL-2) has been used extensively in animal models and clinical trials. It was initially developed based on a chemotherapeutic paradigm and administered to humans at the MTD. However, three major deficiencies were identified: low response rate, severe dependent toxicity, and high cost of delivery. Despite the initially promising results in patients with advanced cancer, no significant improvements or dose-related efficacy have been identified. Thus, high-dose rIL-2 alone has demonstrated some activity in metastatic renal cell cancer and melanoma with objective response rates observed in 15 to 20% of patients. Interleukin-2 is a T-cell proliferative–activating cytokine as well as a potent natural killer (NK) and lymphokine-activated killer (LAK) augmenting agent. In contrast to the initial trials, the current trials focus on the design of well-tolerated schedules with low doses of rIL-2. In some studies, cells are cultured with IL-2 in vitro for 72 h or longer (up to 6 to 8 weeks) and infused back into the patient. As such, rIL-2 is important to all facets of T cell and NK cell augmentation and proliferation. Recombinant interleukin-2 has been approved by the FDA for use as a single agent in the treatment of renal cell carcinoma and, more recently, melanoma. In addition, it is administered in conjunction with LAK or T-cell infiltrating lymphocytes (TILs) in adoptive cellular therapy protocols. The TIL cells are T cells obtained from tumors and are expanded in vitro with lower levels of IL-2 or IL-7, or both, and then used with LAK cells in the presence of tumor antigen. The overall goal is to expand a population of tumor-specific cytotoxic T lymphocytes (CTL), and a more recent objective has been to develop virus-specific CTL, including CTL directed against Cytomegalovirus (CMV), Epstein–Barr Virus (EBV), and human immunodeficiency virus (HIV).

Some researchers have questioned whether the adoptive transfer of LAK cells is necessary or adds to the clinical efficacy of rIL-2. Indeed, there has been little indication of an improved therapeutic effect of rIL-2 plus LAK (or TIL) cells versus rIL-2 alone (Rosenberg et al. 1993; West et al. 1987). When the clinical trials with rIL-2 are rigorously examined, neither strategy has impressive (as opposed to significant) therapeutic activity (Rosenberg et al. 1993; West et al. 1987). The overall response rate with rIL-2 is 7–14% and is associated with considerable toxicity (Lotze et al. 1986); however, it should be noted that these responses are durable. In one of the first clinical studies (Heslop et al. 1989), partial responses were observed in 4 out of 31 patients. Interestingly, these partial responses did not arise in patients with increased LAK or NK cell activity. The antitumor effect of both TIL and LAK cells could be due to either a direct effect or secondary to the generation of other cytokine mediators. The latter mechanism is supported by the observation that rIL-2-stimulated lymphocytes produce IFN-γ and tumor necrosis factor (TNF) as well as other cytokines. Further, the therapeutic activity of rIL-2 may be synergistic with these cytokines (Heslop et al. 1989).

Many of the rIL-2 infusion clinical trials with or without LAK cells in metastatic renal cell carcinoma have used an MTD of rIL-2. Fefer and colleagues (Thompson et al. 1992) compared maintenance rIL-2 therapy at the MTD of 6×10^6 IU/m²/day to 2×10^6 IU/m²/day. They found that it was possible to maintain the patients for a median of 4 days at 6×10^6 IU/m²/day but in the presence of severe hypotension and capillary leak syndrome. In the lower dose protocol, none of the patients experienced severe hypotension or capillary leak syndrome, and the median duration of maintenance rIL-2 therapy was 9 days. This arm of the trial had a total response rate of 41% in contrast to the higher-dose protocol (with a shorter duration of administration), which had a 22% response rate. These investigators suggest that there may be an improved therapeutic activity associated with the longer maintenance protocol at lower doses. In another trial (Lauria et al. 1996) of low-dose rIL-2, which was administered to 11 non-Hodgkin's lymphoma patients, 2 patients with residual disease after autologous bone marrow transplantation (BMT) obtained complete responses following 7 and 10 months of therapy, respectively.

A recent study that examined the transcriptional regulation of cytokine mRNA levels in the peripheral blood leukocytes (PBL) of cancer patients receiving rIL-2 suggested that (1) doses of rIL-2 as low as 3×10^4 IU/day could augment T cell function, and (2) higher doses of rIL-2 ($\geq 1 \times 10^5$ IU/day) increased not only T cell but macrophage function (Hladik et al. 1994). The latter was measured as TNF levels and increased levels of TNF were observed at the higher doses of rIL-2. These higher levels of TNF, when combined with the T cell production of IFN-γ, which occurs at the lower doses of rIL-2, may be responsible for the toxicity of rIL-2 (Mier et al. 1988). Recently, renal cell cancer patients were randomized to receive either a high-dose regimen (FDA approved dose) or one using one-tenth of the dose (72,000 IU/kg/8 h). Both doses of IL-2 were administered by the same schedule (days 1–5 and 15–19, which was repeated every 4–6 weeks). An interim report of this ongoing trial (Kovacs et al. 1995) reported similar response rates in the two groups: 7% complete responses (CR) and 8% partial responses (PR) in the low-dose group versus 3% CR and 17% PR in the high-dose group. However, the toxicity of the low-dose regimen was substantially less than that of the high-dose regimen.

Recently, chronic rIL-2 administration at low doses ($\approx$200,000 IU/m^2/day) has been found to increase CD4$^+$ cell number and the CD4 : CD8 ratio in AIDS patients (Jacobson et al. 1996; Kovacs et al. 1995). The goal of one of these studies (Jacobson et al. 1996) was to give asymptomatic HIV$^+$ individuals rIL-2 without promoting viral replication by using an approach patterned after that described by Ritz and coworkers (Sleijfer et al. 1992). Ritz and colleagues (Soiffer et al. 1992) reported that low doses of rIL-2 could be given to cancer patients for periods up to 3 months with minimal toxicity. The results indicate that extremely low rIL-2 doses are nontoxic and effective in stimulating immune reactivity.

These studies suggest that the rIL-2 augmented activity in vivo, whether by NK or LAK cell, is transient and dependent on recent exposure to IL-2. This is consistent with the requirement for continuous exposure to IL-2 to augment NK cells in culture (Talmadge 1985; Talmadge et al. 1987). The in vitro augmentation of maximal NK cell cytotoxicity by IL-2 requires receptor occupancy for 16 to 24 h at about 100 U/ml (Talmadge 1985; Talmadge et al. 1987). To achieve these parameters clinically, either a continuous IL-2 infusion or multiple daily injections of high IL-2 doses by push administration are required (Moertel 1986). Clinical trials have demonstrated that intravenous bolus injections of rIL-2 at doses sufficient to achieve a continuous low serum level result in appreciable toxicity (Sano et al. 1988). It appears that the pharmacologically appropriate route of administration is continuous infusion and that significantly less protein/m^2 is required (Alper 1995; Clark et al. 1990; Crum and Kaplan 1991; Herberman 1989; Fujiwara et al. 1990; Konrad et al. 1990; Lim et al. 1992; Oldham et al. 1989; Sano et al. 1988; Stevenson et al. 1990; Thompson et al. 1988; Vlasveld et al. 1993; Vlasveld et al. 1992; Crum et al. 1991), although this remains somewhat controversial (Alper 1988). The route of rIL-2 administration is important because of its influence on the bioavailability and serum half-life of the rIL-2. The preclinical observation of therapeutic activity at low doses by continuous infusion (Talmadge 1985; Talmadge et al. 1987) suggests that rIL-2 may be therapeutically effective using this less toxic protocol (Alper 1995; Creekmore et al. 1989; Herberman 1989; Oldham et al. 1989; Sano et al. 1988; Sondel et al. 1988; Stevenson et al. 1990; Thompson et al. 1988). Indeed, the lymphoid hyperplasia and therapeutic activity that are observed with continuous administration of rIL-2 suggest that ex vivo LAK-cell induction, cultivation, and infusion technology may be unnecessary (Kornard et al. 1990; Thompson et al. 1988). Similar therapeutic and immunomodulatory activity has been shown preclinically with low doses of rIL-2 when it was delivered chronically by solid phase (Crum and Kaplan 1991) or minipellet administration (Fujiwara et al. 1990).

Together, these and other studies suggest the need for not only infusion or subcutaneous administration of rIL-2 for therapeutic efficacy in cancer and AIDS but also the necessity for chronic administration. The latter observation echoes the studies with IFN-α in hairy cell leukemia. A similar need for pharmacodynamic considerations has been found with g-CSF. A comparison of sc to iv delivery of g-CSF (Sugiura et al. 1997) found an increase in neutrophil counts in the blood after sc compared with iv administration. This occurred despite a lower area under the curve plasma concentration following sc compared with iv administration. Similar results were found with a comparison of 2- versus 24-h iv infusion of GM-CSF following HDT (Vose et al. 1996). Patients, (historical controls) received 500 μg/m^2/day of GM-CSF by 2-h iv infusion and required 22 days to achieve an absolute neutrophil count (ANC) of 500/μl. This is in contrast to patients who received a continuous infusion of 125 μg/m^2/day of GM-CSF and recovered an ANC of 500 or higher in 12 days. Thus, a rational dosage regimen suggests that a slow, constant infusion may be more useful that a rapid infusion to maintain receptor occupancy.

Abnormal Immunoregulatory and Therapeutic Responses

Interferon-gamma (IFN-γ) Nonlinear Dose Responses

Preclinical studies have suggested that recombinant IFN-γ (rIFN-γ) has significant therapeutic activity in animal models of experimental and spontaneous metastasis that occurs with a reproducible bell-shaped dose response curve (Talmadge et al. 1987). Studies of the immune response in normal animals have revealed the same bell-shaped dose response curve for the augmentation of macrophage tumoricidal activity (Black et al. 1993; Talmadge et al. 1987). Further, optimal therapeutic activity is also observed at the same dose and protocol with significantly less therapeutic activity at lower and higher doses. Thus, a significant correlation between macrophage augmentation and therapeutic efficacy has been reported (Black et al. 1993) in these preclinical models, which suggests that immunological augmentation provides an indirect mechanism for the therapeutic effect of rIFN-γ. In addition, this observation supports the hypothesis that treatment with the MTD of rIFN-γ may not be optimal in an adjuvant setting.

The preclinical hypothesis of a bell-shaped dose response curve for rIFN-γ has been confirmed in numerous clinical studies on the immunoregulatory effects of rIFN-γ that defined an optimal immunomodulatory dose (OID) (Jaffe et al. 1988; Maluish et al. 1988). In general, the OID for rIFN-γ has been found to be between 0.1 and 0.3 mg/m^2 following iv or intramuscular injection. In contrast, the MTD for rIFN-γ may range from 3 to 10 μg/m^2 depending on the source of the rIFN-γ, the clinical center, or both factors. The identification of an OID for rIFN-γ in patients with minimal tumor burden has resulted in the development of clinical trials to test the hypothesis. It has been suggested that immunological enhancement induced by rIFN-γ will result in prolongation of the disease-free period and overall survival of patients in an adjuvant setting (Jaffe and Herberman 1988). In another study (Jett et al. 1994), the OID for IFN-γ (0.2 mg/m^2) subcutaneously daily for 6 months was used as an adjuvant therapy for patients with a clinical complete response or an initial response to combination chemoradiotherapy for small-cell lung cancer. In this trial, 100 patients were randomized to IFN-γ or observation with the finding that IFN-γ had no effect on disease-free or overall survival. However, rIFN-γ was active in this study, for macrophage activation was observed (Pujol et al. 1993).

In contrast, rIFN-γ was found, on an empirical basis, to have therapeutic activity in chronic granulomatous disease (CGD) (The International Chronic Granulomatous Disease Cooperative Study Group 1991), and it was for this indication that the FDA approved rIFN-γ. The studies in CGD suggested that the mechanism of therapeutic activity for rIFN-γ is associated with enhanced phagocytic oxidase activity and increased superoxide production by neutrophils. However, data that are more recent suggest that the majority of CGD patients obtain clinical benefit by prolonging rIFN-γ therapy. It has been suggested that the mechanism of action may not be due to enhanced neutrophil oxidase activity but rather to the correction of a respiratory burst deficiency in a subset of monocytes (Woodman et al. 1992). In addition to being licensed for CGD, IFN-γ has also been approved for the treatment of rheumatoid arthritis in Germany.

Interferon alpha (IFN-α). The Need for Chronic Administration for Therapeutic Activity

The initial, nonrandomized, clinical studies with IFN-α suggested that it has therapeutic activity for malignant melanoma, osteosarcoma, and various lymphomas (Misset et al. 1982). However, subsequent randomized trials demonstrated significant therapeutic activity only against less common tumor histiotypes, including hairy cell leukemia, multiple myeloma, and chronic myelogenous leukemia (CML) (Golomb et al. 1988; Misset et al. 1982; Quesada et al. 1984). Subsequently, the list of responding indications was expanded to include renal cell carcinoma (Muss 1987; Quesada et al. 1985), AIDS and Kaposi's sarcoma (Lane et al. 1995), genital warts (Nieminen et al. 1994), hepatitis (Van Thiel et al. 1996), and, most recently, malignant melanoma (Kirkwood et al. 1997; Rusciani et al. 1997; Sondak and Wolfe 1997). It also appears to have activity for low-grade non-Hodgkin's lymphoma (O'Connell et al. 1986), bladder papillomatosis (Tomao et al. 1995), cutaneous T-cell lymphoma (Bunn et al. 1984), and adult T-cell leukemia and lymphoma (Ezaki 1996).

It has taken almost three decades to translate the concept of IFN-α as an antiviral agent to its routine utility in clinical oncology and infectious diseases. Despite extensive study, the development of IFN-α is still in its early stages and such basic parameters as optimal dose and therapeutic schedule remain to be determined (Golomb et al. 1988; Quesada et al. 1984). The mechanism of activity is also controversial because IFN-α has been shown to have dose-dependent antitumor activities in vitro and yet be active at low doses in hairy cell leukemia (Golomb et al. 1988; Quesada et al. 1984) Immunomodulation as the mechanism of therapeutic activity with IFN-α is perhaps best supported by its action against hairy cell leukemia. Treatment with IFN-α in this disease is associated with a 90–95% response rate; however, this is not fully achieved until the patients have been on the protocol for a year, and it appears that low doses of IFN-α are as active as higher doses (Teichmann et al. 1988). In multiple myeloma, the mechanism(s) by which IFN alfa-2b(Intron A) prolongs remission is also unknown. In clinical studies of IFN-α, 2′,5′-oligoadenylate synthetase (2,5-A synthetase) has been used as an objective indicator of in vivo activity. This enzyme has been assayed in cytosol preparations of peripheral blood mononuclear cells (MNCs) in one trial of 111 patients (Millar et al. 1995) who received IFN alfa-2b and 54 patients who did not. In this study, the level of 2,5-A synthetase activity was compared with the response to intensive therapy and with duration of maintenance therapy. Seventy-three percent of patients had measurable amounts of 2,5-A synthetase during the first 6 months of maintenance therapy. However, there was no difference in the magnitude of enzyme induction among patients who were in complete remission, partial response, or who had no change in disease status following intensive therapy. Thus, the studies to date suggest that immune modulation as measured by the levels of 2,5-A synthetase in patients with multiple myeloma is not indicative of a clinical response to IFN alfa-2b.

Initial dose finding studies by Quesada and coworkers determined that a dose of 12×10^6 U/m^2 of rIFN-α could not be tolerated by patients with hairy cell leukemia (Quesada et al. 1984). Subsequently, they found that a dose of 2×10^6 IU/m^2 was both well tolerated and effective when administered three times per week. Later studies (Smalley et al. 1991) demonstrated that highly purified natural IFN-α at a dose of 2×10^6 IU/m^2, when administered for 28 days, was well tolerated in most patients. However, these studies suggested that this dose might be myelosuppressive as well as neurotoxic or cardiotoxic. In these studies, a lower dose of 2×10^5 IU/m^2 was also administered for 28 days and was found to be better tolerated and to induce improvements in peripheral neutrophil and platelet counts as compared with the standard dose. In this trial, substantial clinical improvement, primarily in terms of increased platelet and neutrophil counts, was observed within the first 4 to 8 weeks of treatment. This resulted in an improved quality of life and a decrease in cardiac and neurologic toxicity, flulike syndromes, myelosuppression, the need for platelet transfusions, and the incidence of bacterial infections. The studies suggested that once improvement is obtained at 2×10^5 U/m^2 and patients become tolerant to IFN-α, the dose can then be increased to 2×10^6 U/m^2 to obtain the greater antileukemic effect. It appears that significant improvements in thrombocytopenia and neutropenia can be induced rapidly in the majority of patients when low and minimally toxic doses of IFN-α are used. However, there is also a therapeutic dose-response effect whereby higher doses of IFN-α induce a quantitatively greater antileukemic response than low doses of IFN-α. Recently, IFN-α was approved by the FDA for the treatment of malignant melanoma. Therapeutic activity was demonstrated by the Eastern Cooperative Oncology Group study (Kirkwood et al. 1997; Sondak and Wolfe 1997), which reported the quality-of-life adjusted survival analysis comparing high-dose IFN alfa-2b treatment for 1 year versus observation in 280 high-risk melanoma patients. After 84 months of median follow-up time, the IFN alfa-2b group experienced an average increase in time before relapse of 8.9 months and a-7 month increase in overall survival compared with the observation group. However, this occurred in the presence of significant treatment-related toxicity and treatment delays and dosage reduction required during the treatment. The most prevalent toxicity was constitutional syndromes that included fatigue and flulike symptoms. Indeed, grade 3 or greater toxicity was found in 78% of the patients on IFN alfa-2b treatment, and 23% could not complete the 1-year therapy regime using 20 million U/m^2 delivered intravenously daily 5 days per week for 4 weeks and then three times weekly at 10 million U/m^2 subcutaneously for 48 weeks. Studies such as this demonstrate the therapeutic utility of IFN-α for melanoma and confirm the studies with CML demonstrating that lower doses can initially be used and may be appropriate (Kirkwood et al. 1997).

GROWTH FACTORS

Four recombinant hematopoietic growth factors: G-CSF (filgrastim), GM-CSF (sargramostim), IL-11 (neumega), and erythropoietin (epoetin: EPO) are currently available for clinical use. Recombinant G-CSF is a lineage-specific growth factor that regulates the production and function of neutrophils. It demonstrates a dose-related increase in the neutrophil counts in preclinical animal models after chemotherapy. In the initial phase I/II human trial, G-CSF administered by intravenous infusion resulted in a dose-dependent increase in the granulocyte count (Gabrilove et al. 1988).

The efficacy of G-CSF was evaluated using patients receiving therapy for small-cell lung cancer. In this phase III trial, 211 patients were randomized to receive 200 μg/m^2 of G-CSF by subcutaneous injection or placebo after chemotherapy with cyclophosphamide, doxorubicin, and etoposide. In this study, G-CSF reduced the duration of neutropenia from 6 to 3 days in cycle 1, the incidence of febrile neutropenia (57 versus 28% in cycle 1), the length of first-cycle hospital stay, and the days of intravenous antibiotic use (Crawford et al. 1991). Predominantly on the basis of this phase III trial, G-CSF was approved for clinical use to reduce the incidence of febrile neutropenia for patients with nonmyeloid malignancies receiving myelosuppressive chemotherapy.

Recombinant GM-CSF is a multipotent growth factor active on progenitors of both the myeloid and monocytic lineage. The functions of mature macrophages and neutrophils, including tumoricidal activity, antibody-dependent cell-mediated cytotoxicity, superoxide production, phagocytosis, and secretion of other cytokines are all enhanced by GM-CSF. Recombinant GM-CSF also enhances chemotactic responses and cell-surface adhesion molecules (Lopez et al. 1986). Preclinical studies in mice (Talmadge et al. 1989) and normal cynomolgus monkeys demonstrated an increase in neutrophils, bands, eosinophils, monocytes, and reticulocytes (Donahue et al. 1986). In myelosuppressed monkeys, GM-CSF accelerated neutrophil and platelet recovery after total-body irradiation (TBI) and autologous bone marrow transplantation (ABMT) (Monroy et al. 1987). In the initial trials, GM-CSF administered intravenously or subcutaneously induced leukocytosis with an increase in neutrophils, eosinophils, and monocytes (Lieschke et al. 1990).

In an earlier clinical trial of GM-CSF (Gerhartz et al. 1993), significant efficacy was observed when it was administered to non-Hodgkin's lymphoma patients. In this trial the patients received cyclophosphamide, vincristine, procarbazine, bleomycin, prednisolone, doxorubicin, and mesna and were randomized in a double-blind prospective trial to receive 400 μg /m^2 of *Escherichia coli* (nonglycosylated) GM-CSF or placebo for 7 days following chemotherapy. The administration of GM-CSF resulted in faster neutrophil recovery, fewer days of intravenous antibiotics, and fewer patients hospitalized for infection. Another randomized trial evaluated *E. coli* GM-CSF versus placebo after chemotherapy in patients with small-cell lung cancer. In this trial the patients receiving GM-CSF had a faster neutrophil recovery, but no significant effect on the incidence of febrile neutropenia, days in the hospital, or days on intravenous antibiotics (Hamm et al. 1994).

Recombinant yeast-derived GM-CSF has also been tested in large, multicenter, randomized trials with patients receiving high-dose chemotherapy and ABMT for lymphoid malignancy. In one such trial, 128 patients were randomized to receive placebo or 250 μg/m^2/day of GM-CSF administered as a 2-h intravenous infusion for 21 days posttransplant (Muss 1987). In this study, the neutrophil recovery occurred 7 days earlier in the GM-CSF group than in the placebo group. The clinical end points reduced were days of antibiotic use, length of hospital stay, and incidence of documented infections in the GM-CSF group. Because of this trial, the yeast-derived glycosylated form of GM-CSF was approved to accelerate myeloid recovery for patients with lymphoid cancer undergoing ABMT. Overall, GM-CSF (glycosylated) appears to exhibit no difference in toxicity or biologic profile between its yeast and *E. coli* GM-CSF (unglycosylated) forms.

The first commercially available hematopoietic growth factor for clinical use was EPO, which is a glycoprotein that regulates production of the erythocytic cell lines—especially mature red blood cells (Krantz 1991). In a series of double-blind, placebo-controlled trials, three different populations of patients with cancer were treated: 124 patients did not receive chemotherapy, 132 received a cisplatin-containing regime, and 157 received chemotherapy that did not contain cisplatin. The mean weekly hematocrit level in EPO-treated patients in all three treatment groups increased progressively (28.6 to 32.1%), whereas the mean weekly hematocrit level remained essentially unchanged in the placebo-treated patients (28.4 to 28.8%). The transfusion requirements of EPO-treated patients were less than for placebo patients, but only during the second and third months of therapy, and did not reach statistical significance (Case-DC et al. 1993; Sharp et al. 1992). Currently, EPO is approved for treatment of anemia associated with chronic renal failure as well as chemotherapy- and zidovudine-induced anemia.

Interleukin-11 (IL-11) is a pleiotropic cytokine that was originally detected in the conditioned medium of an IL-1-stimulated primate bone marrow stromal cell line as a mitogen for an IL-6-responsive murine plasmacytoma cell line. Interleukin-11 has multiple effects on both hematopoietic and nonhematopoietic cells, and many of the biological effects described for IL-11 overlap those of IL-6. In vitro, IL-11 can synergize with IL-3, IL-4, and stem cell factor (SCF) to shorten the Go period of early hematopoietic progenitors. Interleukin-11 also enhances IL-3-dependent, megakaryocyte colony formation. Consistent with the in vitro functions of IL-11, in vivo administration of rIL-11 to normal mice was found to enhance the generation of platelets and to increase the cycling rates of bone–marrow-derived CFU-GM, BFU-E, and CFU-GEMM progenitors.

The hematological response of normal and myelosuppressed nonhuman primates to treatment with rIL-11 (human) has also been studied. In normal cynomolgus monkeys, rIL-11 significantly increases peripheral platelet counts when administered at doses of 10 μg/kg/day to 100 μg/kg/day either by constant intravenous infusion or subcutaneous injection. Treatment with rIL-11 for 4 days is sufficient to increase peripheral platelet counts significantly. In addition, extending the treatment period enhances both the magnitude and the duration of the response. When given to a nonhuman primate myelosuppression model using carboplatin, which causes severe thrombocytopenia with platelet counts of less than 20×10^3 platelets/μl, rIL-11 administered subcutaneously at a dose of 125 μg/kg/day, either concurrently or following chemotherapy, prevented severe thrombocytopenia and accelerated platelet recovery compared with control animals. (Schlerman et al. 1996).

A phase I trial of rIL-11 in women with breast cancer (Gordon et al. 1996) used cohorts of three to five women and five dosage levels of rIL-11 (10, 25, 50, and 75 μg/kg/day). Recombinant interleukin-11 alone was initially administered by a daily subcutaneous injection for 14 days during a 28-day pre-chemotherapy "cycle 0." Patients subsequently received up to four 28-day cycles of cyclophosphamide (1500 mg/m^2) and doxorubicin (60 mg/m^2) chemotherapy followed by rIL-11 at an assigned dose (days 3 through 14). Sixteen patients were accrued, and grade 2 constitutional symptoms (myalgias, arthralgias, and fatigue) were observed at 75 μg/kg/day, which was determined to be the MTD. Administration of rIL-11 alone resulted in a mean 76, 93, 108, and 185% increase in platelet counts at doses of 10, 25, 50, and 75 μg/kg/day, respectively. No significant changes in the total leukocyte count were seen. Compared with patients at the 10 μg/kg/day dose, patients receiving doses less than or equal to 25 μg/kg/day experienced less thrombocytopenia in the first two cycles of chemotherapy. In summary, rIL-11 has thrombopoietic activity at all doses studied, is well tolerated at doses of 10, 25, and 50 μg/kg/day, and at doses of 25 μg/kg/day or higher has the ability to reduce chemotherapy-induced thrombocytopenia. Thus, in November 1997, the FDA approved IL-11 (Neumega by Genetics Institute, Inc., Cambridge, Massachusetts) for the reduction of thrombocytopenia following chemotherapy. In contrast, IL-11 has not shown efficacy for patients undergoing bone marrow transplants. In another primary study of rIL-11, patients who had required platelet transfusions as a result of chemotherapy were evaluated. In the group treated with rIL-11 before receiving another dose of chemotherapy, 28% were able to avoid platelet transfusions compared with only 3% in the placebo.

COMBINATION CHEMOTHERAPY AND CYTOKINE THERAPY

Mammalian blood cells originate from a small population of pluripotential stem cells (SC) that in adults constitute 0.1 to 3% of cells in the bone marrow (BM). The pluripotent SC compartment contains a pool of cells capable of both self-renewal and differentiation to all hematopoietic lineages. Therefore, it contains cell types capable of life-long reconstitution of hematopoiesis in lethally irradiated recipients. The development of the therapeutic potential of pluripotent SC in facilitating stem cell transplantation for patients after treatment with aggressive anticancer regimes has been a major challenge. Researchers have been hampered by an inability to identify surface markers unique to pluripoent SC, and these cells have never been purified to homogeneity. Pluripotent SC-enriched populations can be obtained by several methods, such as negative selection using antibodies and complement or positive selection using antibodies to membrane markers on pluripotent SC (e.g., human $CD34^+CD38^-$ or murine Thy^-1^-1 Lin-Sca$^-+$). However, the pluripotent SC-enriched populations always contain significant numbers of uncharacterized, differentiated cells. Therefore, the true nature and membrane phenotype of the elusive pluripotent SC is still obscure and many questions about its biology remain unanswered.

Moreover, it is still unknown what signals are involved in determining lineage commitment and what factors are responsible for the maintenance of the pluripotent SC pool. In addition, the regulatory mechanisms that direct a dividing pluripotent SC to remain in a pluripotent state or undergo differentiation remain in question, although it has been suggested that asymmetrical versus symmetrical cell division determines the outcome. The progeny of pluripotent SC, the early progenitor cells (often called primitive progenitors) and the more differentiated, lineage-committed progenitor cells, can be distinguished from pluripotent SC on the basis of their limited ability to sustain the repopulation of secondary recipients. In general, as hematopoietic cells become more specialized (differentiated), their capacities for self-renewing proliferation and multipotentiality, as well as their ability to sustain hematopoietic reconstitution, gradually diminish.

Over the past decade, multiple cytokines that stimulate pluripotent SC to divide or differentiate, or both, have been described. In most cases, the activity of these cytokines is measured by the formation of specific colonies; thus, the name colony-stimulating factors (CSFs) has been coined. The CSFs include granulocyte-CSF (G-CSF), macrophage-CSF (M-CSF or CSF-1), granulocyte-macrophage-CSF (GM-CSF), and the multi-CSF, now known as IL-3, which induces colonies of all hematopoietic lineages except lymphoid-origin cells (lymphocytes expand diffusely and do not form colonies). Other lineage-specific cytokines are erythropoietin (EPO) and TPO (c-*mpl* ligand), which are specific for the erythroid and thrombocytic lineages, respectively. These cytokines have been assessed for their clinical usefulness in supporting the hematopoietic system of patients, and some of them, (e.g., EPO, G-CSF, GM-CSF, and IL-11) have already been approved as therapeutic drugs. In contrast, cytokines capable of expanding early progenitor cells are not yet clinically approved. These include the soluble ligands for receptors that are expressed, although not exclusively, on early progenitor cells (i.e. SCF, which is also called steel factor or c-kit ligand) and Flt-3 ligand (the ligand for flt3/flt-2 receptor). Numerous studies have shown that the in vitro growth of the more primitive hematopoietic cells requires activation by multiple cytokine receptors. Thus, many cytokines, including SCF and Flt-3 ligand, have little or no effect on the growth of primitive progenitors, but in combination with CSFs, interleukins, or both, they enhance the cell division and viability of these cells. One example is the activity of TPO on primitive hematopoietic cells (Borge et al. 1997). Although TPO was initially described as a lineage-specific, late-acting hematopoietic cytokine promoting the growth and differentiation of megakaryocyte-committed progenitor cells (Cannizzo et al. 1997; Deutsch et al. 1996), recent studies have shown that TPO can also promote the viability of primitive $CD34^+CD38^-$ BM cells in culture. Further, in combination with SCF, IL-3, or both, but not alone, TPO can support the division of primitive hematopoietic cells before the stage of lineage commitment (Cohen et al. 1997; Deutsch et al. 1996; Ohmizono et al. 1997). Thus, it appears that the synergistic action of multiple cytokines, potentially including ones yet to be discovered, is necessary for the in vitro maintenance and expansion of the most primitive hematopoietic stem cells. The availability of recombinant hematopoietic cytokines as biologically active preparations has facilitated their development, but require further investigation into the mechanism that governs the biology of hematopoietic stem cells.

Because the cytokines have unique mechanisms of action, they are ideal candidates for combination therapy with chemotherapeutic agents. However, increased knowledge and consideration of the potential interactions between these two classes of drugs are necessary for optimal clinical use. The use of high-dose chemotherapy (HDT) and stem-cell rescue provides the ultimate in cytoreductive therapy and posttransplant immunotherapy. As shown later in this chapter, stem-cell transplantation provides one of the few statistically supported demonstrations of therapeutic efficacy by T-cell augmentation (comparison of allogeneic with autologous transplantation). Thus, strategies to up-regulate T-cell function after autologous stem-cell transplantation constitute one focus for posttransplantation cytokine therapy. This is important because the return of immunologic function in transplant patients is slow and is accompanied by depressed numbers of $CD4^+$ T cells, a low $CD4:CD8$ T-cell ratio, and depressed cellular responses (Maraninchi et al. 1987).

The role of T cells in controlling neoplastic disease has been demonstrated in allotransplanted patients and described as a graft-versus-tumor (GVT) reaction. A significantly higher risk of relapse is associated with the use of T-cell depleted bone marrow (BM) cells or the clinical use of cyclosporine A (CSA) to prevent graft-versus-host disease (GVHD) (Horowitz et al. 1990; Maraninchi et al. 1987; Mitsuyasu et al. 1986; Weiden et al. 1979). It has been postulated that T-lymphocyte depletion of the graft increases leukemia relapse by removing the cells responsible for the GVT effect (Hood et al. 1987; Horowitz et al. 1990; Mitsuyasu et al. 1986; Weiden et al. 1979). Similar relapse rates are observed in recipients of non-T-cell-depleted transplants receiving CSA, which suggests that it inhibits the same GVT cells that are removed by T-cell depletion. Clearly, GVHD can also have unfavorable effects on transplant-related mortality. In first remission, the decreased relapse rates with acute or chronic GVHD are offset by the increased risk of death from other causes. Consequently, patients with GVHD have a lower risk of treatment failure but an increased risk of morbidity due to GVHD.

Thus, one approach to improving survival of cancer patients has been to use immunotherapy following HDT and stem cell transplantation to induce an autologous GVT response. On the basis of this strategy, studies using rIL-2 alone following bone marrow transplantation (BMT) to induce a post-infusion lymphocytosis have shown an increase in NK cell phenotype and function (Blaise et al. 1990; Higuchi et al. 1991; Negrier et al. 1991; Soiffer et al. 1992). In one such study (Negrier et al. 1991) with 18 evaluable patients, 3 responses were observed. In another study, rIL-2 was infused following both autologous and allogeneic transplantation for a median of 85 days at a dose of 2×10^5 IU/m^2/day (Soiffer et al. 1992). Toxicity was minimal, and the treatment could be undertaken in the outpatient setting via a Hickman catheter. In this study, no patient developed any signs of GVHD, hypotension, or pulmonary capillary leak syndrome. The treatment did not affect the absolute neutrophil count or hemoglobin level, although eosinophilia was observed. Despite the administration of this low dose of rIL-2, significant immunological changes were noted with a five- to fortyfold increase in NK cell numbers. In a similar study, it was shown that, following continuous infusion of rIL-2 in patients receiving autologous bone marrow transplant, the CD-3$^+$ and CD-16$^+$ cells secreted increased levels of IFN-γ and TNF following in vitro culture and that there was a significant increase in serum levels of IFN-γ but not TNF following the administration of rIL-2 (Soiffer et al. 1992). Massumoto and colleagues (Massumoto et al. 1996) treated non-Hodgkin's lymphoma and acute myeoblastic leukemia patients with rIL-2 at 9×10^6 IU for 5 days beginning a median of 35 days following autologous bone marrow transplantation. The patients were apharesised and the ex vivo expanded LAK cells infused. The patients also received 10 days of maintenance rIL-2 following LAK cell infusion. In this study, skin GVHD was observed that was confirmed by skin biopsy in 85% of patients. However, the association of this autologous GVHD with GVL remains to be addressed in a phase III trial.

Recently, Miller and coworkers (Miller et al. 1997a) undertook low-dose subcutaneous rIL-2 therapy after autologous bone marrow and peripheral stem cell transplantation for advanced breast cancer, Hodgkin's disease, and non-Hodgkin's lymphoma patients who were eligible for chronic rIL-2 therapy following transplantation. In this trial, rIL-2 itself was administered at 0.25×10^6 or 0.5×10^6 IU/m^2/day subcutaneously for 84 consecutive days. The lower dose was better tolerated, and 75% of the planned dose was administered versus 48% of the higher dose. Patients who received at least a month of rIL-2 therapy exhibited a greater than tenfold increase in circulating NK cells. Further, LAK cell function against NK resistant targets, including MCF7, was observed using these cells. Thus, in agreement with the study discussed previously, rIL-2 at approximately 200,000 IU/m^2/day can be administered chronically with acceptable toxicity (Miller et al. 1997b).

Similar posttransplantation strategies with rIFN-α have been undertaken with the observation of a reduced risk of relapse and an increase in myelosuppression (Klingemann et al. 1991; Meyers et al. 1987). The Seattle Group (Meyers et al. 1987) reported in an early study of the prophylactic use of leukocyte interferon following allogeneic BMT that adjuvant treatment with IFN-α had no effect on the probability or severity of cytomegalovirus infections or GVHD in acute lymphocytic leukemia patients who were in remission at the time of transplantation. However, in this large study, there was a significant reduction in the probability of relapse in the rIFN-α recipients ($p = .004$) as compared with transplant patients who did not receive rIFN-α. However, survival rates did not differ between the rIFN-α recipients and control patients. It was noted that the administration of rIFN-α following transplantation reduced the risks of relapse but did not affect CMV infection—perhaps because rIFN-α was not initiated until a median of 18 days following transplantation and was not administered chronically. Recently, Ratanatharathorn and coworkers (Ratanatharathorn et al. 1994) extended the approach of inducing GVT reaction to a combination study that utilized both CSA-induced GVHD and IFN-α augmentation of this effect in autologous transplant patients. Twenty-two patients were enrolled, of which 17 were considered evaluable. Thirteen of the patients who received rHu-IFN-α2a developed GVHD regardless of whether they received CSA, whereas only two of the four patients who received CSA alone developed detectable GVHD. Patients receiving 1×10^6 U/day of rHu-IFN-α2a concomitant with CSA showed a trend towards increased severity of clinical GVHD as compared with patients receiving CSA alone ($p = .06$). These researchers concluded that IFN-α administration can be safely started on day 0 of ABMT and can induce autologous GVHD as a single agent with the potential to improve therapy.

In similar studies (Kennedy et al. 1994) women with advanced breast cancer were treated with combined therapy of CSA for 28 days and 0.025 mg/m^2 of sc rIFN-γ every other day on days 7–28 after high-dose chemotherapy and ABMT. The researchers observed that autologous GVHD developed in 56% of the patients, an incidence comparable with that previously observed with CSA alone. The severity of GVHD was greater with CSA plus rIFN-γ than with CSA alone, for 16 patients required corticosteroid therapy for dermatologic GVHD. Note that strategies to induce an autologous GVT, although conceptually interesting, have not matured sufficiently to allow a discussion of efficiency.

SUMMARY

In the last 20 years, nonspecific immunostimulation has progressed from trials with crude microbial mixtures and extracts to more sophisticated immunopharmacologically active compounds (only a few of which are discussed here) having diverse actions on the immune system. A body of pharmacodynamic knowledge has evolved that shows substantial divergence from conventional pharmacology—particularly in terms of the relationship of dosing schedules to immunopharmacodynamics. This knowledge is important in evaluating agents and predicting appropriate use. Although much remains to be learned and new compounds need to be cloned, the future of immunotherapy seems bright. Several cytokines have been approved, as well as numerous supplemental indications (Gosse et al. 1977), in the United States, Europe, and Asia. However, it is apparent that the combinations of cytokines and biological response modifiers (BRMs) will have optimal activity when used as adjuvants with more traditional therapeutic modalities.

REFERENCES

1993. FDA okays surrogate markers. *Science* 259:A.

Alper, J. 1995. Cetus' proleukin in cancer—The excitement grows. *SCRIP* 1363:24–24.

Bhalla, K., Tang, C., Lbrado, A. M., Grant, S., Tourkina, E., Holladay, C., Hughes, M., Mahoney, M. E., and Huang, Y. 1992. Granulocyte-macrophage colony-stimulating factor/Interleukin-3 fusion protein (PIXY 321) enhances high-dose ara-C-induced programmed cell death or apoptosis in human myeloid leukemia cells. *Blood* 80:2883–2890.

Black, P. L., Phillips, H., Tribble, H. R., Pennington, R. W., Schneider, M., and Talmadge, J. E. 1993. Antitumor response to recombinant murine interferon γ correlates with enhanced immune function of organ-associated, but not recirculating cytolytic T lymphocytes and macrophages. *Cancer Immunol. Immunother.* 37:299–306.

Blaise, D., Olive, D., Stoppa, A. M., Viens, P., Pourreau, C., Lopez, M., Attal, M., Jasmin, C., Monges, G., Mawas, C., Mannoni, P., Palmer, P., Franks, C., and Phillip, T. 1990. Hematologic and immunologic effects of the systemic administration of recombinant interleukin-2 after autologous bone marrow transplantation. *Blood* 76:1092–1097.

Borge, O. J., Ramsfjell, V., Cui, L., and Jacobsen, S. E. 1997. Ability of early acting cytokines to directly promote survival and suppress apoptosis of human primitive CD34 + CD38-bone marrow cells with multilineage potential at the single-cell level: Key role of thrombopoietin. *Blood* 90:2282–2292.

Bruno, E., Briddell, R. A., Cooper, R. J., Brandt, J. E., and Hoffman, R. 1992. Recombinant GM-CSF/IL-3 fusion protein: Its effect on in vitro human megakaryocytopoiesis. *Exp. Hemato.* 20:494–499.

Bunn, P. A., Foon, K. A., Ihde, D. C., Longo, D. L., Eddy, J., Winkler, C. F., Veach, S. R., Zeffren, J., Sherwin, S., and Oldham, R. 1984. Recombinant leukocyte A interferon: An active agent in advanced cutaneous T-cell lymphomas. *Ann. Intern. Medi.* 101:484–487.

Cannizzo, S. J., Frey, B. M., Raffi, S., Moore, M. A., Eaton, D., Suzuki, M., Singh, R., Mack, C. A., and Crystal, R. G. 1997. Augmentation of blood platelet levels by intratracheal administration of an adenovirus vector encoding human thrombopoietin cDNA. *Nat. Biotechnol.* 15:570–573.

Case-DC, Jr., Bukowski, R. M., Carey, R. W., Fishkin, E. H., Henry, D. H., Jacobson, R. J., Jones, S. E., Keller, A. M., Kugler, J. W., Nichols, C. R., Salmon, S. E., Silver, R. T., Storniolo, A. M., Wampler, G. L., Dooley, C. M., Larholt, K. M., Nelson, R. A., and Abels, R. I., 1993. Recombinant human erythropoietin therapy for anemic cancer patients on combination chemotherapy. *J. Natl. Cancer Inst.* 85:801–806.

Clark, J. W., Smith, J. W., II, Steis, R. G., Urba, W. J., Crum, E., Miller, R., McKnight, J., Beman, J., Stevenson, H. C., Creekmore, S., Stewart, M., Conlon, K., Sznol, M., Kremers, P., and Longo, D. L. 1990. Interleukin-2 and lymphokine-activated killer cell therapy: Analysis of a bolus interleukin-2 and a continuous infusion interleukin-2 regimen. *Cancer Res.* 50:7343–7350.

Cohen, S. K., Debili, N., Vainchenker, W., and Wendling, F. 1997. Thombopoietin (Mpl-ligand) and the regulation of platelet production. *Thromb. Haemost.* 78:37–41.

Crawford, J., Ozer, H., Stoller, R., Johnson, D., Lyman, G., Tabbara, I., Kris, M., Grous, J., Picozzi, V., Rausch, G., Smith, R., Gradishar, W., Yahanda, A., Vincent, M., Stewart, M., and Slaspy, J. 1991. Reduction by granulocyte colony-stimulating factor of fever and neutropenia induced by chemotherapy in patients with small-cell lung cancer [see comments]. *N. Eng. J. Med.* 325:164–170.

Creekmore, S. P., Harris, J. E., Ellis, T. M., Braun, D. P., Cohen, I. I., Bhoopalam, N., Jassak, P. F., Cahill, M. A., Canzoneri, C. L., and Fisher, R. I. 1989. A phase I clinical trial of recombinant interleukin-2 by periodic 24-hour intravenous infusions. *J. Clin. Oncol.* 7:276–284.

Crum, E. D., and Kaplan, D. R. 1991. In vivo activity of solid phase interleukin 2. *Cancer Res.* 51: 875–879.

Deutsch, V. R., Eldor, A., Olson, T., Barak, V., Pick, M., and Nagler, A. 1996. Stem cell factor (SCF) synergizes with megakaryocyte colony stimulating activity in post-irradiated aplastic plasma in stimulating human megakaryocytopoiesis. *Med. Oncol.* 13:31–42.

Donahue, R. E., Wang, E. A., Stone, D. K., Kamen, R.,Wong, G. G., Sehgal, P. K., Nathan, D. G., and Clark, S. C., 1986. Stimulation of haematopoiesis in primates by continuous infusion of recombinant human GM-CSF. *Nature* 321:872–875.

Ezaki, K. 1996. Cytokine therapy for hematological malignancies. *Int. J. Hematol.* 65:17–29.

Gabrilove, J. L., Jakubowski, A., Scher, H., Sternberg, C., Wong, G., Grous, J., Yagoda, A., Fain, K., Moore, M. A., Clarkson, B., Oettgen, H. F., Alton, K., Welte, K., and Souza, L. 1988. Effect of granulocyte colony-stimulating factor on neutropenia and associated morbidity due to chemotherapy for transitional-cell carcinoma of the urothelium. *N. Engl. J. Med.* 318:1414–1422.

Gerhartz, H. H., Engelhard, M., Meusers, P., Brittinger, G., Wilmanns, W., Schlimok, G., Mueller, P., Huhn, D., Musch, R., Siegert, W., and et al. 1993. Randomized, double-blind, placebo-controlled, phase III study of recombinant human granulocyte-macrophage colony-stimulating factor as adjunct to induction treatment of high-grade malignant non-Hodgkin's lymphomas [see comments]. *Blood* 82:2329–2339.

Golomb, H. M., Fefer, A., Golde, D. W., Ozer, H., Portlock, C., Silber, R., Rappeport, J., Ratain, M. J., Thompson, J., Bonnem, E., Spiegel, R., Tensen, L., Burke, J. S., and Vardiman, J. W. 1988. Report of a multi-institutional study of 193 patients with hairy cell leukemia treated with interferon-α 2b. *Semin. Oncol.* 15:7–9.

Gordon, M. S., McCaskill-Stevens, W. J., Battiato, L. A., Loewy, J., Loesch, D., Breeden, E., Hoffman, R., Beach, K. J., Kuca, B., Kaye, J., and Sledge-GW, Jr. 1996. A phase I trial of recombinant human interleukin-11 (neumega rhIL-11 growth factor) in women with breast cancer receiving chemotherapy. *Blood* 87:3615–3624.

Gosse, M. E., and Nelson, T. E. 1977. Approval times for supplemental indications for recombinant proteins. Bio/Technol. 15:130.

Hamm, J., Schiller, J. H., Cuffie, C., Oken, M., Fisher, R. I., Shepherd, F., and Kaiser, G. 1994. Dose-ranging study of recombinant human granulocyte-macrophage colony-stimulating factor in small-cell lung carcinoma. *J. Clini. Onol.* 12:2667–2676.

Herberman, R.1989. Clinical cancer therapy with IL-2 [editorial; comment]. *Cancer Invest.* 7:515–516.

Heslop, H. E., Gottlieb, D. J., Bianchi, A. C. M., Meager, A., Prentice, H. G., Mehta, A. B., Hoffbrand, A. V., and Brenner, M. K. 1989. In vivo induction of gamma interferon and tumor necrosis factor by interleukin-2 infusion following intensive chemotherapy or autologous marrow transplantation. *Blood* 74:1374–1380.

Higuchi, C. M., Thompson, J. A., Petersen, F. B., Buckner, C. D., and Fefer, J. 1991. Toxicity and immunomodulatory effects of interleukin-2 after autologous bone marrow transplantation for hematologic malignancies. *Blood* 77:2561–2568.

Hladik, F., Tratkiewicz, J. A., Tilg, H., Vogel, W., Schwulera, U., Kronke, M., Aulitzky, W. E., and Huber, C. 1994. Biologic activity of low dosage IL-2 treatment in vivo. Molecular assessment of cytokine network interaction. *J. Immunol.* 153:1449–1454.

Hood, A. F., Vogelsang, G. B., Black, L. P., Farmer, E. R., and Santos, G. W. 1987. Acute graft-versus-host disease. Development following autologous and syngeneic bone marrow transplantation. *Arch. Dermatol.* 123:745–750.

Horowitz, M. M., Gale, R. P., Sondel, P. M., Goldman, J. M., Kersey, J., Kolb, H. J., Rimm, A. A., Ringden, O., Rozman, C., Speck, B., Truitt, R. L., Swaan, F. E., and Bortin, M. M. 1990. Graft-versus-leukemia reactions after bone marrow transplantation. *Blood* 75:555–562.

Jacobson, E. L., Pilaro, F., and Smith, K. A. 1996. Rational interleukin 2 therapy for HIV positive individuals: Daily low doses enhance immune function without toxicity. *Proc. Natl. Acad. Sci. U.S.A.* 93:10405–10410.

Jaffe, H. S., and Herberman, R. B. 1988. Rationale for recombinant human IFN-α adjuvant immunotherapy for cancer. *J. Nat. Cancer Inst.* 314:1065–1069.

Jett, J. R., Maksymiuk, A. W., Su, J. Q., Mailliard, J. A., Krook, J. E., Tschetter, L. K., Kardinal, C. G., Twito, D. I., Levitt, R., and Gerstner, J. B. 1994. Phase III trial of recombinant interferon gamma in complete responders with small-cell lung cancer. *J. Clin. Onol.* 12:2321–2326.

Fujiwara, T., Sakagami, K., Matsuoka, J., Shiozaki, S., Uchida, S., Fujioka, K., Takada, Y., Onoda, T., and Orita, K. 1990. Application of an interleukin-2 slow delivery system to the immunotherapy of established murine colon 26 adenocarcinoma liver metastases. *Cancer Res.* 50:7003–7007.

Kennedy, M. J., Vogelsang, G. B., Jones, R. J., Farmer, E. R., Hess, A. D., Altomonte, V., Huelskamp, A. M., and Davidson, N. E. 1994. Phase I trial of interferon gamma to potentiate cyclosporine-induced graft-versus-host disease in women undergoing autologous bone marrow transplantation for breast cancer. *J. Clin. Onol.* 1:249–257.

Kirkwood, J. M., Resnick, G. D., and Cole, B. F. 1997. Efficacy, safety, and risk-benefit analysis of adjuvant interferon alfa-2b in melanoma. *Semin. Oncol.* 24:S16–23.

Klingemann, H. G., Grigg, A. P., Wilkie-Boyd, K., Barnett, M. J., Eaves, A. C., Reece, D. E., Shepherd, J. D., and Phillips, G. L. 1991. Treatment with recombinant interferon (alpha-2b) early after bone marrow transplantation in patients at high risk for relapse. *Blood* 78:3306–3311.

Konrad, M. W., Hamstreet, G., Hersh, E. M., Mansell, P. W. A., Mertelsmann, R., Kolitz, J. E., and Bradley, E. C. 1990. Pharmacokinetics of recombinant interleukin-2 in humans. *Cancer Res.* 50:2009–2017.

Kovacs, J. A., Baseler, M., Dewar, R. J., Vogel, S., Davey, R. T., Jr., Falloon, J., Polis, M. A., Walker, R. E., Stevens, R., Salzman, N. P., Metcalf, J. A., Masur, H., and Lane, H. C. 1995. Increases in CD4 T lymphocytes with intermittent courses of interleukin-2 in patients with human immunodeficiency virus infection. A preliminary study [see comments]. *N. Engl. J. Med.* 332:567–575.

Krantz, S. B. 1991. Ery thro poietin. *Blood* 77:419–434.

Kuga, T., Komatsu, Y., Yamasaki, M., Sekine, S., Miyaki, H., Nishi, T., Sato, M., Yokoo, Y., Asano, M., Okabe, M., and Itoh, S. 1989. Mutagenesis of human granulocyte colony stimulating factor. *Biochem. Biophys. Res. Commun.* 159:103–111.

Lane, H. C., Feinberg, J., Davey, V., Deyton, L., Baseler, M., Manischewitz, J., Masur, H., Kovacs, J. A., Herpin, B., Walker, R., Metcalf, J. A., Salzman, N., Quinnan, G., and Fauci, A. S. 1995. Anti-retro-viral effects of interferon-α in AIDS-associated Kaposi's sarcoma. *Lancet* 2:1218–1222.

Lauria, F., Raspadori, D., Ventura, M. A., Rondelli, D., Zinzani, P. L., Gherlinzoni, F., Miggiano, M. C., Fiacchini, M., Rosti, G., Rizzi, S., and Tura, S. 1996. Immunologic and clinical modifications following low-dose subcutaneous administration of rIL-2 in non-Hodgkin's lymphoma patients after autologous bone marrow transplantation. *Bone Marrow Transplant.* 18:79–85.

Lieschke, G. J., Maher, D., O'Connor, M., Green, M., Sheridan, W., Rallings, M., Bonnem, E., Burgess, A. W., McGrath, K., Fox, R. M., Morstyn, G. 1990. Phase I study of intravenously administered bacterially synthesized granulocyte-macrophage colony-stimulating factor and comparison with subcutaneous administration. *Cancer Res.* 50:606–614.

Lim, S. H., Newland, A. C., Kelsey, S., Bell, A., Offerman, E., Rist, C., Gozzard, D., Bareford, D., Smith, M. P., and Goldstone, A. H. 1992. Continuous intravenous infusion of high-dose recombinant interleukin-2 for acute myeloid leukemia-a phase II study. *Cancer Immunol. Immunother.* 34:337–342.

Lopez, A. F., Williamson, D. J., Gamble, J. R., Begley, C. G., Harlan, J. M., Klebanoff, S. J., Waltersdorph, A., Wong, G., Clark, S. C., and Vadas, M. A. 1986. Recombinant human granulocyte-macrophage colony-stimulating factor stimulates in vitro mature human neutrophil and eosinophil function, surface receptor expression, and survival. *J. Clin. Invest.* 78:1220–1228.

Lotze, M. T., Chang, A. E., Seipp, C. A., Simpson, C., Vetto, S. J., and Rosenberg, S. A. 1986. High-dose recombinant interleukin 2 in the treatment of patients with disseminated cancer. Responses, treatment-related morbidity and histologic findings. *JAMA* 256:3117–3124.

Lu, H. S., Boone, T. C., Souza, L. M., and Lai, P. H. 1989. Disulfide and secondary structures of recombinant human granulocyte colony stimulating factor. *Arch. Biochem. Biophys.* 268:81–92.

MacVittie, T. J., Farese, A. M., Herodin, F., Grab, L. B., Baum, C. M., and McKearn, J. P. 1996. Combination therapy for radiation-induced bone marrow aplasia in nonhuman primates using synthokine SC-55494 and recombinant human granulocyte colony-stimulating factor. *Blood* 87:4129–4135.

Maluish, A. E., Urba, W. J., Longo, D. L. O., Overton, W. R., Coggin, D., Crisp, E. R., Williams, R., Sherwin, S. A., Gordon, K., and Steis, R. G. 1988. The determination of an immunologically active dose of interferon-gamma in patients with melanoma. *J. Clin. Oncol.* 6:434–445.

Maraninchi, D., Gluckman, E., Blaise, D., Guyotat, D., Rio, B., Pico, J., Leblond, V., Michallet, M., Dreyfus, F., and Ifrah, N. 1987. Impact of T-cell depletion on outcome of allogeneic bone marrow transplantation for standard-risk leukaemias. *Lancet* 2:175–178.

Massumoto, C., Benyunes, M. C., Sale, G., Beauchamp, M., York, A., Thompson, J. A., Buckner, C. D., and Fefer, A. 1996. Close simulation of acute graft-versus-host disease by interleukin-2 administered after autologous bone marrow transplantation for hematologic malignancy. *Bone Marrow Transplant.* 17:351–356.

Meyers, J. D., Flournoy, N., Sanders, J. E., McGuffin, R. W., Newton, B. A., Fisher, L. D., Lum, L. G., Appelbaum, F. R., Doney, K., Sullivan, K. M., Storb, R., Buckner, C. D., and Thomas, E. D. 1987. Prophylactic use of human leukocyte interferon after allogeneic marrow transplantation. *Ann. Intern. Medi.* 107:809–816.

Mier, J. W., Vachino, G., and Van Der Meet, J. W. M. 1988. Induction of circulating tumor necrosis factor (TNF-alpha) as the mechanism for the febrile response to interleukin-2 (IL-2) in cancer patients. *J. Clin. Immunol.* 8:426–432.

Mihich, E. 1986. Future perspectives for biological response modifiers: A viewpoint. *Semin. Oncol.* 13:234–254.

Millar, B. C., and Bell, J. B. 1995. $2'$, $5'$-Oligoadenylate synthetase levels in patients with multiple myeloma receiving maintenance therapy with interferon alpha 2b do not correlate with clinical response. *Br. J. Cancer* 72:1525–1530.

Miller, J. S., Prosper, F., and McCullar, V. 1997a. Natural killer (NK) cells are functionally abnormal and NK cell progenitors are diminished in granulocyte colony-stimulating factor-mobilized peripheral blood progenitor cell collections. *Blood* 90:3098–3105.

Miller, J. S., Tessmer-Tuck, J., Pierson, B. A., Weisdorf, D., McGlave, P., Blazar, B. R., Katsanis, E., Verfaillie, C., Lebkowski, J., Radford, J., and Burns, L. J. 1997b. Low dose subcutaneous interleukin-2 after autologous transplantation generates sustained in vivo natural killer cell activity. *Biol. Blood Marrow Transplant.* 3:34–44.

Misset, J. L., Mathe, G., Gastiaburu, J., Goutner, A., Dorval, T., Gouveia, J., Schwarzenberg, L., Machover, D., Ribaud, P., and de Vassal, F. 1982. Treatment of leukemias and lymphomas by interferons: II. phase II of the trial treatment of chronic lymphoid leukemia by human interferon a^+. *Biomed. Pharmacother.* 39:112–116.

Mitsuyasu, R. T., Champlin, R. E., Gale, R. P., Ho, W. G., Lenarsky, C., Winston, D., Selch, M., Elashoff, R., Giorgi, J. V., Wells, J., Terasaki, P., Billing, R., and Feig, S. 1986. Treatment of donor bone marrow with monoclonal anti-T-cell antibody and complement for the prevention of graft-versus-host disease. A prospective, randomized, double-blind trial. *Ann. Int. Med.* 105:20–26.

Moertel, C. G. 1986. On lymphokines, cytokines and breakthroughs. *JAMA* 3141–3141.

Monroy, R. L., Skelly, R. R., MacVittie, T. J., Davis, T. A., Sauber, J. J., Clark, S. C., and Donahue, R. E. 1987. The effect of recombinant GM-CSF on the recovery of monkeys transplanted with autologous bone marrow. *Blood* 70:1696–1699.

Muss, H. B. 1987. Interferon therapy for renal cell carcinoma. *Semin. Oncol.* 14:36–42.

Negrier, S., Ranchere, J. Y., Phillip, I., Merrouche, Y., Biron, P., Blaise, D., Attal, M., Rebattu, P., Clavel, M., Pourreau, C., Palmer, P., Favrot, M., Jasmin, C., Maraninchi, D., and Phillip, T. 1991. Intravenous interleukin-2 just after high dose BCNU and autologous bone marrow transplantation. Report of a multicentric French pilot study. *Bone Marrow Transplant.* 8:259–264.

Nieminen, P., Aho, M., Lehtinen, M., Vesterinen, E., Vaheri, A., and Paavonen, J. 1994. Treatment of genital HPV infection with carbon dioxide laser and systemic interferon alpha-2b. *Sex. Transm. Dis.* 21:65–69.

O'Connell, M. J., Colgan, J. P., Oken, M. M., Ritts, R. E., Jr., Kay, N. E., and Itri, L. M. 1986. Clinical trial of recombinant leukocyte A interferon as initial therapy for favorable histology non-Hodgkin's lymphomas and chronic lymphocytic leukemia. An Eastern Cooperative Oncology Group pilot study. *J. Clin. Oncol.* 4:128–136.

Ohmizono, Y., Sakabe, H., Kimura, T., Tanimukai, S., Matsumura, T., Miyazaki, H., Lyman, S. D., and Sonoda, Y. 1997. Thrombopoietin augments ex vivo expansion of human cord blood-derived hematopoietic progenitors in combination with stem cell factor and flt-3 ligand. *Leukemia* 11:524–530.

Oldham, O., Maleckar, J., West, W., and Yannelli, J. 1989. Il-2 and cellular therapy: Lymphokine- activated killer cells and tumor-derived activated cell. Cytokines in hematopoiesis malignant melanoma receiving recombinant interleukin-2. *Int. J. Cancer* 43:410–414.

Oldham, R. K. 1982. Biological response modifiers program. *J. Biol. Resp. Mod.* 1:81–100.

Pujol, J. L., Gibney, D. J., Su, J. Q., Maksymiuk, A. W., and Jett, J. R. 1993. Immune response induced in small-cell lung cancer by maintenance therapy with interferon gamma. *J. Natl. Cancer Inst.* 85:1844–1850.

Quesada, J. R., Reuben, J., Manning, J. T., Hersh, E. M., and Gutterman, J. U. 1984. Alpha interferon for induction of remission in hairy-cell leukemia. *N. Eng. J. Med.* 310:15–18.

Quesada, J. R., Rios, A., Swanson, D., Trown, P., and Gutterman, J. U. 1985. Antitumor activity of recombinant-derived interferon alpha in metastatic renal cell carcinoma. *J. Clin. Oncol.* 3:1522–1528.

Ratanatharathorn, V., Uberti, J., Karanes, C., Lum, L. G., Abella, E., Dan, M. E., Hussein, M., and Sensenbrenner, L. L. 1994. Phase I study of alpha-interferon augmentation of cyclosporine-induced graft versus host disease in recipients of autologous bone marrow transplantation. *Bone Marrow Transplant.* 13:625–630.

Rosenberg, S. A., Grimm, E., McGrogan, M., Doyle, M., Kawasaki, E., Koths, K., and Mark, D. F. 1984. Biological activity of recombinant human interleukin-2 produced in *Eschericia coli*. *Science* 223:1412–1415.

Rosenberg, S. A., Lotze, M. T., Yang, J. C., Topalian, S. L., Chang, A. E., Schwartzentruben, D. J., Aebersold, P., Leitman, S., Linehan, W. M., and Seipp, C. A. 1993. Prospective randomized trial of high-dose interleukin-2 alone or in conjunction with lymphokine-activated killer cells for the treatment of patients with advanced cancer. *J. Natl. Cancer Inst.* 85:622–632.

Rusciani, L., Petraglia, S., Alotto, M., Calvieri, S., and Vezzoni, G. 1997. Postsurgical adjuvant therapy for melanoma. Evaluation of a 3-year randomized trial with recombinant interferon-alpha after 3 and 5 years of follow-up. *Cancer* 79:2354–2360.

Sano, T., Saijo, N., Sasaki, Y., Shinkai, T., Eguchi, K., Tamura, T., Sakurai, M., Takahashi, H., Nakano, H., Nakagawa, K., and Hong, W. S. 1988. Three schedules of recombinant human interleukin-2 in the treatment of malignancy: Side effects and immunologic effects in relation to serum level. *Jpn. J. Cancer Res.* 79:131–143.

Sharp, J. G., Kessinger, A., Vaughan, W. P., Mann, S., Crouse, D. A., Dicke, K., Masih, A., and Weisenberger, D. D. 1992. Detection and clinical significance of minimal tumor contamination of peripheral blood stem cell harvests. *Int. J. Cell. Cloning* 10:92–94.

Sleijfer, D. T., Janssen, R. A., Butler, J., deVries, E. G., Willemse, P. H., and Mulder, N. H. 1992. Phase II study of subcutaneous interleukin-2 in unselected patients with advanced renal cell cancer on an outpatient basis. *J. Clin. Oncol.* 10:1119–1123.

Smalley, R. V., Anderson, S. A., Tuttle, R. L., Connors, J., Thurmond, L. M., Huang, A., Castle, K., Magers, C., and Whisnant, J. K. 1991. A randomized comparison of two doses of human lymphoblastoid Interferon-α in hairy cell leukemia. *Blood* 78:3133–3141.

Soiffer, R. J., Murray, C., Cochran, K., Cameron, C., Wang, E., Schow, P. W., Daley, J. F., and Ritz, J. 1992. Clinical and immunologic effects of prolonged infusion of low-dose recombinant interleukin-2 after autologous and T cell-depleted allogeneic bone marrow transplantation. *Blood* 79:517–526.

Sondak, V. K., and Wolfe, J. A. 1997. Adjuvant therapy for melanoma [see comments]. *Curr. Opin. Oncol.* 9:189–204.

Sondel, P. M., Kohler, P. C., Hank, J. A., Moore, K. H., Rosenthal, N. S., Sosman, J. A., Bechhofer, R., and Storer, B. 1988. Clinical and immunological effects of recombinant interleukin-2 given by repetitive weekly cycles to patients with cancer. *Cancer Res.* 48:2561–2567.

Stevenson, H. C., Creekmore, S., Stewart, M., Conlon, K., Sznol, M., Kremers, P., Cohen, S., and Longo, D. L. 1990. Interleukin-2 and lymphokine-activated killer cell therapy: Analysis of bolus interleukin 2 and a continuous infusion interleukin 2 regimen. *Cancer Res.* 50:7343–7350.

Sugiura, M., Yamamoto, K., Sawada, Y., and Iga, T. 1997. Pharmacokinetic/pharmacodynamic analysis of neutrophil proliferation induced by recombinant granulocyte colony-stimulating factor (rhG-CSF): Comparison between intravenous and subcutaneous administration. *Biol. Pharm. Bull.* 20:684–689.

Talmadge, J. E. 1985. Immunoregulation and immunostimulation of murine lymphocytes by recombinant human interleukin-2. *J. Biol. Response Mod.* 4:18–34.

Talmadge, J. E., and Herberman, R. B. 1986. The preclinical screening laboratory. Evaluation of immunomodulatory and therapeutic properties of biological response modifiers. *Cancer Treat. Res.* 70:171–182.

Talmadge, J. E., Phillips, H., Schindler, J., Tribble, H., and Pennington, R. 1987. Systematic preclinical study on the therapeutic properties of recombinant human interleukin-2 for the treatment of metastatic disease. *Cancer Res.* 47:5725–5732.

Talmadge, J. E., Tribble, H., Pennington, R., Bowersox, O., Schneider, M. A., Castelli, P., Black, P. L., and Abe, F. 1989. Protective, restorative, and therapeutic properties of recombinant colony-stimulating factors. *Blood* 73:2093–2103.

Talmadge, J. E., Tribble, H. R., Pennington, R. W., Phillips, H., and Wiltrout, R. H. 1987. Immunomodulatory and immunotherapeutic properties of recombinant γ-interferon and recombinant tumor necrosis factor in mice. *Cancer Res.* 47:2563–2570.

Teichmann, J. V., Sieber, G., Ludwig, W. D., and Ruehl, H. 1988. Modulation of immune functions by long-term treatment with recombinant interferon-α2 in a patient with hairy-cell leukemia. *J. Interferon Res.* 8:15–24.

The International Chronic Granulomatous Disease Cooperative Study Group. 1991. A controlled trial of interferon gamma to prevent infection in chronic granulomatous disease. *N. Engl. J. Med.* 324:509–516.

Thompson, J. A., Lee, D. J., Lindgren, C. G., Benz, L. A., Collins, C., Levitt, D., and Fefer, A. 1988. Influence of dose and duration of infusion of interleukin-2 on toxicity and immunomodulation. *J. Clin. Oncol.* 6:669–678.

Thompson, J. A., Shulman, K. L., Benyunes, M. C., Lindgren, G., Collins, C., Lange, P. H., Bush, W. H., Jr., Benz, L. A., and Fefer, A. 1992. Prolonged continuous intravenous infusion interleukin-2 and lymphokine-activated killer-cell therapy for metastatic renal cell carcinoma. *J. Clini. Oncol.* 10:960–968.

Tomao, S., Mozzicafreddo, A., Raffaele, M., Romiti, A., Papo, M. A., and Campisi, C. 1995. Interferons in the therapy of solid tumors. *Clin. Ter.* 146:491–502.

Tomlinson, E. 1989. Site-specific drugs and delivery systems: Toxicological and regulatory implications. In *Topics in pharmaceutical sciences*, eds. D. D. Breimer, D. J. A. Crommolin, and K. K. Midha, 661–671. The Hague: International Pharmaceutical Federation.

Tomlinson, E. 1991. Site-specific proteins. In *Polypeptide and protein drugs: Production, characterization and formulation*, eds. R. C. Hider and D. Barlow, 151–364. Chichester: Ellis Horwood Ltd.

Van Thiel, D. H., Friedlander, L., Molloy, P. J., Kania, R. J., Fagiuoli, S., Wright, H. I., Gasbarrini, A., and Caraceni, P. 1996. Retreatment of hepatitis C interferon non-responders with larger doses of interferon with and without phlebotomy. *Hepatogastroenterol.* 43:1557–1561.

Vlasveld, L. T., Hekman, A., Vyth-Dreese, F. A., Rankin, E. M., Scharenberg, J. G. M., Voordouw, A. C., Sein, J. J., Dellemijn, T. A. M., Rodenhuis, S., and Melief, C. J. M. 1993. A phase I study of prolonged continuous infusion of low dose recombinant interleukin-2 melanoma and renal cell cancer. Part II: Immunological aspects. *Br. J. Cancer* 68:559–567.

Vlasveld, L. T., Rankin, E. M., Hekman, A., Rodenhuis, S., Beijnen, J. H., Hilton, A. M., Dubbelman, A. C., Vyth-Dreese, F. A., and Melief, C. J. M. 1992. A phase I study of prolonged continuous infusion of low dose recombinant interleukin-2 in melanoma and renal cell cancer. Part I: Clinical aspects. *Br. J. Cancer* 65:744–750.

Vose, J. M., Bierman, P. J., Ruby, E., Reed, E. C., Bishop, M. R., Tarantolo, S., Kessinger, A., and Armitage, J. O. 1996. Use of continuous infusion granulocyte-macrophage colony-stimulating factor alone or followed by granulocyte colony-stimulating factor to enhance engraftment following high-dose chemotherapy and autologous bone marrow transplantation for lymphoid malignancies. *Bone Marrow Transplant.* 17:951–956.

Weiden, P. L., Flournoy, N., Thomas, E. D., Prentice, R., Fefer, A., Buckner, C. D., and Storb, R. 1979. Antileukemic effect of graft-versus-host disease in human recipients of allogeneic-marrow grafts. *N. Eng. J. Med.* 300:1068–1073.

West, W. H., Tauer, K. W., Yannelli, J. R., Marshall, G. D., Orr, D. W., Thurman, G. B., and Oldham, R. K. 1987. Constant-infusion recombinant interleukin-2 in adoptive immunotherapy of advanced cancer. *N. Eng. J. Med.* 316:898–905.

Woodman, R. C., Richard, W., Rae, J., Jaffe, H. S., and Curnutte, J. T. 1992. Prolonged recombinant interferon-γ therapy in chronic granulomatous disease: Evidence against enhanced neutrophil oxidase activity. *Blood* 79:1558–1562.

Biotechnology and Safety Assessment, 2nd ed.
Edited by John A. Thomas
Copyright © 1998 Taylor & Francis

12

Biotechnology: Therapeutic and Nutritional Products

John A. Thomas

University of Texas Health Science Center, San Antonio, Texas

HISTORY

Although biotechnology has been used for centuries, only recently has it been applied to more complex molecular structures such as in the genetic manipulation of both plant and animal cells. The word may take on somewhat different interpretations, but in a classic sense biotechnology refers to any technique that uses living organisms (or components of these organisms) to modify or create, to improve plants or animals, or to develop microorganisms for specific uses. Biotechnology has been used for centuries in making wine, cheese, yogurt, and bread and in the selective cross-breeding of both animals and plants to enhance specific desirable traits. It is believed that the term *biotechnology* was first coined by a Hungarian, Karl Ereky, toward the end of World War I. Ereky reportedly used the term to refer to intensive agricultural methods.

The cornerstone of applied genetics is genetic engineering. This technology involves the ability to manipulate, modify, or otherwise "engineer" genetic material to produce desired characteristics that have been an integral part of biotechnology (Liberman et al. 1991). Genetic engineering also encompasses recombinant DNA technology. The contemporary era of biotechnology began about 25 years ago with the discovery of endonucleases (restriction enzymes) and their ability to cut and paste segments of DNA. These enzymes act as highly specific chemical scalpels and can be used to obtain particular sequences within genetic material. Thus, restriction enzymes recognize and cut DNA at precise locations. Recombinant DNA and similar genetic engineering procedures led to the transfer of genetic material across species barriers and thus obviated the need for traditional breeding techniques.

Table 1 lists several important milestones in biotechnology beginning with early fermentation techniques (e.g., antibiotics) and extending to present-day recombinant DNA-derived products or materials. Such mileposts include not only important therapeutic proteins but also environmental and agricultural improvements using these technologies.

TABLE 1. *Selected milestones in biotechnology*

Year (Approximately)	Milepost
1940	Commercial production of antibiotics
1953	Double-helical structure of DNA discovered
1965	RNA used to break genetic code
1967	Automatic protein synthesizer developed
1967	Genetically engineered food introduced—potato
1970	Discovery of restriction enzymes (endonucleases)
1972	Splicing of viral DNA
1975	Discovery of monoclonal antibodies (hybridomas)
1976	First commercial company founded to develop recombinant DNA
1977	Expression of gene for human somatostatin
1978	Insulin produced using recombinant DNA
1978	Mammal to mammal gene transplants
1979	Human growth hormone and interferons reproduced
1979	Effect of recombinant BST on milk production in dairy cows
1980	Transgenic animals produced (mouse to mouse)
1982	Recombinant DNA animal vaccine
1982	Production of synthetic growth hormone
1983	Polymerase chain reaction (PCR) technique developed
1989	Bioremediation technology
1990	Clinical trials of human gene therapy
1991	Genetically engineered biopesticide
1993	Genetically engineered tomato
1993	Recombinant DNA erythropoietin
1993	Recombinant DNA blood coagulation factors
1995	Clinical use of recombinant follicle-stimulating hormone (FSH)
1996	Cloning of adult mammal
1997	Plasmid DNA vaccines
1998	Further development of xenotransplants

There are numerous applications of these new biotechnologies. Hormone substitutes, nutritional supplements, and improved food supplies are but a few general examples of how this new biotechnology can impact the quality of life. The biotechnology industry consists of several segments, including therapeutics, diagnostics (e.g., monoclonal antibodies), agriculture, and the environment. Although there is the potential to discover many new drug modalities for the treatment of various diseases, biotechnology will also play a major role in food production to feed a human population that has almost doubled in size in the last 50 years.

TRANSGENE TECHNOLOGIES

Animals

The introduction of transgene technologies, which involve the transfer of a gene from one species to another species, has occurred along with other advances in recombinant DNA methodologies or genetic engineering. The creation of laboratory mouse models for biomedical research has progressed rapidly, and so-called designer mice (produced, for example, through transgenics) are being used to study immune deficiencies, cancer, and developmental biology (Grosveld and Kollias 1992). Transgenic mouse models simulating human diseases include: AIDS, Demyelinating diseases, Diabetes mellitus, Glomerulosclerosis, Hepatocarcinogenesis, Hypertension, Inflammatory diseases, Neonatal hepatitis, Oncogenesis, Osteogenesis imperfecta, and Sickle-cell anemia (Goodnow 1992). Over a decade has elapsed since the development of a transgenic mouse model. Transgenic animals are produced by inserting a foreign gene into an embryo; this gene subsequently becomes an integral part of the host animal's genetic material. Increasingly, there have been improvements in the methods used for gene introduction into animals. Several techniques have been used, including microinjections of DNA into the pronucleus, retrovirus infection, and embryonic stem cells that can be grown in vitro from explanted blastocytes. All of these techniques have been employed to create chimeric animals.

Plants

Generally speaking, progress has been slower in developing transgene species in plant cells than in animal cells. Plant-breeding techniques have evolved over the millennium with the purpose of improving crops. Artificially induced mutations and selection have led to improvements in major cereal crops. Oftentimes, however, artificially induced mutations can also produce undesirable traits. Initially, improvements in plant biotechnology have been directed at enhanced agronomic traits, including pest resistance, herbicide tolerance, and disease resistance (e.g., viruses, fungi).

REGULATORY APPROVAL

Historically, the regulatory approval of genetically engineered products has varied widely. The safety and regulatory approval of therapeutic proteins has generally proceeded without any major impediments. Alternate sources of insulin for the treatment of diabetes mellitus and of growth hormone for treating children with deficiencies have not been controversial and have posed no significant risks. Still other novel therapeutic proteins have undergone rigorous testing to ensure efficacy and safety prior to any regulatory approval. Plant biotechnology involving improvements in food quality has not been so readily accepted by the public nor in every instance by regulatory agencies. Generally, the somewhat slower progress in the regulatory approval of genetically modified foods has not been an issue of safety but rather one involving the environment.

Several countries have well-established regulatory groups for biotechnology-derived-drugs or therapeutic proteins (Table 2). In the United States, the Food and Drug Administration (FDA) through the Center for Biologic Evaluation and Research (CBER) has oversight responsibilities for ensuring the safety and efficacy of biotechnology-derived agents.

Regulatory responsibility for biotechnology-derived foods may involve several U.S. agencies, including the Department of Agriculture, Environmental Protection Agency (EPA), and the FDA. The FDA has oversight of food safety as it relates

TABLE 2. *Regulatory aspects of biotechnology*

Country	Regulatory Committee or Agency
Europe (EU)	Committee on Proprietary Medicinal Products (CPMP)
	Safety Working Party (SWP)
	Biotechnology Quality Working Party (BQWP)
	Operations Working Party (OWP)
	Council Directive 87.22 EEC
	Concertation Procedure
	Notes for Guidance (e.g., preclinical safety [1987])
Japan (MHW)	Pharmaceutical Affairs Bureau (PAB)
	Central Pharmaceutical Affairs Council (CPAC)
	Committee on Drugs
	Committee on Antibiotic Drugs
	Committee on Blood Products
	National Institute of Hygienic Sciences
	Notification No. 243 (1984)
	Notification No. 10 (1988)
United States (FDA)	Center for Drug Evaluation and Research (CDER)
	Center for Biologic Evaluation and Research (CBER)
	Public Health Services Act
	Food, Drug, and Cosmetic Act
	Federal Register: biotechnology notice (1984)
	Federal Register: regulation of biotechnology
	Product-(class-)oriented "Points to Consider" (PTC)

SOURCES: Bass et al. 1992 and Thomas 1995.

to new plant varieties (e.g., dairy products, processing aids). The Department of Agriculture regulates poultry products and meat as well as field tests for genetically modified plants. The EPA, through its regulatory responsibility for pesticides, may oversee aspects of genetically altered plants with traits such as herbicide tolerance and pest resistance.

With the globalization of commerce, many agencies worldwide have become involved in the regulation of genetically modified organisms (GMOs), which may involve both plant and animal biotechnologies. Several of the organizations dealing with GMOs are involved in plant and animal biotechnology:

AcNFP	Advisory Committee on Novel Foods and Processes (UK)
BQWP	Biotechnology Quality Work Party (E.U.)
CBD	Convention on Biological Diversity
CBER	Center for Biological Evaluation and Research (U.S.)
CDER	Center for Drug Evaluation—Research (U.S.)
CPMP	Committee on Proprietary Medicinal Products (E.U.)
GIBiP	Green Industry Biotechnology Platform
ICH	International Conference on Harmonization
NIBSC	National Institute of Biological Standards and Control (UK)
OECD	Organization for Economic Cooperation and Development
UNCED	UN Conference on Environment and Development
UNEP	UN Environmental Program
UNIDO	UN Industrial Development Organization.

Many of these organizations have no regulatory authority but may propose guidelines for the use of GMOs. Early efforts have been devoted to harmonization and standardization of safety protocols for biotechnology-derived products.

There are several advantages and disadvantages to regulatory guidelines. Regulatory guidelines advantages include:

- Formalization of a consensus scientific opinion
- Promotion of consistency of review
- Improvement of quality of studies performed, and
- Assistance to industry.

Regulatory guidelines disadvantages include:

- Nonuniformity in study designs
- Retreat to "check-the-box" approach owing to uncertainties
- Disincentive to industry to be creative, and
- Failure to incorporate new and evolving technologies appropriately (Henck et al. 1996).

It is important to observe that the advantages outweigh the disadvantages. Harmonization or adoption of consistent review protocols is imperative and will lead to better overall assessment of safety and efficacy of new biotechnology-derived products.

BIOPHARMACEUTICS

Introduction

The development of recombinant DNA technologies has enabled the pharmaceutical-related industries to manufacture large quantities of macromolecules in relatively pure form. This has led to the production of such therapeutic proteins as human insulin, human growth hormone (HGH), human tissue plasminogen activator (rTPA), and human erythropoietin. Although hormone substitutes (e.g., insulin and HGH) represented an early entry in biopharmaceutics, many other classes of therapeutic agents have been produced or are otherwise undergoing development by the pharmaceutical industry. Because most of the early biopharmaceuticals were represented by simple hormone replacement therapies (e.g., insulin and HGH), they did not undergo extensive pre-clinical safety and efficacy studies such as those required for the interferons. During the last few years, several biopharmaceutical products have been approved for various clinical indications (Table 3). Several HGH preparations (e.g., Genotropin, Biotropin, and Norditropin) have recently received regulatory approval. New clinical indications have also been approved for some of the interferons (e.g. Intron A for malignant melanoma, Roferon-A for chronic myelogenous leukemia). Most interest has been in the development of therapeutic agents and not necessarily in monoclonal antibody assays used for diagnostic agents. Blood coagulation factors have been an important area of development. The regulatory approval process for HGH was probably accelerated by discovery that some cadaveric HGH was possibly contaminated by the latent neural viruses responsible for Creutzfeldt–Jakob disease.

TABLE 3. *Selected list of biopharmaceuticals*

Generic name	Trade name	Classification	Indication(s)
Alteplase (rTPA)	Activase	Recombinant protein	Acute myocardial infarction Pulmonary embolism
Epoetin alfa (EPO)	Epogen, Procrit	Recombinant protein	Anemia of chronic renal failure Anemia secondary to zidovudine treatment in HIV-infected patients
Antihemophilic factor	Recombinate	Recombinant protein	Hemophilia A
Factor VIII	Kogenate	Recombinant protein	Hemophilia A
Factor VIIIc	Monoclate-P	Recombinant protein	Hemophilia A
Factor IX	L-Nine SC	Recombinant protein	Hemophilia B
Filgrastim (G-CSF)	Neupogen	Recombinant protein	Neutropenia following chemotherapy
Sargramostim	Leukine	Recombinant protein	Myeloid reconstitution after bone marrow transplantation
Interferon alfa-2a	Roferon-A	Recombinant protein	Hairy cell leukemia AIDS-related Kaposi's sarcoma
Interferon alfa-2b	Intron A	Recombinant protein	Hairy cell leukemia AIDS-related Kaposi's sarcoma Hepatitis non-A, non-B/C Condylomata acuminata
Interferon alfa-n3	Alferon N	Human	Genital warts
Interferon beta-1a	Avonex	Recombinant protein	Multiple sclerosis
Interferon beta-1b	Betaseron	Recombinant protein	Multiple sclerosis
Interferon gamma-1b	Actimmune	Recombinant protein	Chronic granulomatous disease
I-IGF-1	Myotropin	Recombinant protein	Amyotrophic lateral sclerosis (ALS)
Aldesleukin (IL-2)	Proleukin	Recombinant protein	Kidney cancer
Dornase-alfa	Pulmozyme	Recombinant protein	Cystic fibrosis
Human insulin	Humulin	Recombinant protein	Diabetes mellitus
Somatrem (rMehGH)	Protropin	Recombinant protein	Growth hormone deficiency
Somatropin (rhGH)	Humatrope Saizen Nutropin	Recombinant protein	Growth hormone deficiency
Follicle-stimulating hormone	Puregon	Recombinant protein	Induces ovulation
Somatotropin (rbGH)[a]	Posilac	Recombinant protein	Increases bovine lactation
Pegaspargase (PEG-L-asparaginase)	Oncaspar	Polyethylene glycol modified protein	Acute lymphocytic leukemia
Hemophilus B conjugate vaccine	HibTITER	Recombinant protein	Prophylaxis against *Hemophilus influenza*

TABLE 3. *Continued.*

Generic name	Trade name	Classification	Indication(s)
Hepatitis B vaccine	Engerix-B Recombivax HB	Recombinant protein	Prophylaxis against Hepatitis B infection
Satumomab	OncoScint CR103	Monoclonal antibody	Colorectal cancer imaging
Satumomab	OncoScint OV103	Monoclonal antibody	Ovarian cancer imaging
Muromonab-CD3 (OKT3)	Orthoclone (OKT3)	Monoclonal antibody	Acute allograft rejection in renal transplant patients

[a]Veterinary product.

Preclinical assessment of potential biotherapeuticals destined for human medical therapies has brought forth the issue of immunogenicity between species. Macromolecules derived through biotechnology are expected to challenge the immune system of the laboratory animal used for toxicological testing protocols. Despite some of these unique and early concerns, guidelines for the preclinical testing of large-molecular-weight biopharmaceuticals have proceeded and have generally been adopted by regulatory agencies. Not only has the advent of macromolecules represented a challenge to the design of toxicological protocols, but it has also led to challenges in the formulation and delivery of biopharmaceuticals.

The New Biologies

Progress in the scientific disciplines of genetics, immunology, and molecular biology has provided a strong impetus to advances in biotechnology—particularly in biopharmaceutics (Tomlinson 1992). Indeed, biotechnology-derived products could not have progressed to their current state had it not been for the coincidentally and independently developed technology of monoclonal antibodies. Certainly, recombinant DNA technologies and monoclonal antibody methodologies are two areas that are inseparably important to the development of new biopharmaceuticals. Monoclonal antibodies are essential in the identification, extraction, and purification of most macromolecules. The term molecular biology has come, in particular, to signify the biochemical study of nucleic acids and has been advanced by the discovery of a series of enzymes (viz., endonucleases) that allow specific manipulations of RNA and DNA.

Several new techniques or approaches are being used in molecular biology (Ausubel et al. 1995). Some molecular cell biology tools include:

Recombinant DNA techniques
- Site-directed mutagenesis
- Single-strand conformational polymorphism (SSCP)
- Ligated gene fusions

Hybridoma technology
- For production of antibodies

Carbohydrate engineering

Novel instrumental techniques
- Polymerase chain reaction (PCR) and reverse transcription PCR (RT–PCR)
- Subtractive hybridization differential display
- Immunoblotting
- Confocal laser microscopy
- Fluorescence-activated cell sorting (FACS)
- Scanning tunneling electron microscopy

Pulsed field electrophoresis

Northern blotting and solution hybridization

In situ hybridization

Southern blotting: gene deletions.

The ability to manipulate the genetic material of a cell to modify gene expression has necessitated a variety of techniques that have led to new biopharmaceuticals, and consequently a broad range of molecular biology techniques have become increasingly routine.

Therapeutic Proteins

Major initiatives in recombinant DNA technologies have focused on mammalian macromolecules, often peptides, or for discovering therapeutically useful proteins. Mammalian biotechnology has advanced more rapidly than plant biotechnology, although the latter field promises to be very fast-moving. Early successes were manifested in the use of prokaryotic cell systems (e.g., *Escherichia coli*) capable of producing rather complex mammalian proteins (e.g., insulin). Subsequently, and with more advances in techniques employed in molecular biology (Table 4), prokaryotic cell systems were found capable of secreting even more complex proteins and glycoproteins. Through improved methods of molecular characterization, it has been possible to create more complex and more highly purified molecular entities that possess therapeutic potential. Mammalian organisms possess an abundance of proteins that modulate many physiologic events.

Modern molecular biology, particularly through the use of polymerase chain reactions (PCRs), has become very important in the discovery and production of recombinant proteins. Acting as a molecular "photocopier," PCRs have been employed extensively to amplify DNA sequences. Subsequently, the design of new biopharmaceuticals has proceeded with sufficient quantities of purified material to fulfill the amounts required for preclinical and clinical protocols or trials.

The majority of new biopharmaceuticals usually must be administered parenterally. Proteins or large peptides are vulnerable to gastric degradation and loss of biological activity. Considerable interest exists in the need to develop drug delivery systems

TABLE 4. *Advances in the characterization of biological products*

Method	Examples
In vitro bioassays	Cell culture techniques for hormones and cytokines
Gene analysis	Polymerase chain reaction analysis (viral contaminations of blood products)
Immunological tests	Enzyme-linked immunosorbent assay
	Epitope mapping by panels of monoclonal antibodies
	Immunoblotting of gels (Western blots)
	Testing for process-related impurities
Advanced chromatographic techniques	Size exclusion and reversed-phase high-performance liquid chromatography
	Capillary zone electrophoresis
Analysis of three-dimensional structure	Circular dichroism
	Nuclear magnetic resonance
Protein analysis by mass spectrometry	Fab-mass spectrometry (antigen binding)
	Electrospray mass spectrometry
Analysis of glycosylation isoforms	Detailed structure–function studies required

SOURCE: Tomlinson 1992.

for these macromolecules. Selective drug targeting is a characteristic that would be highly desirable. Advances have been made with some biopharmaceuticals to modify their biologic dispersion (Table 5). Such approaches have included site-directed mutagenesis and hybrid site-specific proteins.

Toxicologic Testing Protocols

Evaluating the safety of drugs produced by biotechnology resembles the assessment of conventional new chemical entities (NCE) but with certain major differences in testing protocols (Zbinden 1990; Dayan 1995). Several aspects of the new biotechnologies represent a challenge to the classical preclinical safety and efficacy protocols. A host

TABLE 5. *Approaches for modifying the biological dispersion of therapeutic proteins*

Approach	Therapeutic protein
Site-directed mutagenesis	α_1-Antitrypsin
Hybrid site-specific proteins	Immunoglobulin and toxin
Linked synthetically	Fragments
Fused-gene products	Growth factors
Protectants	
Polyethylene glycols	Interleukins
Administration	
Frequency and rate	Growth hormone (effectiveness altered)
Route	Insulin (site of injection)
Staging	Combinations of interferon and tumor necrosis factor

SOURCE: Demain 1991.

of safety considerations of biotechnologically derived products must be incorporated into the preclinical testing protocols (Tomlinson 1992):

Product-related impurities or variants
Genetic variants and mutants
Aggregated forms
Glycosylation patterns
Process-related impurities
Pyrogens
DNA and potential oncogenicity
Host-cell–derived proteins
Viruses (human, simian, murine, bovine).

Each biotechnology-derived product necessitates careful quality control because of concern about immunogenic proteins, peptides, endotoxins released by the harvesting of prokaryotic cell systems that secrete the product, and other possible chemical contaminants emanating from the processing procedures. According to Zbinden (1990), there are at least three areas of concern pertinent to biotechnology-derived products: (1) toxicology issues relating to differences in pharmacodynamic properties, (2) intrinsic toxicity (i.e., adverse effects due to the molecules themselves that are not a direct result of pharmacodynamic actions), and (3) biological toxicity such as responses resulting from the activation of a physiological mechanism (e.g., antigen–antibody reaction).

Several toxicologic or safety issues form the basis of toxicity testing, including the intrinsic actions of the product itself and the cellular system or process that leads to the production or secretion of the macromolecule (Dayan 1995; Thomas and Thomas 1993; Thomas 1995). Some possible product-related impurities or variants are mutants, aggregated forms, and aberrant glycosylation. Process-related impurities may include pyrogens, host-cell derived proteins, viruses, and possibly prions (Jeffcoate 1992). Table 6 is a summary of some of the more important aspects in evaluating the safety of potential biopharmaceuticals. Providing reliable cell systems, expressing the correct gene, and ensuring its proper insertion along with any promoter are of fundamental importance. The secretion or presence of other bioactive moieties must always be considered, and needs to be addressed through proper extraction and purification processes. Protein or peptide contaminants can produce both unwanted immunologic and nonimmunologic actions. Clinical or standard toxicologic protocols may not be relevant when attempts are made to characterize the pharmacodynamics of a macromolecule. Nonimmunologic testing is dependent on the product's potential therapeutic use; at a minimum testing should include biodistribution, developmental and reproductive considerations, and mutagenicity–oncogenicity assessments. It is very important that a list of characteristics (i.e., panel) for detecting possible immune alterations be an integral part of the toxicology testing protocol (Table 7). The first tier consists of screening factors for evaluating immunopathology, cell-mediated immunity, and humoral immunity. Tier I should be used to identify agents that may elicit an immune response. Those agents testing positive in Tier I are subsequently evaluated

TABLE 6. *Selected guidelines for evaluating biopharmaceutics*

Toxicological issues arising from the producing system

- Prokaryotic production system (recombinant DNA)
  Correct gene and promoter, stable expression, contaminating toxins, etc.
- Eukaryotic production system (recombinant DNA)
  Correct gene and promoter, stable expression, presence of antigens and other bioactive peptides, etc.
- Other
  Chemical modifications, infection-free animal vectors, etc.

Toxicological issues arising from the production process

- De- and renaturation of protein(s)
- Presence of other bioactive molecules
- Presence of chemical residues
- Microbial contamination (e.g., endotoxins)

Toxicological issues arising from biopharmaceutics

- Pharmacodynamics
- Pharmacological and toxicological actions
- Immunological
- Other (e.g., live or attenuated virus)

SOURCE: Dayan 1995.

in more detail. On the basis of Tier II, the presence of any immunopathological effects are evaluated.

Regulatory Considerations

There are some difficulties in conceiving and applying guidelines for the safety evaluation of biotechnology-derived products (Claude 1992; Cohen-Haguenauer 1996). Safety is highly dependent on the particular industrial process and its quality control. Pharmaceutical companies and regulatory agencies responsible for public health must establish relevant and meaningful guidelines for the preclinical and clinical evaluation of new biopharmaceuticals. When possible, an effort should be made to compare the new biopharmaceutical to a natural biological entity. Regulatory guidelines must reflect the importance of selecting an appropriate animal model.

Manufacturing procedures for biopharmaceuticals are a very important part of the safety and regulatory process (Jeffcoate 1992; Federici 1994). Certain aspects of the manufacturing process that convert the new biologies into biopharmaceuticals require attention by both manufacturers and regulatory agencies (Table 8).

Biopharmaceuticals are generally classified as biologicals by the regulatory agencies. Such a classification ordinarily includes "any virus, therapeutic serum, toxin, antitoxin, or analogous product applicable to the prevention, treatment or cure of diseases or injuries of man." The quality control of biopharmaceuticals engenders many of the same concepts as those applied to the analysis of conventional or low-molecular-weight pharmaceuticals.

TABLE 7. *Panel for detecting immune alterations*

Tier I
Hematology (e.g., leukocyte counts)
Weights—body, spleen, thymus, kidney, liver
Cellularity—spleen, bone marrow
Histology of lymphoid organ
IgM antibody plaque-forming cells (PFCs)
Lymphocyte blastogenesis
T-cell mitogens (Phytohemagglutinin [PHA], Con A)
T-cell (mixed leukocyte response [MLR])
β-cell (lipopolysaccharide [LPS])
Natural killer (NK) cell activity

Tier II
Quantitation of splenic B and T cells
Lymphocytes (surface markers)
Enumeration of IgG antibody PFC response
Cytotoxic T lymphocyte (CTL)
Cytolysis or delayed hypersensitivity response (DHR)
Host resistance
Syngeneic tumor cells
PYB6 sarcoma (tumor incidence)
B16F10 melanoma (lung burden)
Bacterial models
Listeria monocytogenes (morbidity)
Streptococcus species (morbidity)
Viral models
Influenza (morbidity)
Parasite models
Plasmodium yoelii (parasitemia)

SOURCE: Condensed from Luster et al. 1992.

TABLE 8. *Aspects of regulatory importance*

Manufacturing procedures
Source materials:
Genetically engineered microorganisms
Transformed mammalian cell lines
Hybridoma technology
Manufacture:
Clear production strategy and in-process controls
Validation of virus inactivation and removal
Purification of final product:
Fraction and chromatographic procedures
Affinity purification (e.g., MAb)
Pasteurization
Lyophilization
End-product quality:
Accurate and precise methods
Rigorous specifications

SOURCE: Modified from Jeffcoate 1992, Table 1, p. 192.

AGRIBIOTECHNOLOGY

Food Sources and World Population

Plants are one of many novel hosts that can be used not only in the production of biopharmaceuticals, but more importantly, in the genetic engineering of crops and foods. Since the early discoveries of Mendel, geneticists have been interested in the prospects of directed genetic change. Genetic modification of crop plants to improve many of their qualities or traits has led to a proliferation of recombinant products. Biotechnology has been used for centuries, but its newest form has enabled scientists to transfer traits between different plant species. Agribiotechnology has many potential benefits, including increased crop and livestock productivity, enhanced micronutrient composition, and improved pest control through the development of herbicide-resistant crops. This technology may also improve food processing and even provide diagnostic tools for detecting plant pathogens. Still another important dimension of transgenic plants is the development and production of oral vaccines. Such vaccines could be used in the treatment of diarrheal diseases.

Genetic engineering of plants is not unlike the types of modifications that were used by previous generations of plant breeders, but the techniques are more complex and diverse. In the 1950s, there was a doubling of the world's food supply. Thirty years later, farmers' productivity in the United States exceeded the population growth curve and led to an increase in per capita grain output of about 40%. In the United States, this led to significant agricultural surpluses. Simultaneously, the world's food supply continued to shrink, and stockpiles of grain have been diminishing since the mid-1980s. These supplies continue to dwindle, and the demand for food worldwide has nearly tripled from the 1950s to the 1980s. Currently, the world's population is about 6 billion and growing at a rate of approximately 90 million/year. Because the amount of tillable land remains unchanged, the need for technologies that will improve worldwide food production is becoming increasingly urgent.

Genetic Traits and Plant Biotechnology

Several important agronomic traits can be produced through biotechnology (Stark et al. 1993):

- Insect resistance
- Bacterial and fungal resistance
- Viral resistance
- Herbicide tolerance
- Stress tolerance
- Extended shelf life
- Nutrient modification.

TABLE 9. *Selected biotechnology products*

Product	Altered trait	Source of gene(s)
Canola	High lauric acid	Turnip, oilseed, rape, etc.
Cotton	Herbicide resistance	Bacteria, virus
Potato	Resistance to beetles	Bacteria
Soybean	Herbicide resistance	Bacteria, virus
Squash	Viral resistance	Virus
Tomato	Delayed ripening	Tomato, bacteria, virus
Vaccina virus	Vaccine	Rabies virus

SOURCE: Modified from Paoletti and Pimentel 1996.

Transgenically introduced traits that have received considerable attention include herbicide (e.g., glyphosate) tolerance, virus resistance, and insect resistance. Other traits that are under active investigation or development include resistance to fungal and bacterial infections and tolerances to different stresses such as extremes of salt, heavy metals, temperature, nitrogen, and phytohormones.

Representative products introduced as having altered traits include both foods and vaccines (Table 9). Canola can be modified using an acyl carrier protein thioesterase gene. A glufosinate-tolerant canola can be made using a phosphenothricin acetyltransferase gene obtained from *Streptomyces viridochromogenes*. A glyphosate-tolerant cotton can be made using the enolpyruvylshikimate-3-phosphate gene from *Agrobacterium* species-strain CP4. Likewise, canola and soybeans can be rendered glyphosate tolerant. Insect-protected potatoes and corn can be obtained using the *crylIIA* gene from *Bacillus thuringiensis*. Squash can be rendered virus-resistant by using coat protein genes of watermelon mosaic viruses or zucchini yellow mosaic viruses. A modified fruit-ripening tomato can be produced using several different gene-modifying approaches.

Field Trials

Over 38 different crop species have been tested in field experiments in 31 countries since the first trials with tobacco in 1986 (Dale 1995) (Table 10). The United States leads in the number of field trials, although China reportedly has more acreage under transgenic crops. Several transgenic plant varieties have been approved for commercial use. The plant trait that seems to be most consistently modified is that of herbicide tolerance (Table 11). Insect and viral resistance are two other commonly modified traits that have undergone field releases. Field trials of transgenic crop plants have exploded in number and acreage over the past decade. However, different methods of categorizing and counting procedures make precise comparisons and assessment of worldwide inventories difficult (Giddings 1996). Transgenic plant field trials in the United States are subject to regulatory oversight involving specific areas by the Animal and Plant Health Inspection Service (APHIS) of the U.S. Department of

TABLE 10. *Field trials in various countries*

Country	Field trials (No.)
Belgium	68
Denmark	7
Finland	17
France	138
Germany	5
Hungary	17
Italy	5
Norway	1
Portugal	2
Spain	13
Sweden	11
Switzerland	2
The Netherlands	52
United Kingdom	54
United States	385

SOURCES: Green Industry Biotechnology Platform; Dale, 1995.

Agriculture and by the U.S. Environmental Protection Agency (EPA). If the commercial end use of the product warrants, the U.S. Food and Drug Administration (FDA) may also be involved in regulatory aspects.

Environmental Considerations

Field releases of genetically engineered plants have raised safety issues related to the environment. Consequently, principles or guidelines have been promulgated to ensure environmental monitoring of field experiments. Many scientists have expressed concern regarding the possible environmental risks of genetically engineered organisms (cf. Paoletti and Pimentel 1996). Some have asserted that the release of genetically

TABLE 11. *Plants modified by transformation*

Modified characteristic	Field releases (1986–1993)
Insect resistance	127
Viral resistance	164
Bacterial resistance	11
Fungal resistance	29
Herbicide tolerance	340
Modified quality	184
Resistance to marker genes	93
Multiple traits	26
Unspecified	31

SOURCES: Green Industry Biotechnology Platform; Dale, 1995.

engineered organisms may adversely affect both tropical and temperate biodiversity. Ironically, it has been estimated that between 75 to 100% of agricultural crops contain some degree of host plant resistance—a resistant trait acquired by classical plant breeding (Oldfield 1984).

The gene for the *Bt.* toxin (*Bacillus thuringiensis*) was discovered over a hundred years ago. This soil bacterium has been used extensively in so-called organiç farming. Since 1991, the *B. t.* gene has been inserted into more than 50 plant crops (cf. Paoletti and Pimentel 1996). A concern has arisen about the development of resistance to *B. t.* toxin in field and laboratory tests (Lambert and Peferoen 1992; Stone et al. 1991; Tabashnik et al. 1992). At this time, the long-range environmental consequences of the overzealous use of *B. t.* toxin in major crop plants remain uncertain or unknown. It is noteworthy that some of these modified crop traits, such as herbicide tolerance, are environmentally friendly because their use results in less chemical spraying.

Food Safety and Allergenicity

It has been estimated that there are thousands of different proteins in the body. However, only a few proteins in food cause allergies. Because genetically engineered crop plants introduce new proteins, there is always a possibility that a protein encoded into this new genetic material may produce an allergic response. Many foods that are not genetically modified, including milk, eggs, tree nuts, shell fish, and certain legumes (e.g., peanuts and soybeans), are commonly allergenic. Thus, a food crop (e.g., legume) that is genetically modified is apt to be just as allergenic in a sensitive person as the same legume from a nongenetically modified version of the same crop.

There is no evidence that recombinant proteins in newly developed foods are more allergenic than traditional proteins. In fact, evidence suggests that the vast majority of these proteins are safe to the consumer (Lehrer et al. 1996). The stability (or instability) of food allergens to gastric digestion represents another factor in their potential immunogenicity (Astwood et al. 1996). Furthermore, the temperature extremes that may be used in the processing of the final food product and the effect it may have on allergenicity are often overlooked.

It is possible to construct a decision tree for evaluating recombinant food proteins for their immunogenicity (Fig. 1). Recombinant proteins from known allergens can readily be tested using a tier of in vitro immunochemical assays. Several biochemical and immunotoxicity tests can be used to evaluate allergenicity. Unfortunately, recombinant proteins from unknown allergenic sources are not easily recognized or necessarily anticipated. Thus, a main problem is assaying recombinant proteins from sources with undetermined allergenic activity. Finally, there is no evidence that recombinant proteins are any more allergenic than traditional proteins. In fact, the vast majority of recombinant proteins present in new food products are safe for consumption. Nevertheless, it is important that new food products, regardless of source, be prudently developed and tested for any potential immunogenicity.

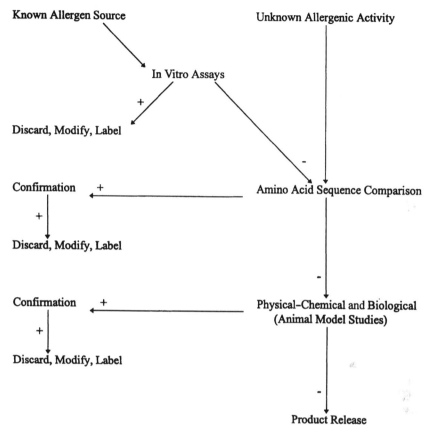

FIG. 1. Testing recombinant proteins for their immunogenicity (from Lehrer et al. 1996, Figure 3, p. 561).

SUMMARY

The next millennium will witness the use of an increasing number of an complex molecules as biotherapeutic agents. Recombinant DNA-derived products have already become established in a variety of medical therapies and hormone replacement regimens. These products have gained widespread public acceptance, and new biotherapies will continue to be novel and creative sources of drugs.

Agribiotechnology, although slower to gain public acceptance, will also experience major scientific advances in the twenty-first century. These advances will be driven, in part, by the burgeoning world population and the need to provide adequate food supplies.

REFERENCES

Ahh Goy, P., and Ducsing, J. H. 1995. From pots to plots: Genetically modified plants on trial. *Biotechnology*. 13:454–448.

Astwood, J. D., Leach, J. N., and Fuchs, R. L. 1996. Stability of food allergens to digestion *in vitro*. *Nature Bio/Technology* 14:1269–1273.

Ausubel, F. M., Brent, R., and Kingston, R. E. 1995. *Current protocols in molecular biology*. New York: John Wiley & Sons.

Bass, R., Kleeberg, U., Schrode, R. H., and Scheibner, E. 1992. Current guidelines for the preclinical safety assessment of therapeutic proteins. *Toxicol. Lett.* 64/65:339–347.

Claude, J-R. 1992. Difficulties in conceiving and applying guidelines for the safety evaluation of biotechnologically-produced drugs: Some examples. *Toxicol. Lett.* 64/65:349–355.

Cohen-Haguenauer, O. 1996. Safety and regulation at the leading edge of biomedical biotechnology. *Curr. Opin. Biotechnol.* 7:265–272.

Dale, P. J. 1995. R & D regulation and field trialling of transgenic crops. *Trends Biotechnol.* 13:398–403.

Davis, J. R. E. 1996. Molecular biology techniques in endocrinology. *Clin. Mol. Endocrinol.* 45:125–133.

Dayan, A. D. 1995. Safety evaluation of biological and biotechnology-derived medicines. *Toxicol.* 105:59–68.

Demain, A. L. 1991. An overview of biotechnology. *Occupa. Med.: State Art Rev.* 6:157–168.

Federici, M. M. 1994. The quality control of biotechnology products. *Biologicals* 22:151–159.

Giddings, L. V. 1996. Transgenic plants on trial in the USA. *Curr. Opin. Biotechnol.* 7:275–280.

Goodnow, C. C. 1992. Transgenic mice and analysis of β-cell tolerance. *Ann. Rev. Immunol.* 10:489–518.

Grosveld, F., and Kollias, G., eds. 1992. *Transgenic animals*. San Diego: Academic Press.

Henck, J. W., Hilbish, K. G., Serabian, M. A., Cavagnaro, J. A., Hendrickx, A. G., Agnish, N. D., Kung, A. H. C., and Mordenti, J. 1996. Reproductive toxicity testing of therapeutic biotechnology agents. *Teratol.* 53:185–195.

Jeffcoate, S. L. 1992. New biotechnologies: Challenges for the regulatory authorities. *J. Pharm. Pharmacol.* 44:191–194.

Lambert, B., and Peferoen, M. 1992. Insecticidal promise of *Bacillus thuringiensis*: Acts and mysteries about a successful biopesticide. *Biosci.* 42:112–122.

Lehrer, S. B., Horner, W. E., and Reese, G. 1996. Why are some proteins allergenic? Implications for biotechnology. *Crit. Rev. Food Science Nutri.* 36:553–564.

Liberman, D. F., Israeli, E., and Fink, R. 1991. Risk assessment of biological hazards. *Occupa. Med.: State Art Rev.* 6:285–299.

Luster, M. I., Portier, C., and Pait, D. G. 1992. Risk assessment in immunotoxicology; sensitivity and predictability of immune tests. *Fundam. Appl. Toxicol.* 18:200–210.

Oldfield, M. L. 1984. *The value of conserving genetic resources*. Washington, D. C.: U.S. Department of Interior, National Park Service.

Paoletti, M. G., and Pimentel, D. 1996. Genetic engineering in agriculture and the environment. *Biosci.* 46:665–673.

Stark, D. M., Barry, G. F., and Kishore, G. M. 1993. Impact of plant biotechnology on food and food ingredient production. in Chap. 8 *Science for the food industry of the 21st century*, edited by M. Yalpani, 115–132. Mount Prospect, NY: ATL Press, Inc, Science Publishers.

Stone, T., Sims, S. R., MacIntosh, S. C., Fuch, R., and Marrone, P. G. 1991. Insect resistance to *Bacillus thuringiensis*. In *Biotechnology of biological control of pests and vector*, edited by K. Maramorosch, 53–68. Boca Raton, FL: CRC Press.

Tabashnik, B. A., Schwartz, J. M., Finson, N., and Johnson, M. W. 1992. Inheritance of resistance to *Bacillus thuringiensis* in diamondback moth (*Lepidoptera plutellidae*). *J. Econ. Entomol.* 85:1046–1055.

Thomas, J. A., and Thomas, M. J. 1993. New biologics: Their development, safety, and efficacy. In *Biotechnology and Safety Assessment*, edited by J. A. Thomas and L. A. Myers, 1–22. New York: Raven Press.

Thomas, J. A. 1995. Recent developments and perspectives of biotechnology-derived products. *Toxicol.* 105:7–22.

Tomlinson, E. 1992. Impact of the new biologies on the medical and pharmaceutical sciences. *J. Pharm. Pharmacol.* 44:147–159.

Zbinden, G. 1990. Safety evaluation of biotechnology products. *Drug Safety* 5:58–64.

Biotechnology and Safety Assessment, 2nd ed.
Edited by John A. Thomas
Copyright © 1998 Taylor & Francis

13

Clinical Toxicity of Interferons

Thierry Vial and Jacques Descotes

INSERM 498-X, Claude Bernard University, Lyon, France·

The interferons (IFNs), first described in 1957, are a group of at least five natural human glycoproteins (alpha, beta, gamma, omega and, tau) with marked antiviral, antiproliferative, and immunomodulatory activity. Only the first three types are currently used clinically. Interferon-α (IFN-α) is an heterogeneous family of more than 20 subspecies of proteins and glycoproteins. Natural lymphoblastoid IFN-α (IFN α-N1) is a mixture of nine IFN subtypes produced from a human B lymphoblastoid cell line, whereas recombinant IFN-α (IFN alfa-2a and IFN alfa-b) is produced by DNA techniques using a strain of *Escherichia coli.* More recently, researchers developed a so-called consensus IFN-α (CIFN) by assigning the most frequently observed amino acids at each position to generate a consensus molecule (Keeffe and Hollinger 1997). Consensus IFN-α was later produced in a recombinant system. Interferon-β (IFN-β) has antiviral and antiproliferative properties very similar to those of IFN-α and is being used as a natural fibroblast or in recombinant preparations. Two recombinant forms of IFN-β (i.e., IFN beta-1b and, more recently, IFN beta-1a), have been approved for the treatment of multiple sclerosis. Interferon-γ (IFN-γ) is produced essentially by activated T lymphocytes and shares no similarities with the other IFNs. Interferon-γ primarily exerts immunoregulatory properties and is available only as a recombinant form for the treatment of chronic granulomatous disease.

This chapter will review the available literature on the clinical adverse effects associated with IFN therapy in humans.

INTERFERON-α

Various forms of IFN-α have been approved worldwide in the treatment of a variety of diseases, including solid and hematological malignancies and viral disorders, such as hepatitis B virus- or hepatitis C virus-related chronic hepatitis. Interferon-α is currently the only approved drug for the treatment of chronic viral hepatitis. Thousands of patients have now been treated, and thus considerable clinical experience has been gained. The optimal dose and duration of treatment is still under investigation. From a recent metanalysis of randomized trials using various preparations of IFN-α, the authors concluded that a 1-year treatment course seems to offer the best efficacy-to-risk ratio in the treatment of chronic hepatitis C (Poynard et al. 1996). Patients are now exposed to treatment for relatively long periods as compared with the first trials using a 6-month course of treatment. Data on the safety and incidence of IFN-α largely come from clinical trials and retrospective surveys. Almost all patients treated with IFN-α will experience mostly mild-to-moderate adverse effects during treatment. They occur early after the initiation of therapy and are easily manageable without discontinuation of treatment. The profile of adverse effects observed with the various available types of IFN-α is very similar, as shown by randomized, double-blind, controlled studies comparing IFN alfa-2a and IFN alfa-N1 (Rumi et al. 1996) or IFN alfa-2b and CIFN (Keeffe and Hollinger 1997).

The incidence of IFN-α-associated severe adverse effects is more difficult to assess, for it largely depends on the dose, the schedule of treatment, and the disease to treat. Data from IFN-α trials with hepatitis C showed that at least 4–5% of patients had to stop treatment because of adverse effects (Poynard et al. 1996). Dosage reduction was required in 9% of those receiving 3 million units (MU) and in 22% of those receiving 5 MU. Delayed or irreversible toxicities can be troublesome, but very severe or life-threatening adverse effects associated with IFN-α are rarely encountered. In a large retrospective evaluation of 11,241 patients treated for chronic viral hepatitis, fatal or life-threatening toxicites were noted in 5 (including 4 cases of irreversible liver failure) and 8 patients (including attempted suicide in 2 and severe bone marrow depression in 6), respectively, which corresponds to an incidence close to 1 in 1000 (Fattovitch et al. 1996). In addition, severe but usually reversible nonhepatic adverse effects occurred in 131 (1%) patients and mostly consisted of thyroiditis (54%), neuropsychiatric manifestations (13%), dermatological adverse effects (11%), or diabetes (8%). Data from Japanese investigators showed that higher doses are associated with a higher rate of adverse effects in patients treated for chronic hepatitis C. Among approximately 1000 patients initially receiving 42 to 70 MU daily, dosage reduction or treatment withdrawal was required in 31% of these cases (Okanoue et al. 1996). Flu-like symptoms and hematological disorders were the most common findings. Significant but reversible adverse effects were noted in 12% of the remaining 677 patients, and depression or thyroid disorders were again the most frequent side effects. The delayed occurrence of several adverse effects associated with IFN-α treatment is sometimes difficult to interpret because the underlying disease may itself be the cause of the observed disorders. This is particularly true with chronic hepatitis C and the increased evidence of its pathogenic role in a variety of extrahepatic diseases.

Although hundreds of reports have dealt with a wide spectrum of IFN-α-induced adverse effects, the pathogenesis is still poorly understood. Two major mechanisms are usually postulated; namely, a direct toxic effect of IFN-α or an indirect effect via immunologic pathways, but clear experimental evidence is still lacking in most instances.

General Toxicity

Whatever the dose, nearly 90% of patients experience a flu-like syndrome during the first days of treatment (Vial and Descotes 1994). This reaction typically includes fever and chills, sweating, malaise, tachycardia, headache, arthralgias, and myalgias persisting a few hours following injection. The tolerance to this syndrome greatly varies among patients and is not influenced by age. Although the severity tends to increase with the dose, the syndrome is rarely treatment-limiting. The severity is reduced by preadministration of acetaminophen (paracetamol). In addition, tachyphylaxis resulting in less frequent and less severe symptoms usually develops within 2 weeks after repeated injections. Most of the cytokines currently used clinically (e.g., interleukins [IL-1, IL-2, IL-3] or granulocyte-macrophage colony-stimulating factor) have been associated with such first-dose reactions (Vial and Descotes 1995). The mechanism of flu-like symptoms remains unclear, but the acute release of fever-promoting factor (i.e., several eicosanoids, IL-1 or TNF-α) secondary to cytokine administration has been suggested.

Psychiatric Disorders

A wide range of neuropsychological disturbances have been described in IFN-α-treated patients. Symptoms such as cephalalgia, malaise, somnolence, or mnesic disorders are typically associated with the flu-like syndrome or occur shortly thereafter. Weight loss with a mean decrease of 2 to 5 kg is also common.

After several weeks of treatment, patients often complain of various cognitive and behavioral changes as well as intellectual and mental impairment (Meyers and Valentine 1995). Although usually mild-to-moderate in intensity, these later adverse effects are sometimes considered unacceptable and prompt patients to seek temporary dosage reduction or treatment withdrawal. The most severe and acute neuropsychiatric manifestations are usually associated with the higher doses. In patients treated with low-dose IFN-α, the onset of these disorders is more insidious and not always easy to recognize. These neuropsychiatric manifestations are now considered as one of the more frequent and major late complications of IFN-α treatment, and the reported incidence ranges from 7 to 35% (Prasad et al. 1992; Meyers and Valentine 1995). In the initial description of patients treated for chronic viral hepatitis, these neuropsychiatric disorders occurred in 17% of patients and were classified into three main categories: organic personality syndrome with irritability and personality changes; organic affective syndrome with depressive disorders and emotional lability; and psychotic manifestations with agitation, delirium, paranoia, and suicidal behavior (Renault et al. 1987). Although the underlying treated disease can undoubtedly play a role in the emergence of such symptoms, psychiatric morbidity was found to be significantly higher in two open label trials of treated versus control patients, either in the setting of chronic hepatitis C (McDonald et al. 1987) or chronic myelogenous leukemia (Pavol et al. 1995). In addition, symptoms typically lessened after IFN-α dosage reduction or withdrawal. Both findings strongly suggest that interferon per se is involved in these psychiatric manifestations. Severe depressive disorders and psychosis are the most troublesome neuropsychiatric complications, and the occurrence of suicidal ideation requires early intervention. Although reported as a major outcome, the incidence or excess risk of suicide associated with IFN-α treatment is not clearly established (Janssen et al. 1994). It could be higher than previously expected, as suggested by the results of a French prospective study that found 2 suicides and 2 attempted suicides among 219 patients treated for chronic hepatitis C (Rifflet et al. 1996).

Most patients recovered from psychiatric adverse effects after dosage reduction or treatment withdrawal. The possibility for persistent neurotoxicity has emerged from a single retrospective evaluation of only 14 cancer patients after completion of 26 weeks of treatment on average with IFN-α (Meyers et al. 1991). Neuropsychological evaluation found personality changes, impaired memory, deficits in motor coordination with ataxia and akathisia, decrease in frontal lobe executive functions, parkinsonian-like tremor, and moderate dementia.

A previous history of psychiatric disorders or addictive behavior was sometimes found in patients with the most severe IFN-α neurotoxicity. Despite early findings, only a few neuropsychological changes were observed in asymptomatic HIV-1 positive patients receiving IFN-α (Mapou et al. 1996). Unfortunately, no predictive factors for the development of psychiatric adverse effects have yet been clearly identified, and thus all patients require careful examination when psychiatric manifestations occur. In an uncontrolled open trial of 31 patients suffering from active psychiatric disorders, only 4 experienced worsening of the psychiatric disease, and treatment withdrawal was necessary in 2 (Van Thiel et al. 1995). Nevertheless, IFN-α is usually contraindicated in patients with preexisting psychiatric conditions.

As suggested by electroencephalographic findings with slowing of dominant alpha rhythm and occasional emergence of intermittent delta and theta activities in the frontal lobes, the mechanism of IFN-α-associated neurotoxicity could involve a direct effect on fronto-subcortical functions (Renault et al. 1987; Meyers and Valentine 1995). Indeed, although regional differences exist, the ability of IFN-α to cross the blood–brain or blood–spinal cord barrier was demonstrated in several central nervous system areas, at least in animals (Pan et al. 1997). However, the pathophysiological mechanisms of IFN neurotoxicity are still unknown, and a direct vascular effect, the alteration of neuroendocrine hormone production, change in dopaminergic pathways, or an effect mediated by the secondary induction of another cytokine were all suggested to be involved (Meyers and Valentine 1995). An excitatory effect of IFN-α on opioid receptors is also possible, for naltrexone improved cognitive dysfunction in IFN-α-treated patients (Valentine et al. 1995).

Neurological and Neurosensorial Disorders

Various severe neurological symptoms have been reported. A 1.3% incidence of generalized tonic-clonic seizures was found in a retrospective study of 311 patients (Shakil et al. 1996). This rate is far higher than the 0.03–0.05% annual rate of new seizure onset in the general population. Seizures were also found to be relatively common adverse effects in children treated for chronic hepatitis B (Woynarowski and Socha 1997). Overall, 10 of 225 children developed tremors or seizures. In children aged under 5, associated fever, and perinatal central nervous system injury may clearly play a contributory role. Whereas dose-related paresthesias have been reported in as much as 7% of patients (Vial and Descotes 1994), frank peripheral neuropathy seems to occur in a limited number of patients. Such isolated case reports were mostly described in patients who received high cumulative doses of IFN-α (Cudillo et al. 1990), but induction or exacerbation of axonal polyneuropathy is increasingly reported following long-term low-dose treatment (La Civita et al. 1996; Negoro et al. 1994; Rutkove 1997). Although the underlying disease may be involved in the occurrence of peripheral nerve dysfunction, IFN-α should reasonably be considered as a possible cause.

An increasing number of reports have addressed the risk of developing myasthenia gravis during IFN-α therapy. The diagnostic criteria for myasthenia gravis were clearly fulfilled in these patients, and an autoimmune pathogenesis was the most probable because each patient had positive serum antiacetylcholine receptor antibodies (Batocchi et al. 1995; Rohde et al. 1996). A familial genetic predisposition was sometimes found (Mase et al. 1996). The triggering of an underlying silent myasthenia gravis can be suggested, for these patients still required permanent anticholinesterase drugs long after withdrawal of treatment and had persistent low titers of antiacetylcholine receptor antibodies.

Ophthalmic disorders were found in more than half of patients (Hayasaka et al. 1995; Kawano et al. 1996). Retinal abnormalities included mild ischemic lesions with cotton wool exudate, capillary occlusion, or retinal hemorrhage. The lesions were usually asymptomatic and reversible upon IFN-α discontinuation or despite continuation of treatment. Optic neuritis (Manesis et al. 1994) or acute visual loss (Yamada et al. 1994) were reported in isolated patients Diabetes mellitus and hypertension were both found as significant risk factors. An increased C5a plasma level was suggested to predict the occurrence of retinal hemorrhage (Sugano et al. 1994).

Prospective investigations in 49 chronic-viral-hepatitis-treated patients showed frequent but spontaneously reversible otological impairment. Tinnitus, sensorineural hearing loss of at least 20 dB, or both disorders, were respectively found in 8, 16, and 20% after IFN-α or IFN-β administration (Kanda et al. 1994). The occurrence of these disorders partly depended on the cumulative dose and were considered treatment-limiting in only 2 of 22 patients.

The pathophysiology of these neurosensorial damages is unknown and may involve microvascular damages resulting from vasculitis or vasospasm. Recent experimental data have suggested that retinal microinfarction might be the result of IFN-α-induced increase in leukocyte adherence to the vascular endothelium (Nishiwaki et al. 1996).

Cardiovascular Complications

Although no clear dose-related effect can be found in the available data, severe and life-threatening cardiac toxicities (i.e., angina pectoris or myocardial infarction, severe ventricular dysfunction, arrhythmias) have mostly been reported in patients receiving high-dose IFN-α or concomitant cardiotoxic drugs in the setting of solid cancer or hematological malignancies (Vial and Descotes 1994). Severe cardiac dysfunction in patients without evidence of previous cardiac disease or in those receiving low-dose IFN-α are indeed anecdotal (Mateo et al. 1996; Sonnenblick and Rosin 1991). Severe cardiac manifestations were considered exceptional in patients treated for chronic viral hepatitis, and only 7 experienced myocardial infarction, angina, or dysrhythmias among 11,241 patients (Fattovitch et al. 1996). Other investigators found a higher incidence, with 6 reversible cardiovascular adverse effects (namely arrhythmia, ischemic heart disease, and myocardial disease) among 295 patients with chronic hepatitis C (Teragawa et al. 1996). A careful prospective evaluation of the cardiac adverse effects of IFN-α in only 31 chronic viral hepatitis patients treated for at least 6 months found no significant clinical changes in cardiovascular tests—even in patients with a previous cardiac history (Kadayifci et al. 1997). In sharp contrast to these results, a reversible reduction of more than 10% in left ventricular ejection fraction, which might be critical in patients with preexisting cardiac disease, was evidenced in 5 of 11 patients (Sartori et al. 1995). The mechanism of IFN-α-induced cardiac effects is unknown. Whatever the dose, a flu-like syndrome with fever, tachycardia, and an increase in oxygen consumption can precipitate acute cardiac dysfunction—particularly in patients with patent underlying cardiac disorders. Interestingly, arrhythmias and myocardial ischemia have been produced in an animal model following repeated IFN-α infusions (Zbinden 1990).

Severe Raynaud's phenomenon, sometimes leading to digital necrosis, has been described in a few patients (Arslan et al. 1994; Liozon et al. 1997). Although probably underrecognized, a moderate form of Raynaud's phenomenon was observed in 13 of 25 patients receiving long-term IFN-α treatment for chronic myelogenous leukemia (Creutzig et al. 1996). Different mechanisms probably account for these vascular abnormalities. An autoimmune etiology is unlikely, for these patients had no symptoms of autoimmune disease. A vasospasm is presumably the mechanism involved. Repeated episodes of chest pain at rest, which occurred after IFN-α injection only in a patient with a predisposition to coronary vasospasm (Tanaka et al. 1995), as well as reports of ischemic coronary manifestations (Mansat-Krzyzanowska et al. 1991) or ocular damage (see previous section) are in accord with this hypothesis. A direct toxicity of IFN-α on the vascular endothelium cannot be ruled out. Interferon-α has also been associated with digital vasculitis (Reid et al. 1992)

Pulmonary Toxicity

Reports of bronchiolitis obliterans organizing pneumonia (Ogata et al. 1994) or interstitial pneumonia were initially described in Japenese patients also using the popular "Sho-saiko-to" herbal preparation (Chin et al. 1994; Ishizaki et al. 1996; Kamisako et al. 1993). Because the latter preparation is sometimes involved in the occurrence of pulmonary infiltrates, the role of IFN-α could not be determined. More recently, similar cases of pneumonitis reversible upon IFN-α discontinuation were described in Western patients, which indicates that IFN-α can be considered as a causal factor in some patients (Lutsman et al. 1995; Nouri et al. 1996). Pulmonary symptoms were typically observed within 1–3 months of treatment. A drug lymphocyte stimulation test was found positive to IFN-alpha, suggesting that an immune-mediated mechanism, rather than a direct toxic effect, was involved (Ishizaki et al. 1996). In addition, an elevated CD8$^+$/CD4$^+$ ratio in the bronchoalveolar lavage fluid and a reversibly increased soluble interleukin-2 receptor indicated possible T-cell activation.

Renal Disorders

A wide range of renal changes have been noted in IFN-α-treated patients. Whereas a significant number of patients presented asymptomatic and transient proteinuria, microscopic hematuria, increased enzymuria, or a mild increase in serum creatinine levels (Quesada et al. 1986), only a few patients experienced severe nephrotoxicity with marked nephrotic syndrome, acute renal insufficiency, or both. The most severe renal complications, sometimes irreversible, have been described in patients treated for malignancies, in those receiving high-dose IFN-α, or in patients with underlying renal disease (Dimitrov et al. 1997). When available, various pathological findings were evidenced; namely, a nephrotic syndrome with minimal or severe glomerulous nephropathy (Lederer and Truong 1992), an extracapillary glomerulonephritis with crescents (Durand et al. 1993), a membrano-proliferative glomerulonephritis (Kimmel et al. 1994), a focal segmental glomerulosclerosis (Coroneos et al. 1996), an acute interstitial nephritis (Averbuch et al. 1984), or an acute tubular necrosis (Dimitrov et al. 1997).

Because of the variety of renal lesions, the mechanism of renal dysfunction is probably multifactorial and is currently poorly understood. A direct nephrotoxic effect with a possible additive effect of nonsteroidal anti-inflammatory drugs, an immune-mediated renal disease involving activated T lymphocytes, or an immune-complex glomerulonephritis have been suggested to occur. In other cases, slowly progressive renal insufficiency was attributed to hemolytic-uremic syndrome (Schlaifer et al. 1994) or renal thrombotic microangiopathy (Honda et al. 1997), but the respective roles of the underlying hematological malignancy and IFN treatment were not clear. Only chronological events with the appearance of symptoms under treatment, prompt resolution following IFN-α withdrawal, and rare but convincing cases with further positive rechallenge (Selby et al. 1985), argue for a potential role of the cytokine.

Metabolic Disorders

Interferon-α can induce changes in serum lipids. Complete investigations of the effect of IFN-α on serum and triglyceride concentrations have been performed (Malaguarnera et al. 1996; Shinohara et al. 1997). A significant increase in triglyceride levels and a reduction in total cholesterol, HDL-cholesterol, LDL-cholesterol, and apoprotein A1 were noted, suggesting an increased cardiovascular risk. Very large increases in serum triglyceride concentrations that reversed upon the cessation of treatment have been found (Sunderkötter et al. 1993; Yamagishi et al. 1994). The clinical relevance of these findings is unknown but probably limited, and pancreatic or cardiac complications have not been reported so far. An inhibiting effect of IFN-α on lipoprotein lipase, triglyceride lipase, or an increased hepatic lipogenesis were suggested to account for these metabolic disorders (Shinohara et al. 1997; Yamaghishi et al. 1994). The treated disease does not seem to account for this effect, for hypertriglyceridemia has been noted both in patients treated for chronic hepatitis (Shinohara et al. 1997) or hematological malignancies (Sgarabotto et al. 1997).

Hematological Toxicity

Interferon-α can inhibit stem-cell proliferation and differentiation and acts as a myelosuppressive agent. This largely explains dose-related neutropenias and thrombocytopenias, which are the most common hematological complications of IFN-α treatment. Thrombopenia may also result from increased hepatic uptake, but the platelet survival time was not significantly modified by IFN-α treatment (Sata et al. 1997). Serum thrombopoietin levels were found to be increased as a result of IFN-induced thrombocytopenia (Shiota et al. 1997). Moderate anemia with a progressive decrease in hemoglobin and hematocrit was also observed. The average decrease in platelets and white blood cells was in the range of 30–50% as compared with baseline values, and patients with severe liver disease and hypersplenism were more prone to severe complications (Vial and Descotes 1994). Hematological toxicity is usually not treatment-limiting in patients with no hematological disorders before treatment, but frequent monitoring of blood cell counts for dosage adjustment or drug discontinuation is required. An increased risk for severe hematological toxicity has also been found in patients receiving angiotensin-converting enzyme inhibitors together with IFN-α (Casato et al. 1995). In a large randomized trial of patients receiving low-dose IFN-α for chronic viral hepatitis, thrombocytopenia (platelets lower than 49,000/mm^3) and leukopenia (white blood cells lower than 900/mm^3) were observed in 10 and 20% of patients, respectively (Poynard et al. 1995).

Hematological toxicity can also be mediated by immune blood cell destruction. Reversible autoimmune hemolytic anemia (Braathen and Stavem 1989) or asymptomatic positive direct Coombs' test (Akard et al. 1986) were sometimes reported. Both acute exacerbation shortly after the initiation of treatment or delayed de novo appearance of immune hemolytic anemia can be observed (Andriani et al. 1996). In chronic myelogenous leukemia, immune-mediated hemolysis was identified in 1% of patients (Sacchi et al. 1995). Interferon-α-induced autoimmune hemolytic anemia was suggested to mimic that described with α-methyldopa (Barbolla et al. 1993). However, Coombs' test was found negative in several patients (Hirashima et al. 1994), indicating that nonimmune hemolysis is also possible. Immune thrombocytopenia is sometimes associated with chronic hepatitis C (Pawlotsky et al. 1995), but IFN-α is also a possible cause of thombocytopenia with features of autoimmune thrombocy-topenic purpura (Dourakis et al. 1996; Khan et al. 1996). That autoimmune-mediated thrombocytopenia recurred following IFN readministration argues for a direct causal role of IFN-α (Zuffa et al. 1996). Interestingly, IFN-β may be a safe alternative in patients who experienced IFN-α-induced immune thrombo-cytopenia (Tappero et al. 1996). Preexisting autoimmune thrombocytopenia may also be worsened by IFN-α administration (Bacq et al. 1996; Stern et al. 1994). However, IFN-α does not appear to be harmful in patients with chronic hepatitis C previously positive for platelet-associated immunoglobulin G (Taliani et al. 1996).

Clinically insignificant coagulation disorders were sometimes described (Mirro et al. 1985). More recently, IFN-α was associated with antiphospholipid antibodies in cancer (Becker et al. 1994) or chronic hepatitis C (Doutre et al. 1997) patients. The clinical relevance of these findings was confirmed by the high prevalence of throm-botic complications noted in four of five antiphospholipid antibody positive melanoma patients (Becker et al. 1994). The underlying cancer may have played a role in this outcome, for thrombosis resulting from antiphospholipid antibodies has not been yet reported in chronic hepatitis patients. Interferon-α treatment was also suspected of inducing antifactor VIII autoantibody in one patient without hemophilia (Stricker et al. 1994), but a small uncontrolled study was unable to detect increased antibodies in hemophilia A patients treated with IFN-α (Mauser-Bunschoten et al. 1996).

Digestive Toxicity

Mild and transient gastrointestinal adverse effects, including nausea and diarrhea, are common during treatment. More severe symptoms; namely ischemic colitis (Tada et al. 1996) and reversible acute pancreatic disorder (Sotomatsu et al. 1995), are rarely encountered.

Hepatic Adverse Effects

A mild-to-moderate reversible increase in liver transaminases is commonly reported in cancer patients receiving high-dose IFN-α. In patients treated for chronic hepatitis B, Hbe antigen seroconversion is sometimes preceded by a transient worsening of transaminase levels. In the most severe cases, this was associated with a fatal decom-pensation observed in less than 0.5% of patients (Fattovitch et al. 1996; Janssen et al. 1993). Patients with active cirrhosis or a previous history of decompensated cirrhosis, are particularly prone to hepatic decompensation during treatment (Krogsgaard et al. 1996).

The most troublesome hepatic consequence of IFN-α treatment is the possible acute exacerbation of a latent autoimmune hepatitis. On the basis of a false-positive enzyme immunoassay for hepatitis C virus antibodies, several patients were initially misdiagnosed as having hepatitis C, but further biological investigations later proved that autoimmune hepatitis rather than chronic viral hepatitis was the correct diagnosis

(Papo et al. 1992). More recently, it has been clearly evidenced that serological markers frequently positive in autoimmune hepatitis (namely antinuclear, anti-smooth–muscle, and antiliver and antikidney microsomal autoantibodies) and unequivocal evidence of chronic hepatitis C may coexist in the same patient before any treatment (Bayraktar et al. 1997; Calleja et al. 1996; Muratori et al. 1994; Todros et al. 1995). In addition, the induction or exacerbation of autoimmune hepatitis may occur in patients previously negative for serological markers of autoimmune hepatitis. In a large study involving 144 chronic hepatitis C patients treated with IFN-α, the acute flare-up of hepatitis experienced in seven patients was subsequently attributed to a classical autoimmune hepatitis (Garcia-Buey et al. 1995). Although only three patients had positive nonorgan-specific autoantibodies before treatment, the authors proposed that IFN-α may have triggered a latent autoimmune hepatitis. Unfortunately, the distinction between autoimmune hepatitis and chronic hepatitis C virus infection cannot readily be done on the basis of biological or histological data. A possible association with several human leukocyte antigen haplotypes (DR-3, DR-52, DQ-2, DQ-6) has been suggested, but this was based on a small number of patients (Garcia-Buey et al. 1995). Furthermore, the systematic detection of these autoantibodies failed to predict the risk of overt autoimmune hepatitis. There is no agreement on the management of patients with features of both chronic viral and autoimmune hepatitis. Corticosteroids may increase the level of viremia, whereas IFN-α may exacerbate an autoimmune liver disease. Several investigators proposed IFN-α as a valuable and safe treatment as compared with corticosteroids (Calleja et al. 1996; Todros et al. 1995), whereas others considered corticosteroids as a better first alternative option (Muratori et al. 1994; Tran et al. 1997). In every case, a very close monitoring of liver test functions should be performed to detect an acute worsening of transaminase levels.

Overall, de novo induction rather than exacerbation of a latent autoimmune hepatitis seems to be a very rare event reported in only isolated patients treated for chronic viral hepatitis or cancer. This is in keeping with the usual lack of autoantibodies specific for autoimmune hepatitis in IFN-α-treated patients. Other hepatic events associated with INF-α treatment included primary biliary cirrhosis (D'Amico et al. 1995) and granulomatous hepatitis resolving after IFN-α discontinuation and despite persistence of hepatitis C infection (Veerabagu et al. 1997).

Dermatological Disorders

Skin lesions reported with IFN-alpha are manifold (Asnis and Gaspari 1995). The most frequent and nontreatment-limiting adverse cutaneous effects consisted of skin dryness, pruritus, or nonspecific cutaneous lesions; namely, rash, diffuse erythema, or urticaria. Interestingly, no cases of immediate-type allergic reactions have so far been reported. Moderate and reversible telogen effluvium alopecia is a common finding (Tosti et al. 1992) and was observed in 9–24% of patients with chronic hepatitis C (Davis et al. 1990). Several recent reports suggested that interferons can induce vitiligo in patients with metastatic melanoma (Le Gal et al. 1996) or chronic hepatitis

C (Simsek et al. 1996). An IFN-induced increased susceptibility of melanoma cells to cytolytic T lymphocytes is a possible mechanism.

The attribution of skin disorders to IFN-α treatment should be regarded cautiously because there is increasing evidence that dermatological diseases, such as cryoglobulin-associated vasculitis, lichen planus, or porphyria cutanea tarda are dermatologic manifestations of hepatitis C infection (Schwaber and Zlotogorski 1997). This is particularly true for lichen planus, which is believed to be autoimmune. Reports on chronic hepatitis C patients receiving IFN-α showed both the development or exacerbation of lichen planus, or, in contrast, a complete reversibility of the lesions (Areias et al. 1996; Dupin et al. 1994; Hildebrand et al. 1995). However, several cases of lichen planus in cancer patients (Aubin et al. 1995) or reports with recurrence of the lesions after IFN-α readministration or increased dosage (Strumia et al. 1993) provided evidence for a causal relationship with IFN-α treatment.

The pathogenesis of psoriasis probably involves a TH1-mediated process. Because IFN-α seems to influence the cytokine network toward increased TH1 response, the association between IFN-α and psoriasis is not totally unexpected. Since the first reports in patients receiving very high doses (Quesada et al. 1986), subsequent cases have largely confirmed the involvement of IFN-α in the development or acute exacerbation of psoriasis (Georgeston et al. 1993; Nguyen et al. 1996; Wolfe et al. 1995). The causal relationship was particularly compelling in patients who experienced a rapid recurrence of lesions after IFN-α readministration (Funk et al. 1991; Jucgla et al. 1991). In the most severe instances, exacerbation of psoriasis was associated with psoriatic arthritis (Georgeston et al. 1993; Makino et al. 1994).

Injection site reactions following subcutaneous injections, namely local erythema and skin induration, are sometimes observed and may require regular changes in the area of injection. More severe reactions with inflammatory nodules, ulcerations, and typical cutaneous necrosis have seldom been reported whatever the dose used or the indication of treatment (Azagury et al. 1996; Kontochristopoulos et al. 1996; Oeda and Shinohara 1993). The pathophysiology is still mysterious, and a variety of mechanisms have been discussed, such as a local procoagulant effect of IFN-α with an additional role for an underlying coagulation deficiency, a local immune-mediated inflammatory process, a direct toxic effect of IFN-α, or an unintentional intra-arterial injection.

Although reported is several patients, the involvement of IFN-α in other dermatological immune or autoimmune disorders, such as bullous pemphigus (Parodi et al. 1993) or pemphigus foliaceus (Niizeki et al. 1994) awaits further confirmation.

Autoantibodies and Autoimmune Disorders

The spectrum of IFN-α-induced organ- and nonorgan-specific autoimmune diseases has expanded year after year (Vial and Descotes 1995; Bacq et al. 1996; Braathen and Stavem 1989). In this respect, it is usually contraindicated to treat patients with overt autoimmune diseases because of possible exacerbation.

Interferon-α treatment has been associated with a high prevalence of various organ-specific- or nonorgan-specific autoantibodies. An increased titer or the new occurrence of various autoantibodies (i.e., antinuclear antibodies, antithyroid antibodies, parietal cell antibodies, smooth muscle antibodies, rheumatoid factors) was noted in 4–30% of previously negative patients with malignant tumours or chronic hepatitis (Vial and Descotes 1995). Preexisting autoantibodies or their development during IFN-treatment do not appear to influence the response to treatment (Wada et al. 1997). The exact contribution of IFN-α treatment is still not clarified because the underlying disease can also be associated with immunological disorders. This is particularly true for chronic hepatitis C (Lunel 1995). However, the frequent development of at least one de novo autoantibody in patients treated for chonic hepatitis B strongly argues for a direct role of IFN-α (Mayet et al. 1989).

In most instances, the pathogenicity of autoantibodies is unclear, and there is no consensus on the management of patients previously positive for nonorgan-specific autoantibodies frequently associated with autoimmune diseases. It is generally accepted that the presence of these autoantibodies, at least in low titers, is not predictive of the further occurrence of autoimmune disease. For example, no patients developed clinical systemic lupus erythematosus as a result of positivity for antinuclear antibodies in large series (Imagawa et al. 1995; Noda et al. 1996; Watanabe et al. 1994), and no difference in the development of autoimmune diseases was observed between patients who had, or did not have, preexisting autoantibodies (Bayraktar et al. 1997). The presence of serum autoantibodies in patients with no clinical signs of co-existing autoimmune disease cannot be considered a contraindication to treatment. Patients who are positive for organ-specific autoantibodies can be more predisposed to develop an autoimmune disease, but this has been convincingly shown only for thyroid disorders.

Only very few cases of IFN-α-associated systemic lupus erythematosus have been reported (Flores et al. 1994; Tolaymat et al. 1992; Yoshida et al. 1994). The clinical and biological features were rather different from those usually observed in drug-related lupus and more in accordance with the unmasking of an idiopathic disease. Polymyositis was also reported (Solis et al. 1996). The induction or exacerbation of inflammatory rheumatologic disorders also probably falls within the scope of IFN-α-associated autoimmune disease. In the largest available study, which included 137 patients treated for myeloproliferative disorders, 27 experienced arthralgias, myalgias, and Raynaud's phenomenon, which suggested lupoid or rheumatic-like disease (Rönnblom et al. 1992). Most of these patients had positive antinuclear antibodies. Isolated cases of arthritis and polyarthritis were also described, but serological markers were inconsistently found (Chazerain et al. 1992; Chung and Older 1997; Jumbou et al. 1995). Rheumatoid factor was one of the most frequently positive autoantibodies before treatment, and decreased or unchanged titers were observed after IFN-α introduction. The clinical significance of newly occurring rheumatoid factor during treatment is difficult to interpret. Nonorgan-specific autoimmune diseases were confirmed as very rare complications of IFN-α treatment with only 1 case of lupus-like syndrome and 2 cases of polyarthritis among 677 patients treated for chronic hepatitis C (Okanoue et al. 1996).

In addition to its antiviral activity, IFN-α exerts immunoregulatory and anti-inflammatory properties (Tilg 1997). The idea that IFN-α can play a role in the pathogenesis of autoimmune diseases came from the positive association between IFN-α serum levels and disease activity in autoimmune disorders such as systemic lupus erythematosus and rheumatoid arthritis. As reviewed elsewhere (Miossec 1997), the association between autoimmune disorders and IFN could have been predicted from our knowledge of the role of cytokines in immune regulation—in particular in the control of the TH1/TH2 balance—in which IFN-α favors the development of TH1 clones and enhances T-cell mediated immune reactions with a TH1 and proinflammatory pattern. However, it is not yet clearly known whether interferon-α can induce or unmask a silent autoimmune disease. A genetic predisposition for autoimmune phenomena induced by IFN-α remains to be studied carefully. The A2, B7, and DR2 haplotypes were suggested to be associated with the occurrence of autoimmune diseases in IFN-α treated patients, but this was based on isolated reports only (Pittau et al. 1997; Rönnblom et al. 1991).

The immune system is also involved in the pathogenesis of Behçet's disease. Three chronic myelogenous leukemia (CML) patients without previous symptoms of Behçet disease experienced characteristic manifestations of Behçet disease during IFN-α treatment (Budak et al. 1997; Segawa et al. 1995). In addition, 2 of 11 CML patients who had a negative skin pathergy test before treatment developed positive test reactions while receiving IFN-α for at least 6 months (Budak et al. 1997). Other rare immune side effects included the induction of subcutaneous sarcoid nodules (Blum et al. 1993) and the occurrence (Nakajima et al. 1996) or the recurrence of generalized sarcoidosis (Teragawa et al. 1996).

Endocrinological Disorders

Thyroid Disorders

Thyroid disorders are the most common organ-specific diseases associated with IFN-α. Since the first report of hypothyroidism induced by IFN-α treatment for breast cancer (Fentiman et al. 1985), numerous investigators have reported their experience in patients treated for chronic viral hepatitis (Marazuella et al. 1996; Preziati et al. 1995; Watanabe et al. 1994), solid tumors (Rönnblom et al. 1991), or hematological malignancies (Gisslinger et al. 1992; Vallisa et al. 1995), and it is beyond the scope of this chapter to review all available studies. Abnormalities ranged from asymptomatic antithyroid autoantibodies to clinical thyroid disorders with hypothyroidism, hyperthyroidism, or biphasic thyroiditis marked by transient hyperthyroidism followed by sustained hypothyroidism. Most cases appear within the first 4 months of treatment, but few patients experienced thyroid disorders several weeks after the completion of IFN-α treatment (Lisker-Melman et al. 1992; Preziati et al. 1995), which suggests that long-term surveillance is required following treatment withdrawal.

Reversal of thyroid dysfuntion is usually observed after interferon withdrawal (Baudin et al. 1993; Marazuella et al. 1996). However, sustained hypothyroidism requiring long-term substitutive hormone therapy has sometimes been reported (Lisker-Melman et al. 1992; Marazuella et al. 1996; Watanabe et al. 1994), and the long-term outcome of thyroid function awaits further investigations. Severe initial symptoms and elevated thyroid antibodies were more frequently found in patients who developed sustained or irreversible hypothyroidism (Mekkakia-Benhabib et al. 1996). Interestingly, the presence of thyroid abnormalities was not associated with the clinical response to treatment (Imagawa et al. 1995; Watanabe et al. 1994). The management of patients who developed clinical thyroid disorders during treatment is not clearly established. The discontinuation of IFN-α treatment is not always required, and each situation should be discussed case-by-case depending on the initial response to treatment and the severity of thyroid disorders.

The incidence of thyroid disorders is now well-characterized in the setting of chronic viral hepatitis treatment. From a recent review of 11 published prospective studies involving 494 chronic hepatitis C patients treated with natural or recombinant IFN-α, clinical or subclinical thyroid abnormalities were found in 7% of patients (Vial et al. 1996). Hypothyroidism and hyperthyroidism were equally frequent in these patients. Overall, the incidence was 1–2% in patients treated for hepatitis B, 5–10% in patients treated for chronic hepatitis C, and 10–45% in cancer patients (Vial and Descotes 1995). Interestingly, this incidence was even higher in patients receiving both IFN-α and IL-2.

Several risk factors predisposing thyroid dysfunction have been investigated. As expected from available data in the general population, IFN-α-induced thyroid disorders are more frequent in female patients, but no differences between sex were found when using a logistic multiple regression model (Watanabe et al. 1994). The prevalence of thyroid dysfunction is not influenced by the origin of IFN-α or the dose. In contrast, the available data clearly suggest that pretreatment positivity for thyroid antibodies or their occurrence during treatment is strongly associated with the further development of thyroid abnormalities. Approximately 60% of previously antithyroid-antibody-positive patients developed thyroid abnormalities, and 70% of patients who developed antibodies during treatment experienced biochemical or clinical thyroid disorders (Vial et al. 1996). In addition, patients with previous autoimmune thyroid diseases are more prone to develop a severe form of hypothyroidism with TSH receptor antibodies (Chen et al. 1996). The possible influence of hepatitis C virus genotype Ib should be further explored (Sachithanandan et al. 1997).

The mechanism of thyroid dysfunction in IFN-α-treated patients has not been fully clarified. Owing to the immunomodulatory properties of interferon, an autoimmune mechanism is the most attractive hypothesis. In autoimmune thyroid diseases, thyrocytes express major histocompatibility complex (MHC) class II antigens. This aberrant expression is involved in the initiation and maintenance of autoimmune thyroid disorders and can be caused by IFN-γ. Although IFN-α does not induce MHC class II antigen expression, it has been reported to increase the release of several cytokines, such as IFN-γ and IL-6, which supports the role of cytokine network activation by IFN-α in the exacerbation of thyroid autoimmunity. Thyroid antibodies were detected in about two-thirds of patients with a diagnosis of thyroid disease during interferon therapy, the majority of whom previously had thyroid antibodies with other newly developed antibodies, suggesting an exacerbation of a previous or latent autoimmune thyroiditis. However, although the immuno-genetic background probably plays a pivotal role, no predictive genetic predisposition has yet been found in patients who developed thyroid antibodies during treatment. Another piece of evidence for an autoimmune phenomenon has emerged from a recent study that found that other autoantibodies, such as antinuclear and anti-dsDNA antibodies, as well as clinical signs of autoimmune disorders, were significantly more frequent in patients with thyroid disorders (Marazuella et al. 1996). Several patients also developed both hypothyroidism and seronegative rheumatoid-like arthritis or systemic–lupus–erythematosus-like syndrome (Pittau et al. 1997). Finally, the clinical course of thyroid disease after IFN-α treatment was suggested to mimic postpartum autoimmune thyroiditis closely (Chen et al. 1996).

However, autoimmunity is not universal mechanism, and other pathophysiological mechanisms are probably involved in the 20–30% of negative thyroid antibody patients developing thyroid diseases. A direct toxicity of interferon-α on thyroid gland or thyroid-stimulating hormone release, or a cytotoxic effect on thyrocytes possibly mediated by natural killer cells, may also be involved. For example, the acute administration of IFN-α in volunteers decreased thyroid-stimulating hormone levels (Wiedermann et al. 1991). More recently, in vitro studies showed that IFN-α directly inhibits thyrocyte function at clinically observable levels (Yamazaki et al. 1993). On the basis of a prospective evaluation using the iodide–perchlorate discharge test, Roti et al. (1996) also found reversible defects in the intrathyroidal organification of iodine in seven (22%) antithyroid-antibody-negative patients treated with IFN-α.

Because thyroid disorders are relatively frequent, the routine assessment of thyroid-stimulating hormone before and during treatment is strongly recommended.

Diabetes

Although only sporadic reports are available, IFN-α has been repeatedly associated with type 1 diabetes. Once again, the treated disease should be taken into consideration, and thus a higher prevalence of diabetes mellitus is observed in chronic hepatitis C, but not in chronic hepatitis B patients, as compared with the general population (Fraser 1996). Nevertheless, IFN-α-induced diabetes has also been described in cancer patients, and prompt amelioration or complete recovery has sometimes been noted after treatment withdrawal (Gori et al. 1995; Guerci et al. 1994). In addition, transgenic mice expressing IFN-α in pancreatic beta-cells develop Type 1 diabetes, which is an effect prevented by neutralizing IFN-α antibodies (Stewart et al. 1993), and more recently an increased expression of IFN-α was found in the pancreas of patients with Type 1 diabetes (Huang et al. 1995). Although the positivity for islet cell antibody was inconsistently detected in patients with IFN-α-induced diabetes, it is tempting to speculate that IFN-α can trigger an autoimmune phenomenon or latent diabetes. Interestingly, the DR4-HLA haplotype, which is positively correlated with the occurrence of diabetes, was found in several patients. The rapid occurrence of Type 2 diabetes in predisposed patients (Lopes et al. 1994) is also in keeping with an IFN-induced insulin resistance and a subsequent accelerated autoimmune destruction of stimulated β-cells (Koivisto et al. 1989). Finally, the appearance of islet cell antibodies in IFN-α-treated patients has never been demonstrated (Di Cesare 1996; Imagawa et al. 1995; Mayet et al. 1989). In contrast, IFN-α was shown to increase the prevalence of insulin autoantibodies (3.3% before treatment versus 13.3% after 6 months of treatment) (Di Cesare et al. 1996), which suggests that IFN-α can elicit a pancreatic autoimmune phenomenon. Overall, these data suggest that IFN-α-induced diabetes should probably be expected in a limited number of patients—in particular in patients with predisposing factors such as obesity or a previous personal or familial history of glucose intolerance.

Immunodeficency and Interferon-α Treatment

Earlier experimental reports suggested that long-term administration of low-dose IFN-α might be associated with immunosuppression (Teichmann et al. 1989). However, the available clinical data on the immunosuppressive effects of IFN-α are still limited, and no firm conclusions can be drawn. The sporadic occurrence of severe *Candida* esophagitis (Hassanein et al. 1994) or *Pneumocystis carinii* pneumopathy (Dhôte et al. 1996) in immunocompetent patients, and a decrease in CD4[+] T cells sometimes associated with the development of opportunistic infections in several HIV-infected patients (Pesce et al. 1993; Soriano et al. 1994) are in keeping with possible immunosuppressive effects of IFN-α. Autoantibodies against a sequence of human leukocyte antigen (HLA) class II antigens were evidenced in the sera of CD4[+] cell-depleted patients. Because these autoantibodies were homologous to a sequence found on HIV-1 gp41, an autoimmune destruction of CD4[+] cells was postulated as a result of autoantibody production in HIV-infected patients with particular HLA haplotypes (Vento et al. 1993).

Interferon-α and Transplantation

An increased incidence or severity of graft-versus-host reaction following IFN-α treatment in recipients of allogeneic bone marrow transplantation has been reported (Browett et al. 1994; Samson et al. 1996). Earlier reports found an increased incidence of steroid-resistant rejection in transplant patients receiving prophylactic IFN-treatment for viral infection (Kovarik et al. 1988). More recent trials in renal transplant patients with chronic viral hepatitis also evidenced a deterioration in renal function after the adjunction of IFN-α (Magnone et al. 1995; Rostaing et al. 1995)—particularly in patients who had a previous episode of renal rejection in the 6 months before IFN-α treatment (Durlik et al. 1995). However, no definite evidence of increased renal graft rejection induced by IFN-α emerged from other studies (Gayowski et al. 1997). Similarly, a possible increased risk of acute or chronic rejection in liver transplant recipients (Dousset et al. 1994; Féray et al. 1995) is debated, for other studies failed to find a deleterious influence of IFN-α in this setting (Min and Bodenheimer 1995).

Interferon-α and Pregnancy

Experimental data argue against a teratogenic or mutagenic effect of IFN-α. Different ex vivo assay techniques were unable to demonstrate a transfer of IFN-α across the human placenta (Waysbort et al. 1993), and very low concentrations of IFN-α were found immediately after delivery in the serum of two newborns as compared with maternal serum concentrations (Haggstrom et al. 1996). Several uncomplicated and successful pregnancies in patients treated with IFN-α during the first trimester or the whole pregnancy have been reported, but the number of reported pregnancies is too small to exclude an increased risk. In addition, a long-term follow-up of newborns is not yet available. In the light of these limitations, IFN-α appears to be a possible, valuable, and safe treatment in selected pregnant patients with severe thrombocythemia or chronic myelogenous leukemia.

Drug Interactions

Most of our knowledge on the effects of IFNs on hepatic drug-metabolizing enzymes is derived from experimental studies in animals or cultured human hepatocytes (Cribb et al. 1994; Donato et al. 1993; Okuno et al. 1993). The observation of a decreased activity in several cytochrome P-450 isoenzymes may translate to a potential alteration in the metabolism of several selected drugs metabolized via this pathway. However, data obtained in vivo in humans are scarce and inconclusive. Using antipyrine or aminopyrine as probes of hepatic microsomal drug metabolism in patients treated with long-term IFN-α, different investigators found a decrease in hepatic clearance (Horsmans et al. 1994) or no effect (Echizen et al. 1990). Results obtained with theophylline consistenly evidenced a significant but widely variable decrease in systemic theophylline clearance (Israel et al. 1993). Clinical experience is even less conclusive, and only one isolated case of increased serum warfarin concentration and prothrombin time following IFN-α treatment has been reported so far (Adachi et al. 1995).

Antibodies to Interferon-α

A large amount of data have now accumulated on the occurrence of both binding and neutralizing antibodies to IFN-α. Neutralizing antibodies were usually detected within 2–4 months of treatment (Antonelli et al. 1996). Available IFN-α preparations appear to differ in antigenicity (Antonelli 1994). In cancer or chronic viral hepatitis patients treated with a similar route of administration and treatment schedule, IFN alfa-2a was repeatedly found to induce more frequent neutralizing antibodies using the same anti-IFN-α antibody assay. The incidence of such antibodies ranged from 20 to 50% for IFN alfa-2a, 6–10% for IFN alfa-2b, and 1–6% for IFN alfa-N1 (McKenna and Oberg 1997). A higher prevalence of anti-IFN-α antibodies was found in patients with hepatitis C virus genotype 3a (Hanley et al. 1996), but such findings await further confirmation. A complete disappearance of IFN antibodies is consistently observed after treatment discontinuation.

The clinical significance of neutralizing antibodies is a critical and strongly debated issue because they are a possible cause of interpatient variability in response to treatment. Several reports comparing the clinical response between antibody-positive and antibody-negative patients showed that failure to respond to treatment or secondary disease reactivation after initial response was significantly more frequent in antibody-positive patients treated for chronic hepatitis C or hematological disorders (Antonelli et al. 1996; Russo et al. 1996). In addition, several investigators demonstrated a complete restored response after replacement of recombinant IFN-α by natural lymphoblastoid human IFN-α (Milella et al. 1995; Russo et al. 1996), which suggests that natural IFN-α can overcome the neutralizing activity of antibodies directed against recombinant IFN. However, a recent analysis of clinical trials using standardized and comparable methods for the detection of neutralizing anti-IFN antibodies suggested that treatment failure was only rarely associated with anti-IFN antibodies and was independent of the type of IFN-α (Bonino et al. 1997).

Interferon-β

The general toxicity of IFN beta-1a or -1b is very similar to that of IFN-α in that fever, myalgia, and a flu-like syndrome are transiently observed in approximately 60% of patients during the first weeks of treatment (Jacobs et al. 1996; The IFNB Multiple Sclerosis Group, 1995). These symptoms are usually prevented, cured, or managed with paracetamol, and, more recently, oral pentoxifylline, which can reduce the production of proinflammatory cytokines, was proposed to minimize the early adverse effects of IFN-β (Rieckmann et al. 1996). In an open-label trial, adverse effects resulting in the discontinuation of IFN-β treatment were noted in 18% of patients—essentially those with chronic progressive disease (Neilley et al. 1996). Transient abnormalities included mild anemia and leukopenia, and moderate increases in liver enzymes. Hypereosinophilia, supposedly caused by the inhibitory effect of IFN-β on IFN-γ production with subsequent preferential induction of TH2 cells, was recently reported in an atopic patient (Gattoni et al. 1997).

Injection-site reactions with swelling and redness occur in 65% of patients following subcutaneous IFN beta-1b injections (The IFNB Multiple Sclerosis Group 1995) but in only 5% of those receiving IFN beta-1a (Jacobs et al. 1996). The clinical features of these cutaneous reactions were sometimes more severe and included sclerotic dermal plaques, painful erythematous plaques, and deep ulcers with necrosis (Elgart et al. 1997). Exacerbation of quiescent psoriasis has also been noted (Webster et al. 1996). A mild perivascular infiltrate and deep dermal vessel thrombosis were the most common histological findings. In contrast to previous claims, even the use of low-dose IFN-β

can produce local cutaneous necrosis (Bérard et al. 1995; Webster et al. 1996). As with IFN-α, the mechanism of IFN-β-induced skin necrosis is unknown and may involve an inflammatory process localized in the blood vessels or platelet-dependent thrombosis.

Interferon-β-associated neurotoxicity was deemed less frequent as compared with IFN-α (Liberati et al. 1990). An increased incidence of depressive disorders was noted after several years of treatment in patients with multiple sclerosis (Neilley et al. 1996). However, severe depressive disorders and suicidal ideation are also frequently found in untreated multiple sclerosis patients. This issue was a matter of debate and several investigators concluded that severe depression is related to the disease rather than to the treatment. However, a previous history of suicidal ideation or current depressive disorders should be carefully evaluated before treatment.

As compared with IFN-α, very few studies carefully investigated the potential of IFN-β to induce autoimmune diseases. Thyroid disorders and antithyroid antibodies were not identified in patients receiving a 24-week course of IFN-β (Pagliacci et al. 1991). However, thyroid disorders associated with antithyroid antibodies during long-term IFN-β treatment for multiple sclerosis were reported in two patients (Schwid et al. 1997). Although not fully classified as an autoimmune disease, induction of sarcoidosis has been described (Bobbio-Pallavicini et al. 1995).

One interesting case of immediate-type hypersensitivity reaction confirmed by positive rechallenge has been reported (Young and Otis 1996).

Neutralizing antibodies were detected in 22 to 38% of patients treated after 2 to 3 years with IFN beta-1a (Jacobs et al. 1996) or IFN beta-1b (The IFNB Multiple Sclerosis Group 1996), respectively. An attenuation of the treatment effect and magnetic resonance imaging findings were observed in 35% of patients with neutralizing antibodies against IFN beta-1b. The clinical consequences of IFN beta-1a antibodies is currently being studied.

Interferon-γ

Although IFN-γ is being investigated in several disorders, the only approved indication is the treatment of chronic granulomatous disease, and thus experience on adverse effects is still limited. In this setting, constitutional symptoms with fever, headache, digestive, and flu-like symptoms, and moderate injection site reaction are commonly observed (ICGDCSG 1991). Neuropsychiatric disturbances are not consistently found in IFN-γ-treated patients. The most significant adverse effects are a dose-related proteinuria (Weiss 1992) with rare acute renal failure (Tashiro et al. 1996) and cardiovascular toxicity upon the administration of high doses in cancer patients (Matsson et al. 1991). Interestingly, no consequences on growth and development have been identified in children after a mean follow-up of 2.5 years (Bemiller et al. 1995).

Psoriatic lesions at the injection site have been reported in patients treated for psoriatic arthritis (Fierlbeck et al. 1990), and erythema nodosum frequently occurred in patients receiving long-term treatment for lepromatous leprosy (Sampaio et al. 1992).

Other indications based on the immune effects of IFN-γ came from a few prospective studies that investigated the appearance of several autoantibodies during treatment. In a 6-month study of chronic hepatitis B, all 11 patients developed at least one new autoantibody (Weber et al. 1994). Most antibodies disappeared after treatment withdrawal, and clinical signs of autoimmune disease were not observed. Induction of antinuclear antibodies and a clinical exacerbation of the disease were also found in several rheumatoid patients (Seitz et al. 1988), and exacerbation of systemic lupus erythematosus has been reported (Machold and Smolen 1990). Interferon-γ was also involved in the induction or reactivation of seronegative arthritis in patients treated for cutaneous psoriasis (O'Connell et al. 1992) or in the exacerbation of multiple sclerosis (Panitch et al. 1987).

Neutralizing antibodies are usually not found in patients treated with IFN-γ (Jaffe et al. 1987).

REFERENCES

Adachi, Y., Yokoyama, Y., Nanno, T., and Yamamoto, T. 1995. Potentiation of warfarin by interferon. *Brit. Med. J.* 311:292.

Akard, L. P., Hoffman, R., Elias, L., and Saiers, J. H. 1986. Alpha-interferon and immune hemolytic anemia. *Ann. Intern. Med.* 105:306.

Andriani, A., Bibas, M., Callea, V., Derenzo, A., Chiurazzi, F., Marceno, R., and Musto, P. 1996. Autoimmune hemolytic anemia during alpha interferon treatment in nine patients with hematological diseases. *Haematologica* 81:258–260.

Antonelli, G. 1994. Development of neutralizing and binding antibodies to interferon (IFN) in patients undergoing IFN therapy. *Antiviral Res.* 24:235–244.

Antonelli, G., Giannelli, G., Currenti, M., Simeoni, E., Del Vecchio, S., Maggi, F., Pistello, M., Roffi, L., Pastores, G., Chemello, L., and Dianzani, F. 1996. Antibodies to interferon (IFN) in hepatitis C patients relapsing while continuing recombinant IFN-alpha 2 therapy. *Clin. Exper. Immunol.* 104:384–387.

Areias, J., Velho, G. C., Cerquiera, R., Barbedo, C., Amaral, B., Sanches, M., Massa, A., and Saraiva, A. M. 1996. Lichen planus and chronic hepatitis C: Exacerbation of the lichen under interferon-alpha-2a therapy. *Eur. J. Gastroenterol. Hepatol.* 8:825–828.

Arslan, M., Ozyilkan, E., Kayhan, B., and Telatar, H. 1994. Raynaud's phenomenon associated with alpha-interferon therapy. *J. Intern. Med.* 235:503.

Asnis, L. A., and Gaspari, A. A. 1995. Cutaneous reactions to recombinant cytokine therapy. *J. Am. Acad. Dermatol.* 33:393–410.

Aubin, F., Bourezane, Y., Blanc, D., Voltz, J. M., Faivre, B., and Humbert, P. 1995. Severe lichen planus-like eruption induced by interferon-alpha therapy. *Eur. J. Dermatol.* 5:296–299.

Averbuch, S. D., Austin, H. A., Sherwin, S. A., Antonovych, T., Bunn, P. A., and Longo, D. L. 1984. Acute interstitial nephritis with the nephrotic syndrome following recombinant leukocyte A interferon therapy for mycosis fungoides. *N. Eng. J. Med.* 310:32–35.

Azagury, M., Pauwels, C., Kornfeld, S., Bataille, N., and Perie, G. 1996. Severe cutaneous reactions following interferon injections. *Eur. J. Cancer* 32A:1821.

Bacq, Y., Sapey, T., Gruel, Y., Fimbel, B., Degenne, D., Barin, F., and Metman, E. H. 1996. Exacerbation d'un purpura thrombopénique auto-immun au cours du traitement par l'interféron alpha chez une femme atteinte d'une hépatite chronique virale C. *Gastroenterologie Clinique et Biologique* 20:303–306.

Barbolla, L., Paniagua, C., Outeirino, J., Prieto, E., and Sanchez Fayos, J. 1993. Haemolytic anaemia to the alpha-interferon treatment: A proposed mechanism. *Vox Sang.* 65:156–157.

Batocchi, A. P., Evoli, A., Servidei, S., Palmisani, M. T., Apollo, F., and Tonali, P. 1995. Myasthenia gravis during interferon alfa therapy. *Neurology* 45:382–383.

Baudin, E., Marcellin, P., Pouteau, M., Colas-Linhart, N., Le Floch, J. P., Lemmonier, C., Benhamou, J. P., and Bok, B. 1993. Reversibility of thyroid dysfunction induced by recombinant alpha interferon in chronic hepatitis C. *Clin. Endocrinol.* 39:657–661.

Bayraktar, Y., Bayraktar, M., Gurakar, A., Hassanein, T. I., and van Thiel, D. H. 1997. A comparison of the prevalence of autoantibodies in individuals with chronic hepatitis C and those with autoimmune hepatitis: The role of interferon in the development of autoimmune diseases. *Hepato-Gastroenterol.* 44:417–425.

Becker, J. C., Winkler, B., Klingert, S., and Brücker, E. B. 1994. Antiphospholipid syndrome associated with immunotherapy for patients with melanoma. *Cancer* 73:1621–1624.

Bemiller, L. S., Roberts, D. H., Starko, K. M., and Curnutte, J. T. 1995. Safety and effectiveness of long-term interferon gamma therapy in patients with chronic granulomatous disease. *Blood Cells, Molecules, and Diseases* 21:239–247.

Bérard, F., Canillot, S., Balme, B., and Perrot, H. 1995. Nècrose cutanée locale après injections d'interféron-béta. *Ann. de Derm. et de Vener.* 122:105–107.

Blum, L., Serfaty, L., Wattiaux, M. J., Picard, O., Cabane, J., and Imbert, J. C. 1993. Nodules hypodermiques sarcoïdosiques au cours d'une hépatite virale C traitée par interféron alpha 2b. *Rev. de Med. Int.* 14:1161.

Bobbio-Pallavicini, E., Valsecchi, C., Tacconi, F., Moroni, M., and Porta, C. 1995. Sarcoidosis following beta-interferon therapy for multiple myeloma. *Sarcoidosis* 12:140–142.

Bonino, F., Baldi, M., Negro, F., Oliveri, F., Colombatto, P., Bellati, G., and Brunetto, M. R. 1997. Clinical relevance of anti-interferon antibodies in the serum of chronic hepatitis C patients treated with interferon-alpha. *J. Interferon Cytokine Res.* 17 (suppl.1):35–38.

Braathen, L. R., and Stavem, P. 1989. Autoimmune haemolytic anaemia associated with interferon alpha-2a in patients with mycosis fungoides. *Brit. Med. J.* 298:1713.

Browett, P. J., Nelson, J., Tiwari, S., Van de Water, N. S., May, S., and Palmer, S. J. 1994. Graft-versus-host disease following interferon therapy for relapsed chronic myeloid leukaemia post-allogeneic bone marrow transplantation. *Bone Marrow Transplant.* 14:641–644.

Budak Alpdogan, T., Demircay, Z., Alpdogan, O., Direskeneli, H., Ergun, T., Bayik, M., and Akoglu, T. 1997. Behçet's disease in patients with chronic myelogenous leukemia: Possible role of interferon-alpha treatment in the occurence of Behçet's symptoms. *Ann. Hematol.* 74:45–48.

Calleja, J. L., Albillos, A., Cacho, G., Iborra, J., Abreu, L., and Escartin, P. 1996. Interferon and prednisone therapy in chronic hepatitis C with non-organ-specific antibodies. *J. Hepatol.* 24:308–312.

Casato, M., Pucillo, L.P., Leoni, M., di Lullo, L., Gabrielli, A., Sansonno, D., Dammacco, F., Danieli, G., and Bonomo, L. 1995. Granulocytopenia after combined therapy with interferon and angiotensin-converting enzyme inhibitors: Evidence for a synergistic hematologic toxicity. *Am. J. Med.* 99:386–391.

Chazerain, P., Meyer, O., Ribard, P., de Bandt, M., Mechelany, C., Marcellin, P., Bernard, J. F., Grossin, M., and Kahn, M. F. 1992. Trois cas de polyarthrite survenant au cours d'un traitement par interféron-alpha recombinant. *Rev. du Rhum. et des Mal. Ostéo.* 59:303–309.

Chen, F. Q., Okamura, K., Sato, K., Kuroda, T., Mizokami, T., Fujikawa, M., Tsuji, H., Okamura, S., and Fujishima, M. 1996. Reversible primary hypothyroidism with blocking or stimulating type TSH binding inhibitor immunoglobulin following recombinant interferon-alpha therapy in patients with pre-existing thyroid disorders. *Clin. Endocrinol.* 45:207–214.

Chin, K., Tabata, C., Satake, N., Nagai, S., Moriyasu, F., and Kuno, K. 1994. Pneumonitis associated with natural and recombinant interferon alpha therapy for chronic hepatitis C. *Chest* 105:939–941.

Chung, A., and Older, S. A. 1997. Interferon-alpha associated arthritis. *J. Rheumatol.* 24:1844–1845.

Creutzig, A., Caspary, L., and Freund, M. 1996. The Raynaud phenomenon and interferon therapy. *Ann. Intern. Med.* 125:423.

Cribb, A. E., Delaporte, E., Kim, S. G., Novak, R. F., and Renton, K. W. 1994. Regulation of cytochrome P-4501A and cytochrome P-4502E induction in the rat during the production of interferon α/β. *J. Pharmacol. Exp. Ther.* 268:487–494.

Cudillo, L., Cantonetti, M., Venditti, A., Lentini, R., Rossini, P. M., Caramia, M., Masi, M., and Papa, G. 1990. Peripheral polyneuropathy during treatment with alpha-2 interferon. *Haematologica* 75:485–486.

D'Amico, E., Paroli, M., Fratelli, V., Palazzi, C., Barnaba, V., Callea, F., and Consoli, G. 1995. Primary biliary cirrhosis induced by interferon-alpha therapy for hepatitis C virus infection. *Dig. Dis. Sci.* 40:2113–3116.

Davis, G. L., Balart, L. A., Schiff, E. R., Lindsay, K., Bodenheimer, H. C., Perillo, R. P., Carey, W., Jacobson, I. M., Payne, J., Dienstag, J. L., van Thiel, D. H., Tamburro, C., Lefkowitch, J., Albrecht, J., Meschievitz, C., Ortego, T. J., and Gibas, A. 1990. Treatment of chronic hepatitis C with recombinant α-interferon: A multicentre randomized controlled trial. *J. Hepatol.* 11 (suppl. 1):S31–S35.

Dhôte, R., Calmus, Y., Belenfant, X., Bachmeyer, C., and Christoforov, B. 1996. Pneumopathie à Pneumocystis carinii au cours d'une hépatite virale C traitée par interféron alpha2 et corticoïdes. *Presse Medicale* 25:2047.

Di Cesare, E., Previti, M., Russo, F., Brancatelli, S., Ingemi, M. C., Scoglio, R., Mazzu, N., Cucinotta, D., and Raimondo, G. 1996. Interferon-alpha therapy may induce insulin autoantibody development in patients with chronic viral hepatitis. *Dig. Dis. Sci.* 41:1672–1677.

Dimitrov, Y., Heibel, F., Marcellin, L., Chantrel, F., Moulin, B., and Hannedouche, T. 1997. Acute renal failure and nephrotic syndrome with alpha interferon therapy. *Nephrol. Dialysis Transplant.* 12:200–203.

Dourakis, S. P., Deutsch, M., and Hadziyannis, S. J. 1996. Immune thrombocytopenia and alpha-interferon therapy. *J. Hepatol.* 25:972–975.

Dousset, B., Conti, F., Houssin, D., and Calmus, Y. 1994. Acute vanishing bile duct syndrome after interferon therapy for recurrent HCV infection in liver-transplant recipients. *N. Eng. J. Med.* 330: 1160–1161.

Doutre, M. S., Baquey, A., Bernard, P., Couzigou, P., Bernard, N., Lacoste, D., and Morlat, P. 1997. Apparition d'anticorps anti-phospholipides chez les patients présentant une hépatite à virus C traitée par interféron-alpha. *Ann. de Med. Int.* 148:99–112.

Dupin, N., Chosidow, O., Francès, C., Boisnic, S., and Lunel-Fabiani, F. 1994. Lichen planus after alpha-interferon therapy for chronic hepatitis C. *Eur. J. Dermatol.* 4:535–536.

Durand, J. M., Retornaz, F., Cretel, E., Kaplanski, G., and Soubeyrand, J. 1993. Glomérulonéphrite extra-capillaire au cours d'un traitement par interféron alpha. *Rev. de Med. Int.* 14:1138.

Durlik, M., Gaciong, Z., Rancewicz, Z., Rowinska, D., Wyzgal, J., Kozlowska, B., Gradowska, L., Lao, M., Nowaczyk, M., Korczak-Kowalska, G., and Gorski, A. 1995. Renal allograft function in patients with chronic viral hepatitis B and C treated with interferon alpha. *Transplant. Proc.* 27:958–959.

Elgart, G. W., Sheremata, W., and Ahn, Y. S. 1997. Cutaneous reactions to recombinant human interferon beta-1b: the clinical and histological spectrum. *J. Am. Acad. Dermatol.* 37:553–558.

Fattovich, G., Giustina, G., Favarato, S., and Ruol, A. 1996. A survey of adverse events in 11.241 patients with chronic viral hepatitis treated with alfa interferon. *J. Hepatol.* 24:38–47.

Fentiman, I. S., Thomas, B. S., Balkwill, F. R., Rubens, R. D., and Hayward, J. L. 1985. Primary hypothyroidism associated with interferon therapy of breast cancer. *Lancet* i; 8438:1166.

Féray, C., Samuel, D., Gigou, M., Paradis, V., David, M. F., Lemonnier, C., Reynès, M., and Bismuth, H. 1995. An open trial of interferon alfa recombinant for hepatitis C after liver transplantation: Antiviral effects and risk of rejection. *Hepatol.* 22:1084–1089.

Fierlbeck, G., Rassner, G., and Muller, C. 1990. Psoriasis induced at the injection site of recombinant interferon gamma. Results of immunohistologic investigations. *Arch. Dermatol.* 126:351–355.

Flores, A., Oliv, A., Feliu, E., and Tena, X. 1994. Systemic lupus erythematosus following interferon therapy. *Brit. J. Rheumatol.* 33:787.

Fraser, G. M., Harman, I., Meller, N., Niv, Y., and Porath, A. 1996. Diabetes mellitus is associated with chronic hepatitis C but not chronic hepatitis B infection. *Israel J. Med. Sci.* 32:526–530.

Funk, J., Langeland, T., Schrumpf, E., and Hanssen, L. E. 1991. Psoriasis induced by interferon-alpha. *Brit. J. Dermatol.* 125:463–465.

Garcia-Buey, L., Garcia-Monzon, C., Rodrigue, S., Borque, M. J., Garcia-Sanchez, A., Iglesias, R., DeCastro, M., Mateos, F. G., Vicario, J. L., Balas, A., and Moreno-Otero, R. 1995. Latent autoimmune hepatitis triggered during interferon therapy in patients with chronic hepatitis C. *Gastroenterol.* 108:1770–1777.

Gattoni, A., Romano, C., Cecere, A., and Caiazzo R. 1997. Eosinophilia triggered by beta-interferon therapy for chronic hepatitis C. *Eur. J. Gastroenterol. Hepatol.* 9:909–911.

Gayowski, T., Singh, N., Marino, I. R., Vargas, H., Wagener, M., Wannstedt, C., Morelli, F., Laskus, T., Fung, J. J., Rakela, J., and Starzl, T. E. 1997. Hepatitis C virus genotypes in liver transplant recipients. Impact on posttransplant reccurence, infections, response to interferon-alpha therapy and outcome. *Transplant.* 64:422–426.

Georgeston, M. J., Yarze, J. C., Lalos, A. T., Webster, G. F., and Martin, P. 1993. Exacerbation of psoriasis due to interferon-α treatment of chronic active hepatitis. *Am. J. Gastroenterol.* 88:1756–1758.

Gisslinger, H., Gilly, B., Woloszczuk, W., Mayr, W. R., Havelec, L., and Linkesch, W. 1992. Thyroid autoimmunity and hypothyroidism during long-term treatment with recombinant interferon-alpha. *Clin. Exp. Immunol.* 90:363–367.

Gori, A., Caredda, F., Franzetti, F., Ridolfo, A., Rusconi, A., and Moroni, M. 1995. Reversible diabetes in patients with AIDS-related Kaposi's sarcoma treated with interferon α-2a. *Lancet* 345:1438–1439.

Guerci, A. P., Guerci, B., Lévy-Marchal, C., Ongagna, J., Ziegler, O., Candiloros, H., Guerci, O., and Drouin, P. 1994. Onset of insulin-dependent diabetes mellitus after interferon-alfa therapy for hairy cell leukemia. *Lancet* 343:1167–1168.

Haggstrom, J., Adriansson, M., Hybbinette, T., Hamby, E., and Thorbert, G. 1996. Two cases of CML treated with alpha-interferon during second and third trimester of pregnancy with analysis of the drug in the new-born immediately postpartum. *Eur. J. Hematol.* 57:101–102.

Hanley, J. P., Jarvis, L. M., Simmonds, P., and Ludlam, C. A. 1996. Development of anti-interferon antibodies and breakthrough hepatitis during treatment for HCV infection in haemophiliacs. *Brit. J. Haematol.* 94:551–556.

Hassanein, T., Schade, R., Lasky, S., and Van Thiel, D. 1994. Is interferon an immunosuppressant? *Gastroenterol.* 106 (suppl):905.

Hayasaka, S., Fujii, M., Yamamoto, Y., Noda, S., Kurome, H., and Sasaki, M. 1995. Retinopathy and subconjunctival haemorrhage in patients with chronic viral hepatitis receiving interferon alfa. *Brit. J. Ophthalmol.* 79:150–152.

Hildebrand, A., Kolde, G., Luger, T. A., and Schwarz, T. 1995. Successful treatment of generalized lichen planus with recombinant interferon alfa-2b. *J. Am. Acad. Dermatol.* 33:880–883.

Hirashima, N., Mizokami, M., Orito, E., Yamauchi, M., and Narita, M. 1994. Chronic hepatitis C complicated by Coombs-negative hemolytic anemia during interferon treatment. *Intern. Med.* 33:300–302.

Honda, K., Ando, A., Endo, M., Shimizu, K., Higashihara, M., Nitta, K., and Nihei, H. 1997. Thrombotic microangiopathy associated with alpha-interferon therapy for chronic myelocytic leukemia. *Am. J. Kidney Dis.* 30:123–130.

Horsmans, Y., Brenard, R., and Geubel, A. P. 1994. Short report: Interferon-α decreases 14C-aminopyrine breath test values in patients with chronic hepatitis C. *Aliment. Pharmacol. Ther.* 8:353–355.

Huang, X., Yuan, J., Goddard, A., Foulis, A., James, R. F. L., Lernmark, A., Pujol-Borrell, R., Rabinovitch, A., Somoza, N., and Stewart, T. A. 1995. Interferon expression in the pancreases of patients with type I diabetes. *Diabetes* 44:658–664.

Imagawa, A., Itoh, N., Hanafusa, T., Oda, Y., Waguri, M., Miyagawa, J. I., Kono, N., Kuwajima, M., and Matsuzawa, Y. 1995. Autoimmune endocrine disease induced by recombinant interferon-α therapy for chronic active type C hepatitis. *J. Clin. Endocrinol. Metab.* 80:922–926.

International Chronic Granulomatous Disease Cooperative Study Group (ICGDCSG). 1991. A controlled trial of interferon gamma to prevent infection in chronic granulomatous disease. *N. Eng. J. Med.* 324: 509–516.

Ishizaki, T., Sasaki, F., Ameshima, S., Shiozaki, K., Takahashi, H., Abe, Y., Ito, S., Kuriyama, M., Nakai, T., and Kitagawa, M. 1996. Pneumonitis during interferon and/or herbal drug therapy in patients with chronic active hepatitis. *Eur. Resp. J.* 9:2691–2696.

Israel, B. C., Blouin, R. A., McIntyre, W., and Shedlofsky, S. I. 1993. Effect of interferon-α monotherapy on hepatic drug metabolism in cancer patients. *Brit. J. Clin. Pharmacol.* 36:229–235.

Jacobs, L. D., Cookfair, D. L., Rudick, R. A., Herndon, R. M., Richert, J. R., Salazar, A. M., Fischer, J. S., Goddkin, D. E., Granger, C. V., Simon, J. H., Alam, J. J., Bartoszak, D. M., Bourdette, D. N., Braiman, J., Brownscheidle, C. M., Coats, C. E., Cohan, S. L., Dougherty, D. S., Kinkel, R. P., Mass, M. K., Munschauer, F. E., Priore, R. L., Pullicino, P. M., Scherokman, B. J., Weinstock-Guttman, B., and Witham, R. H. 1996. Intramuscular interferon beta-1a for disease progression in relapsing multiple sclerosis. *Ann. Neurol.* 39:285–294.

Jaffe, H. S., Chen, A. B., Kramer, S., and Sherwin, S. A. 1987. The absence of interferon antibody formation in patients receiving recombinant human interferon-gamma. *J. Biolog. Response Mod.* 6:576–580.

Janssen, H. L. A., Brouwer, J. T., Nevens, F., Sanchez-Tapias, J. M., Craxi, A., and Hadziyannis, S. 1993. Fatal hepatic decompensation associated with interferon alfa. *Lancet* 306:107–108.

Janssen, H. L. A., Brouwer, J. T., van der Mast, R. C., and Schalm, S. W. 1994. Suicide associated with alfa-interferon therapy for chronic viral hepatitis. *J. Hepatol.* 21:241–243.

Jucgla, A., Marcoval, J., Curco, N., and Servitje, O. 1991. Psoriasis with articular involvement induced by interferon alfa. *Arch. Dermatol.* 127:910–911.

Jumbou, O., Berthelot, J. M., French, N., Bureau, B., Litoux, P., and Dréno, B. 1995. Polyarthritis during interferon alpha therapy: 3 cases and a review of the literature. *Eur. J. Dermatol.* 5:581–584.

Kadayifci, A., Aytemir, K., Arslan, M., Aksoyek, S., Sivri, B., and Kabakci, G. 1997. Interferon-alpha does not cause significant cardiac dysfunction in patients with chronic active hepatitis. *Liver* 17:99–102.

Kamisako, T., Adachi, Y., Chihara, J., and Yamamoto, T. 1993. Intersitial pneumonitis and interferon-alfa. *Brit. Med. J.* 306:896.

Kanda, Y., Shigeno, K., Kinoshita, N., Nakao, K., Yano, M., and Matsuo, H. 1994. Sudden hearing loss associated with interferon. *Lancet* 343:1134–1135.

Kawano, T., Shigehira, M., Uto, H., Kato, J., Hayashi, K., Maruyama, T., Kuribayashi, T., Chuman, T., Futami, T., and Tsubouchi, H. 1996. Retinal complications during interferon therapy for chronic hepatitis C. *Am. J. Gastroenterol.* 91:309–313.

Keeffe, E. B., Hollinger, F. B., and the Consensus Interferon Study Group. 1997. Therapy of hepatitis C: Consensus interferon trials. *Hepatol.* 26 (suppl.1):101–107.

Khan, H. A., Khawaja, F. I., and Mahrous, A. R. S. 1996. Life-threatening severe immune thrombocytopenia after alpha-interferon therapy for chronic hepatitis C infection. *Am. J. Gastroenterol.* 91:821–822.

Kimmel, O. L., Abraham, A. A., and Phillips, T. M. 1994. Membrano-proliferative glomerulo-nephritis in a patient treated with interferon-α for human immunodeficiency virus infection. *Am. J. Kidney Dis.* 24:858–863.

Koivisto, V. A. Pelkonen, R., and Cantell, K. 1989. Effect of interferon on glucose tolerance and insulin sensitivity. *Diabetes* 38:641–647.

Kontochristopoulos, G., Stavrinos, C., Aroni, K., and Tassopoulos, N. C. 1996. Cutaneous necrosis by subcutaneous injection of α-interferon in a patient with chronic type B hepatitis. *J. Hepatol.* 25:271.

Kovarik, J., Mayer, G., Pohanka, E., Schwarz, M., Traindl, O., Graf, H., and Smolen, J. 1988. Adverse effect of low-dose prophylactic human recombinant leucocyte interferon-alpha treatment in renal transplant patients. *Transplant.* 45:402–405.

Krogsgaard, K., Marcellin, M., Trepo, C., Berthelot, P., Sanchez-Tapias, J. M., Bassendine, M., Tran, A., Ouzan, D., Ring-Larsen, H., Lindberg, J., Enriquez, J., Benhamou, J. P., and Bindslev, N. 1996. Prednisolone withdrawal therapy enhances the effect of human lymphoblastoid interferon in chronic hepatitis B. *J. Hepatol.* 25:803–813.

La Civita, L., Zignego, A. L., Lombardini, F., Monti, M., Longombardo, G., Pasero, G., and Ferri, C. 1996. Exacerbation of peripheral neuropathy during alpha-interferon therapy in a patient with mixed cryoglobulinemia and hepatitis B virus infection. *J. Rheumatol.* 23:1641–1643.

Le Gal, F. A., Paul, C., Chemaly, P., and Dubertret, L. 1996. More on cutaneous reactions to recombinant cytokine therapy. *J. Am. Acad. Dermatol.* 35:650–651.

Lederer, E., and Truong, L. 1992. Unusual glomerular lesion in a patient receiving long-term interferon alpha. *Am. J. Kidney Dis.* 20:516–518.

Liberati, A. M., Biagini, S., Perticoni, G., Ricci, S., D'Alessandro, P., Senatore, M., and Cinieri, S. 1990. Electrophysiological and neuropsychological functions in patients treated with interferon-β. *J. Interferon Res.* 10:613–619.

Liozon, E., Delaire, L., Lacroix, P., Labrousse, F., Ly, K., Fauchais, A. L., Loustaud-Ratti, V., Vidal, J., Liozon, F., and Vidal, E. 1997. Syndrome de Raynaud compliqué de gangrène digitale au cours d'un traitement par l'interféron alfa. *Rev. de Med. Int.* 18:316–319.

Lisker-Melman, M., Di Bisceglie, A. M., Usala, S. J., Weintraub, B., Murray, L. M., and Hoofnagle, J. H. 1992. Development of thyroid disease during therapy of chronic viral hepatitis with interferon alfa. *Gastroenterol.* 102:2155–2160.

Lopes, E. P. A., Oliveira, P. M., Silva, A. E., Ferraz, M. L., Costa, C. H. R. M., Miranda, W., and Dib, S. A. 1994. Exacerbation of type 2 diabetes mellitus during interferon-alpha therapy for chronic hepatitis B. *Lancet* 343:244.

Lunel, F. 1995. Hepatitis C virus and autoimmunity: Fortuitous association or reality? *Gastroenterol.* 107:1550–1555.

Lutsman, F., Salhadin, A., Nouwynck, C., and Hanson, B. 1995. Pneumonie interstitielle à la suite d'un traitement par interféron-alpha. *Presse Medicale* 24:1910.

Machold, K. P., and Smolen, J. S. 1990. Interferon-gamma induced exacerbation of systemic lupus erythematosus. *J. Rheumatol.* 17:831–832.

Magnone, M., Holley, J. L., Shapiro, R., Scantlebury, V., McCauley, J., Jordan, M., Vivas, C., Starzl, T., and Johnson, J. P. 1995. Interferon-α-induced acute renal allograft rejection. *Transplant.* 59:1068–1070.

Makino, Y., Tanaka, H., Nakamura, K., Fujita, M., Akiyama, K., and Makino, I. 1994. Arthritis in a patient with psoriasis after alpha-interferon therapy for chronic hepatitis C. *J. Rheumatol.* 21:1771–1772.

Malaguarnera, M., Giugno, I., Ruello, P., Pistone, G., Restuccia, S., and Trovato, B. A. 1996. Effect of interferon on blood lipids. *Clin. Drug Invest.* 11:43–48.

Manesis, E. K., Petrou, C., Brouzas, D., and Hadziyannis, S. 1994. Optic tract neuropathy complicating low-dose interferon treatment. *J. Hepatol.* 21:474–477.

Mansat-Krzyzanowska, E., Dréno, B., Chiffoleau, A., and Litoux, P. 1991. Manifestations cardio-vasculaires associées à l'interféron a-2a. *Annales de Médecine Interne* 142:576–581.

Mapou, R. L., Law, W. A., Wagner, K., Malone, J. L., and Skillman, D. R. 1996. Neuropsychological effects of interferon alfa-n3 treatment in asymptomatic human immunodeficiency virus-1-infected individuals. *J. Neuropsychi. Clin. Neurosci.* 8:74–81.

Marazuella, M., Garcia-Buey, L., Gonzalez-Fernandez, B., Garcia-Monzon, C., Arranz, A., Borque, M. J., and Moreno-Otero, R. 1996. Thyroid autoimmune disorders in patients with chronic hepatitis C before and during interferon-alpha therapy. *Clin. Endocrinol.* 44:635–642.

Mase, G., Zorzon, M., Biasutti, E., Vitrani, B., Cazzato, G., Urban, F., and Frezza, M. 1996. Development of myasthenia gravis during interferon-alpha treatment for anti-HCV positive chronic hepatitis. *J. Neurol. Neurosurg. Psychi.* 60:348–349.

Mateo, R., Jethmalani, S., Angus, D. C., Gorcsan, J. III, Uretsky, B., and Fung, J. 1996. Interferon-associated left ventricular dysfunction in a liver transplant recipient. *Dig. Dis. Sci.* 41:1500–1503.

Mattson, K., Niiranen, A., Pyrhonen, S., Farkkila, M., and Cantell, K. 1991. Recombinant interferon gamma treatment in non-small cell lung cancer. Antitumour effect and cardiotoxicity. *Acta Oncologica* 30:607–610.

Mauser-Bunschoten, E. P., Damen, M., Reesink, H. W., Roosendaal, G., Chamuleau, R. A. F. M., and van den Berg, H. M. 1996. Formation of antibodies to factor VII in patients with hemophilia A who are treated with interferon for chronic hepatitis C. *Ann. Intern. Med.* 125:297–299.

Mayet, W. J., Hess, G., Gerken, G., Rossol, S., Voth, R., Manns, M., and Meyer Zum, Büschenfelde K. H. 1989. Treatment of chronic type B hepatitis with recombinant α-interferon induces autoantibodies not specific for autoimmune chronic hepatitis. *Hepatol.* 10:24–28.

McDonald, E. M., Mann, A. H., and Thomas, H. C. 1987. Interferons as mediators of psychiatric morbidity. *Lancet* 2:1175–1177.

McKenna, R., and Oberg, K. E. 1997. Antibodies to interferon-alpha in treated cancer patients: Incidence and significance. *J. Interferon Cytokine Res.* 17:141–143.

Mekkakia-Benhabib, C., Marcellin, P., Colas-Linhart, N., Castel Nau, C., Buyck, D., Erlinger, S., and Bok, B. 1996. Natural course of dysthyroidism in patients treated with interferon for chronic hepatitis C. *Annales d'endocrinologie* 57:419–427.

Meyers, C. A., and Valentine, A. D. 1995. Neurological and psychiatric adverse effects of immunological therapy. *CNS Drugs* 3:56–68.

Meyers, C. A., Scheibel, R. S., and Forman, A. D. 1991. Persistent neurotoxicity of systemically administered interferon-alpha. *Neurol.* 41:672–676.

Milella, M., Antonelli, G., Santantonio, T., Giannelli, G., Currenti, M., Monno, L., Turriziani, O., Pastore, G., and Dianzani, F. 1995. Treatment with natural IFN of hepatitis C patients with or without antibodies to recombinant IFN. *Hepato-Gastroenterol.* 42:201–204.

Min, A. D., and Bodenheimer, H. C. 1995. Does interferon precipitate rejection of liver allografts? *Hepatol.* 22:1333–1335.

Miossec, P. 1997. Cytokine-induced autoimmune disorders. *Drug Safety* 17:93–104.

Mirro, J., Kalwinsky, D., Whisnant, J., Weck, P., Chesney, C., and Murphy, S. 1985. Coagulopathy induced by continuous infusion of high doses of human lymphoblastoid interferon. *Cancer Treat. Rep.* 69: 315–317.

Muratori, L., Lenzi, M., Cataleta, F., Giostra, F., Cassani, F., Ballardini, G., Zauli, D., and Bianchi, F. B. 1994. Interferon therapy in liver/kidney microsomal antibody type-1 positive patients with chronic hepatitis C. *J. Hepatol.* 21:199–203.

Negoro, K., Fukusako, T., Morimatsu, M., and Liao, C. M. 1994. Acute axonal polyneuropathy during interferon α-2a therapy for chronic hepatitis C. *Muscle and Nerve* 17:1351–1352.

Neilley, L. K., Goodin, D. S., Goodkin, D. E., and Hauser, S. L. 1996. Side effect profile of interferon beta-1b in MS: Results of an open label trial. *Neurol.* 46:552–554.

Nguyen, C., Misery, L., Tigaud, J. D., Petiot, A., Fière, D., Faure, M., and Claudy, A. 1996. Psoriasis induit par l'interféron alpha. A propos d'une observation. *Ann. de Med. Int.* 147:519–521.

Niizeki, H., Inamoto, N., Nakamura, K., Tsuchimoto, K., Hashimoto, T., and Nishikawa, T. 1994. A case of pemphigus foliaceus after interferon alpha-2a therapy. *Dermatol.* 189 (suppl. 1):129–130.

Nishiwaki, H., Ogura, Y., Miyamoto, K., Matsuda, N., and Honda, Y. 1996. Interferon alfa induces leukocyte capillary trapping in rat retinal microcirculation. *Arch. Ophtalmol.* 114:726–730.

Noda, K., Enomoto, N., Arai, K., Masuda, E., Yamada, Y., Suzuki, K., Tanaka, M., and Yoshihara, H. 1996. Induction of antinuclear antibody after interferon therapy in patients with type-C chronic hepatitis: Its relation to the efficacy of therapy. *Scand. J. Gastroenterol.* 31:716–722.

Nouri, K., Valor, R., Rodriguez, M., and Kerdel, F. A. 1996. Interferon alfa-induced interstitial pneumonitis in a patient with cutaneous T-cell lymphoma. *J. Am. Acad. Dermatol.* 35:269–270.

O'Connell, P. G., Gerber, L. H., DiGiovanna, J. J., and Peck, G. L. 1992. Arthritis in patients with psoriasis treated with gamma-interferon. *J. Rheumatol.* 19:80–82.

Oeda, E., and Shinohara, K. 1993. Cutaneous necrosis caused by injection of α-interferon in a patient with chronic myelogeneous leukemia. *Am. J. Hematol.* 44:213–214.

Ogata, K., Koga, T., and Yagawa, K. 1994. Interferon-related bronchiolitis obliterans organizing pneumonia. *Chest* 106:612–613.

Okanoue, T., Sakamoto, S., Itoh, Y., Minami, M., Yasui, K., Sakamoto, M., Nishioji, K., and Katagishi, T. 1996. Side effects of high-dose interferon therapy for chronic hepatitis C. *J. Hepatol.* 25:283–291.

Okuno, H., Takasu, M., Kano, H., Seki, T., Shiozaki, Y., and Inoue, K. 1993. Depression of drug-metabolizing activity in the human liver by interferon-beta. *Hepatol.* 17:65–69.

Pagliacci, M. C., Pelicci, G., Schippa, M., Liberati, A. M., and Nicoletti, I. 1991. Does interferon-beta therapy induce thyroid autoimmune phenomena. *Hormon. Metab. Res.* 23:196–197.

Pan, W., Banks, W. A., and Kastin, A. J. 1997. Permeability of the blood-brain and blood-spinal cord barriers to interferons. *J. Neuroimmunol.* 76:105–111.

Panitch, H. S., Hirsch, R. L., Schindler, J., and Johnson, K. P. 1987. Treatment of multiple sclerosis with gamma interferon: Exacerbations associated with activation of the immune system. *Neurol.* 37: 1097–1102.

Papo, T., Marcellin, P., Bernuau, J., Durand, F., Poynard, T., and Benhamou, J. P. 1992. Autoimmune chronic hepatitis exacerbated by apha-interferon. *Ann. Intern. Med.* 116:51–53.

Parodi, A., Semino, M., Gallo, R., and Rebora, A. 1993. Bullous eruption with circulating pemphigus-like antibodies following interferon-alpha therapy. *Dermatol.* 186:155–157.

Pavol, M. A., Meyers, C. A., Rexer, J. L., Valentine, A. D., Mattis, P. J., and Talpaz, M. 1995. Pattern of neurobehavorial deficits associated with interferon alfa therapy for leukemia. *Neurol.* 45:947–950.

Pawlotsky, J. M., Bouvier, M., Fromont, P., Deforges, L., Duval, J., Dhumeaux, D., and Bierling, P. 1995. Hepatitis C virus infection and autoimmune thrombocytopenic purpura. *J. Hepatol.* 23:635–639.

Pesce, A., Taillan, B., Rosenthal, E., Garnier, G., Vinti, H., Dujardin, P., and Cassuto, J. P. 1993. Opportunistic infections and CD4 lymphocytopenia with interferon treatment in HIV-1 infected patients. *Lancet* 341:1597.

Pittau, E., Bogliolo, A., Tinti, A., Mela, Q., Ibba, G., Salis, G., and Perpignano, G. 1997. Development of arthritis and hypothyroidism during alpha-interferon therapy for chronic hepatitis C. *Clin. Exp. Rheumatol.* 15:415–419.

Poynard, T., Bedossa, P., Chevallier, M., Mathurin, P., Lemonnier, C., Trepo, C., Couzigou, P., Payen, J. L., Sajus, M., Costa, J. M., Vidaud, M., and Chaput, J. C. 1995. A comparison of three interferon alfa-2b regimens for the long-term treatment of chronic non-A, non-B hepatitis. *N. Eng. J. M.* 332:1457–1462.

Poynard, T., Leroy, V., Cohard, M., Thevenot, T., Mathurin, P., Opolon, P., and Zarski, J. P. 1996. Meta-analysis of interferon randomized trials in the treatment of viral hepatitis C: Effects of dose and duration. *Hepatol.* 24:778–789.

Prasad, S., Waters, B., Hill, P. B., Portera, F. A., and Riely, C. A. 1992. Psychiatric side effects of interferon alfa-2b in patients treated for hepatitis C. *Clin. Res.* 40:840A.

Preziati, D., La Rosa, L., Covini, G., Marcelli, R., Rescalli, S., Persani, L., Del Ninno, E., Meroni, P. L., Colombo, M., and Beck-Peccoz, P. 1995. Autoimmunity and thyroid function in patients with chronic active hepatitis treated with recombinant interferon alpha-2a. *Eur. J. Endocrinol.* 132: 587–593.

Quesada, J. R., Talpaz, M., Rios, A., Kurzrock, R., and Gutterman, J. U. 1986. Clinical toxicity of interferons in cancer patients: A review. *J. Clin. Oncol.* 4:234–243.

Reid, I. T., Lombardo, F. A., Redmond, I. J., Hammond, S. L., Coffey, J. A., and Ozer, H. 1992. Digital vasculitis associated with interferon therapy. *Am. J. Med.* 92:702–703.

Renault, P. F., Hoofnagle, J. H., Park, Y., Mullen, K. D., Peters, M., Jones, B., Rustgi, V., and Jones, E. A. 1987. Psychiatric complications of long-term interferon alfa therapy. *Arch. Intern. Med.* 147:1577–1580.

Rieckmann, P., Weber, F., Günther, A., and Poser, S. 1996. The phosphodiesterase inhibitor pentoxifylline reduces side effects of interferon-beta1b treatment in patients with multiple sclerosis. *Neurol.* 47:604.

Rifflet, H., Vuillemin, E., Oberti, F., Laine, P., and Calès, P. 1996. Interféron et suicide au cours des hépatites virales chroniques (abstract). *Gastr. Clin. et Biol.* 20:68.

Rohde, D., Sliwka, U., Schweizer, K., and Jakse, G. 1996. Oculo-bulbar myasthenia gravis induced by cytokine treatment of a patient with metastasized renal cell carcinoma. *Eur. J. Clin. Pharmacol.* 50: 471–473.

Rönnblom, L. E., Alm, G. V., and Oberg, K. E. 1991. Autoimmunity after alpha-interferon therapy for malignant carcinoid tumors. *Ann. Intern. Med.* 115:178–183.

Rostaing, L., Izopet, J., Baron, E., Duffaut, M., Puel, M., and Durand, D. 1995. Treatment of chronic hepatitis C with recombinant interferon-alpha in kidney transplant recipients. *Transplant.* 59:1426–1431.

Roti, E., Minelli, R., Giuberti, T., Marchelli, S., Schianchi, C., and Gardini, E. 1996. Multiple changes in thyroid function in patients with chronic active HCV hepatitis treated with recombinant interferon-alpha. *Am. J. Med.* 101:482–487.

Rumi, M., DelNinno, E., Parravicini, M. L., Romeo, R., Soffredini, R., Donato, M. F., Wilber, J., Russo, A., and Colombo, M. 1996. A prospective, randomized trial comparing lymphoblastoid to recombinant interferon alfa 2a as therapy for chronic hepatitis C. *Hepatol.* 24:1366–1370.

Russo, D., Candoni, A., Zuffa, E., Minisini, R., Silvestri, F., Fanin, R., Zaja, F., Martinelli, G., Tura, S., Botta, G., and Baccarani, M. 1996. Neutralizing anti-interferon-alpha antibodies and response to treatment in patients with Ph(+) chronic myeloid leukaemia sequentially treated with recombinant (alpha 2a) and lymphoblastoid interferon-alpha. *Brit. J. Haematol.* 94:300–305.

Rutkove, S. B. 1997. An unusual axonal polyneuropathy induced by low-dose interferon alfa-2a. *Arch. Neurol.* 54:907–908.

Sacchi, S., Kantarjian, H., O'Brien, S., Cohen, P. R., Pierce, S., and Talpaz, M. 1995. Immune-mediated and unusual complications during interferon alfa therapy in chronic myelogenous leukemia. *J. Clin. Oncol.* 13:2401–2407.

Sachithanandan S., Clarke, G., Crowe, J., and Fielding, J. F. 1997. Interferon-associated thyroid dysfunction in anti-D-related chronic hepatitis. *J. Interferon Cytokine Res.* 17:409–411.

Sampaio, E. P., Moreira, A. L., Sarno, E. N., Malta, A. M., and Kaplan, G. 1992. Prolonged treatment with recombinant interferon-γ induces erythema nodosum leprosum in lepromatous leprosy patients. *J. Exp. Med.* 175:1729–1737.

Samson, D., Volin, L., Schanz, U., Bosi, A., and Gahrton, G. 1996. Feasibility and toxicity of interferon maintenance therapy after allogeneic BMT for multiple myeloma: A pilot study of the EBMT. *Bone Marrow Transplant.* 17:759–762.

Sartori, M., Andorno, S., La Terra, G., Pozzoli, G., Rudoni, M., Sacchetti, G. M., Inglese, E., and Aglietta, M. 1995. Assessment of interferon cardiotoxicity with quantitative radionuclide angiocardiography. *Eur. J. Clin. Invest.* 25:68–70.

Sata, M., Yano, Y., Yoshiyama, Y., Ide, T., and Kumashiro, R. 1997. Mechanisms of thrombocytopenia induced by interferon therapy for chronic hepatitis B. *J. Gastroenterol.* 32:206–210.

Schlaifer, D., Dumazer, P., Spenatto, N., Mignon-Conte, M., Brousset, P., Lumbroso, C., Cooper, M., Muller, C., Huguet, F., Attal, M., Laurent, G., and Pris, J. 1994. Hemolytic-uremic syndrome in a patient with chronic myelogenous leukemia treated with interferon alpha. *Am. J. Hematol.* 47:254–255.

Schwaber, M. J., and Zlotogorski, A. 1997. Dermatologic manifestations of hepatitis C infection. *Int. J. Dermatol.* 36:251–254.

Schwid, S. R., Goodman, A. D., and Mattson, D. H. 1997. Autoimmune hyperthyroidism in patients with multiple sclerosis treated with interferon beta-1b. *Arch. Neurol.* 54:1169–1170.

Segawa, F., Shimizu, Y., Saito, E., and Kinoshita, M. 1995. Behçet's disease induced by interferon therapy for chronic myelogenous leukemia. *J. Rheumatol.* 22:1183–1184.

Seitz, M., Kranke, M., and Kirchner, H. 1988. Induction of antinuclear antibodies in patients with rheumatoid arthritis receiving treatment with recombinant human gamma interferon. *Ann. Rheum. Dis.* 47:642–644.

Selby, P., Kohn, J., Raymond, J., Judson, I., and McElwain, T. 1985. Nephrotic syndrome during treatment with interferon. *Brit. Med. J.* 290:1180.

Sgarabotto, D., Vianello, F., Stefani, P. M., Scano, F., Sartori, R., Caenazzo, A., and Girolami, A. 1997. Hypertriglyceridemia during long-term interferon-alpha therapy in a series of hematologic patients. *J. Interferon Cytokine Res.* 17:241–244.

Shakil, A. O., Di Bisceglie, A. M., and Hoofnagle, J. H. 1996. Seizures during alpha interferon therapy. *J. Hepatol.* 24:48–51.

Shinohara, E., Yamashita, S., Kihara, S., Hirano, K., Ishigami, M., Arai, T., Nozaki, S., Kameda-Takemura, K., Kawata, S., and Matsuzawa, Y. 1997. Interferon alpha induces disorder of lipid metabolism by lowering postheparin lipases and cholesteryl ester transfer protein activities in patients with chronic hepatitis C. *Hepatol.* 25:1502–1506.

Shiota, G., Okubo, M., Kawasaki, H., and Tahara, T. 1997. Interferon increases serum thrombopoietin in patients with chronic hepatitis C. *Brit. J. Haematol.* 97:340–342.

Simsek, H., Savas, C., Akkiz, H., and Telatar, H. 1996. Interferon-induced vitiligo in a patient with chronic viral hepatitis C infection. *Dermatol.* 193:65–66.

Solis, R. A., Pomales, S. Y., and Torres, E. A. 1996. Polymyositis induced by interferon alpha-2B in a patient with chronic hepatitis C. *Am. J. Gastroenterol.* 91:2041.

Sonnenblick, M., and Rosin, A. 1991. Cardiotoxicity of interferon: A review of 44 cases. *Chest* 99:557–561.

Soriano, V., Bravo, R., Samaniego, J. G., Gonzalez, J., Odriozola, P. M., Arroyo, J. L., Vicario, J. L., Castro, A., Colmenero, M., Carballo, E., and Pedreira, J. 1994. CD4+ T-lymphocytopenia in HIV-infected patients receiving interferon therapy for chronic hepatitis C. *AIDS* 8:1621–1622.

Sotomatsu, M., Shimoda, M., Ogawa, C., and Morikawa, A. 1995. Acute pancreatitis associated with interferon-a therapy for chronic myelogenous leukemia. *Am. J. Hematol.* 48:211–212.

Stewart, T. A., Hultgren, B., Huang, X., Pitts-Meek, S., Hully, J., and MacLachlan, N. J. 1993. Induction of type I diabetes by interferon-α in transgenic mice. *Science* 260:1942–1946.

Strumia, R., Venturini, D., Boccia, S., Gamberini, S., and Gullini, S. 1993. UVA and interferon-alfa therapy in a patient with lichen planus and chronic hepatitis C. *Int. J. Dermatol.* 32:386.

Sugano, S., Yanagimoto, M., Suzuki, T., Sato, M., and Onmura, H. 1994. Retinal complications with elevated circulating plasma C5a associated with interferon-alpha therapy for chronic active hepatitis C. *Am. J. Gastroenterol.* 89:2054–2056.

Sunderkötter, C., Luger, T., and Kolde, G. 1993. Severe hypertriglyceridaemia and interferon-alpha. *Lancet* 342:1111–1112.

Tada, H., Saitoh, S., Nakagawa, Y., Hirana, H., Morimoto, L., Shima, T., Shimamoto, K., Okanoue, T., and Kashima, K. 1996. Ischemic colitis during interferon-alpha treatment for chronic active hepatitis C. *J. Gastroenterol.* 31:582–584.

Taliani, G., Duca, F., Clementi, C., and De Bac, C. 1996. Platelet-associated immunoglobulin G, thrombocytopenia and response to interferon treatment in chronic hepatitis C. *J. Hepatol.* 25:999.

Tanaka, H., Yamakado, T., Emi, Y., Nabeshima, K., Itoh, S., and Nakano, T. 1995. Interferon-induced coronary-vasospasm: A case history. *Angiol.* 46:1139–1143.

Tappero, G., Negro, F., Farina, M., Gallo, M., Angelo, A., and Hadengue, A. 1996. Safe switch to β-interferon treatment of chronic hepatitis C after α-interferon-induced autoimmune thrombocytopenia. *J. Hepatol.* 24.

Tashiro, M., Yokoyama, K., Nakayama, M., Yamada, A., Ogura, Y., Kawaguchi, Y., and Sakai, O. 1996. A case of nephrotic syndrome developing during postoperative gamma interferon therapy for renal cell carcinoma. *Nephron* 73:685–688.

Teichmann, J. V., Sieber, G., Ludwig, W. D., and Ruehl, H. 1989. Immunosuppressive effects of recombinant interferon-α during long-term treatment of cancer patients. *Cancer* 63:1990–1993.

Teragawa, H., Hondo, T., Amano, H., Hino, F., and Ohbayashi, M. 1996. Adverse effects of interferon on the cardiovascular system in patients with chronic hepatitis C. *Jap. Heart J.* 37:905–915.

Teragawa, H., Hondo, T., Takahashi, K., Watanabe, H., Ohe, H., Hattori, N., Watanabe, Y., Amano, H., Hino, F., Ohbayashi, M., Urushihara, T., and Yonehara, S. 1996. Sarcoidosis after interferon therapy for chronic active hepatitis C. *Intern. Med.* 35:19–23.

The IFNB Multiple Sclerosis Study Group and the University of British Columbia MS/MRI Analysis Group. 1995. Interferon beta-1b in the treatment of multiple sclerosis: Final outcome of the randomized controlled trial. *Neurol.* 45:1277–1285.

The IFNB Multiple Sclerosis Study Group and the University of British Columbia MS/MRI Analysis Group. 1996. Neutralizing antibodies during treatment of multiple sclerosis with interferon beta-1b: Experience during the first three years. *Neurol.* 47:889–894.

Tilg, H. 1997. New insights into the mechanisms of interferon alfa: An immunoregulatory and anti-inflammatory cytokine. *Gastroenterol.* 112:1017–1021.

Todros, L., Saracco, G., Durazzo, M., Abate, M. L., Touscoz, G., Scaglione, L., Verme, G., and Rizzetto, M. 1995. Efficacy and safety of interferon alfa therapy in chronic hepatitis C with autoantibodies to liver–kidney microsomes. *Hepatol.* 22:1374–1378.

Tolaymat, A., Leventhal, B., Sakarcan, A., Kashima, H., and Monteiro, C. 1992. Systemic lupus erythematosus in a child receiving long-term interferon therapy. *J. Pediatr.* 120:429–432.

Tosti, A., Misciali, C., Bardazzi, F., Fanti, P. A., and Varotti, C. 1992. Telogen effluvium due to recombinant interferon alpha-2b. *Dermatol.* 184:124–125.

Tran, A., Benzaken, S., Yang, G., Schneider, S., Doglio, A., Rampal, A., and Rampal, P. 1997. Chronic hepatitis C and autoimmunity: Good response to immunosuppressive treatment. *Dig. Dis. Sci.* 42:778–780.

Valentine, A. D., Meyers, C. A., and Talpaz, M. 1995. Treatment of neurotoxic side effects of interferon-alpha with naltrexone. *Cancer Invest.* 13:561–566.

Vallisa, D., Cavanna, L., Berté, R., Merli, F., Ghisoni, F., and Buscarini, L. 1995. Autoimmune thyroid dysfunctions in hematologic malignancies treated with alpha-interferon. *Acta Haematologica* 93:31–35.

Van Thiel, D. H., Friedlander, L., Molloy, P. J., Fagiuoli, S., Kania, R. J., and Caraceni, P. 1995. Interferon-alpha can be used successfully in patients with hepatitis C virus-positive chronic hepatitis who have a psychiatric illness. *Eur. J. Gastroenterol. Hepatol.* 7:165–168.

Veerabagu, M. P., Finkelstein, S. D., and Rabinovitz, M. 1997. Granulomatous hepatitis in a patient with chronic hepatitis C treated with interferon-alpha. *Dig. Dis. Sci.* 42:1445–1448.

Vento, S., Di Perri, G., Cruciani, M., Garofano, T., Concia, E., and Bassetti, D. 1993. Rapid decline of CD4+ cells after IFN-α treatment in HIV-1 infection. *Lancet* 341:958–959.

Vial, T., and Descotes, J. 1994. Clinical toxicity of the interferons. *Drug Safety* 10:115–150.

Vial, T., and Descotes, J. 1995. Immune-mediated side effects of cytokines in humans. *Toxicol.* 105:31–57.

Vial, T., Bailly, F., Descotes, J., and Trepo, C. 1996. Effets secondaires de l'interféron alpha (Adverse effects of interferon alpha). *Gastr. Clin. et Biol.* 20:462–489.

Wada, M., Kang, K. B., Kinugasa, A., Shintani, S., Sawada, K., Nishigami, T., and Shimoyama, T. 1997. Does the presence of serum autoantibodies influence the responsiveness to interferon-alpha2a treatment in chronic hepatitis C? *Intern. Med.* 36:248–254.

Watanabe, U., Hashimoto, E., Hishamitsu, T., Obata, H., and Hayashi, N. 1994. The risk factors for development of thyroid disease during interferon-α therapy for chronic hepatitis C. *Am. J. Gastroenterol.* 89:399–403.

Waysbort, A., Giroux, M., Mansat, V., Teixeira, M., Dumas, J. C., and Puel, J. 1993. Experimental study of transplacental passage of alpha interferon by two assay techniques. *Antimicrob. Agents Chemother.* 37:1232–1237.

Weber, P., Wiedmann, K. H., Klein, R., Walter, E., Blum, H. E., and Berg, P. A. 1994. Induction of autoimmune phenomena in patients with chronic hepatitis B treated with gamma-interferon. *J. Hepatol.* 20:321–328.

Webster, G. F., Knobler, R. L., Lublin, F. D., Kramer, E. M., and Hochman, L. R. 1996. Cutaneous ulcerations and pustular psoriasis flare caused by recombinant interferon beta injections in patients with multiple sclerosis. *J. Am. Acad. Dermatol.* 34 (Part 2):365–367.

Weiss, K. S. 1992. Nephrotic range of proteinuria and interferon therapy. *Ann. Intern. Med.* 116:347.

Wiedermann, C. J., Vogel, W., Tilg, H., Wiedermann, F. J., Herold, M., Zilian, U., Wohlfarter, T., Gruber, M., and Braunsteiner, H. 1991. Suppression of thyroid function by interferon-alpha2 in man. *Naunyn-Schmiedeberg's Arch. Pharmacol.* 343:665–668.

Wolfe, J. T., Singh, A., Lessin, S. R., Jaworsky, C., and Rook, A. H. 1995. De novo development of psoriatic plaques in patients receiving interferon alfa for treatment of erythrodermic cutaneous T-cell lymphoma. *J. Am. Acad. Dermatol.* 32:887–893.

Woynarowski, M., and Socha, J. 1997. Seizures in children during interferon alpha therapy. *J. Hepatol.* 26:956–957.

Yamagishi, S. I., Abe, T., and Sawada, T. 1994. Human recombinant interferon alpha-2a (rIRFN alpha-2a) therapy suppresses hepatic triglyceride lipase, leading to severe hypertriglyceridemia in a diabetic patient. *Am. J. Gastroenterol.* 89:2280.

Yamazaki, K., Amishima, M., Fujita, J., Aida, A., and Aoi, K. 1993. A case of chronic hepatitis C complicated by ischemia-like changes seen on the electrocardiogram during interferon treatment. *Kokyu to Junkan* 41:805–809.

Yoshida, A., Takeda, A., Koyama, K., Morozumi, K., and Oikawa, T. 1994. Systemic lupus erythematosus with nephropathy after interferon alpha 2A therapy. *Clin. Rheumatol.* 13:382.

Young, M. C., and Otis, J. 1996. Interferon beta-1b hypersensitivy and desensitization. *J. Allergy Clin. Immunol.* 97 (part 3):345.

Zbinden, G. 1990. Effects of recombinant human alpha-interferon in a rodent cardiotoxicity model. *Toxicol. Lett.* 50:25–35.

Zuffa, E., Vianelli, N., Martinelli, G., Tazzari, P., Cavo, M., and Tura, S. 1996. Autoimmune mediated thrombocytopenia associated with the use of interferon-alpha in chronic myeloid leukemia. *Haematologica* 81:533–535.

Index

Printed and bound by CPI Group (UK) Ltd, Croydon, CR0 4YY

24/10/2024

01778278-0013